Design Principles for the Immune System and Other Distributed Autonomous Systems

Editors

Lee A. Segel
Weizmann Institute, Rehovot, Israel

Irun R. Cohen
Weizmann Institute, Rehovot, Israel

Santa Fe Institute
Studies in the Sciences of Complexity

OXFORD
UNIVERSITY PRESS

2001

OXFORD
UNIVERSITY PRESS

Oxford New York
Athens Auckland Bangkok Bogotá Buenos Aires Calcutta
Cape Town Chennai Dar es Salaam Delhi Florence Hong Kong Istanbul
Karachi Kuala Lumpur Madrid Melbourne Mexico City Mumbai Nairobi
Paris São Paulo Shanghai Singapore Taipei Tokyo Toronto Warsaw

and associated companies in
Berlin Ibadan

Published by Oxford University Press, Inc.
198 Madison Avenue, New York, New York 10016

Library of Congress Cataloging-in-Publication Data
Design principles for the immune system and other distributed
autonomous system / editors, Lee A. Segel, Irun R. Cohen.
p. cm. — (Santa Fe Institute studies in the sciences of complexity. Proceedings)
Includes bibliographical references.
ISBN 0-19-513699-3; ISBN 0-19-513700-0 (pbk.)
1. Immune system—Congresses. 2. System theory—Congresses.
3. Biological systems—Congresses. I. Segel, Lee A. II. Cohen, Irun R.
III. Proceedings volume in the Santa Fe Institute studies in the sciences of complexity
QR182 .D38 2001
616.07'9'01—dc21 00-045657

9 8 7 6 5 4 3 2 1

Printed in the United States of America
on acid-free paper

Design Principles for the Immune System

and Other Distributed Autonomous Systems

About the Santa Fe Institute

The *Santa Fe Institute* (SFI) is a private, independent, multidisciplinary research and education center, founded in 1984. Since its founding, SFI has devoted itself to creating a new kind of scientific research community, pursuing emerging science. Operating as a small, visiting institution, SFI seeks to catalyze new collaborative, multidisciplinary projects that break down the barriers between the traditional disciplines, to spread its ideas and methodologies to other individuals, and to encourage the practical applications of its results.

All titles from the *Santa Fe Institute Studies in the Sciences of Complexity* series will carry this imprint which is based on a Mimbres pottery design (circa A.D. 950–1150), drawn by Betsy Jones. The design was selected because the radiating feathers are evocative of the outreach of the Santa Fe Institute Program to many disciplines and institutions.

Contributors List

Johan Bollen, *Los Alamos National Laboratory, Computer Research & Applications Group, CIC-3, MS B265, Los Alamos, NM 87545; e-mail: jbollen@lanl.gov*

Eric Bonabeau, *Eurobios, 9, rue de Grenelle, 75007 Paris, France; e-mail: eric.bonabeau@eurobios.com*

José A. M. Borghans, *Utrecht University, Theoretical Biology, Padualaan 8, 3584 CH Utrecht, The Netherlands; e-mail: j.a.m.borghans@bio.uu.nl*

Eugene Butcher, *Department of Pathology, Stanford University School of Medicine, Stanford, CA 94305-5324; e-mail: ebutcher@stanford.edu*

Irun R. Cohen, *Department of Immunology, Weizmann Institute of Science, Rehovot, Israel; e-mail: irun.cohen@weizmann.weizmann.ac.il*

Rob J. De Boer, *Utrecht University, Theoretical Biology, Padualaan 8, 3584 CH Utrecht, The Netherlands; e-mail: r.j.deboer@bio.uu.nl*

Thomas N. Denny, *UMD-New Jersey Medical School, 185 South Orange Avenue, Newark, NJ 07103; e-mail: dennytn@umdnj.edu*

Stephanie Forrest, *Santa Fe Institute, 1399 Hyde Park Road, Santa Fe, NM 87505, and Department of Computer Science, University of New Mexico; e-mail: steph@santafe.edu*

Ellen F. Foxman, *Department of Pathology, Stanford University School of Medicine, Stanford, CA 94305-5324*

Deborah Gordon, *Stanford University, Department of Biological Sciences, Stanford, CA 94305-5020; e-mail: gordon@ants.stanford.edu*

Steven A. Hofmeyr, *University of New Mexico, Department of Computer Science, FEC357, Albuquerque, NM 87131; e-mail: steveah@cs.unm.edu*

Dragana Jankovic, *NIH Allergy and Infectious Diseases, Lab of Parasitic Diseases, Bldg. 4 Rm 126, Bethesda, MD 20892-0425; e-mail: djankovic@niaid.nih.gov*

Eric J. Kunkel, *Department of Pathology, Stanford University School of Medicine, Stanford, CA 94305-5324*

Melanie Mitchell, *Los Alamos National Laboratory, Biophysics P-21, Mail Stop D454, Los Alamos, NM 87545; e-mail: mm@biophysics.lanl.gov*

André Noest, *Utrecht University, Theoretical Biology, Padualaan 8, 3584 CH Utrecht, The Netherlands; e-mail: a.j.noest@bio.uu.nl*

Charles G. Orosz, *The Ohio State University, Departments of Surgery, Pathology, Medicine Microbiology & Immunology, College of Medicine, 357 Means Hall, 1654 Upham Drive, Columbus, OH 43210; e-mail: orosz01@medctr.osu.edu*

Junliang Pan, *Department of Pathology, Stanford University School of Medicine, Stanford, CA 94305-5324*

Luis M. Rocha, *Los Alamos National Laboratory, Computer Research & Applications Group, CIC-3, MS B265, Los Alamos, NM 87545; e-mail: rocha@santafe.edu*

John Ross, *Stanford University, Department of Chemistry, Stanford, CA 94305; e-mail: ross@chemistry.stanford.edu*

Lee A. Segel, *Department of Computer Science and Applied Mathematics and Computer Science, Weizmann Institute, Rehovot, Israel; e-mail: lee@wisdom.weizmann.ac.il*

Eli Sercarz, *La Jolla Institute for Allergy & Immunology, 10355 Science Center Drive, San Diego, CA 92121; e-mail: eli@liai.org*

Alan Sher, *NIH Allergy and Infectious Diseases, Lab of Parasitic Diseases, Bldg. 4 Rm. 126, Bethesda, MD 20892-0425; e-mail: alan_sher@nih.gov*

Marcel O. Vlad, *Stanford University, Department of Chemistry, Stanford, CA 94305, and Romanian Academy, Center of Mathematical Statistics, Casa Academiei, Bd 13 Septembrie, Bucharest, Romania; e-mail: marceluc@stanford.edu*

Howard L. Weiner, *Brigham & Womens Hospital, Center for Neurologic Diseases, 77 Avenue Louis Pasteur, HIM 730, Boston, MA 02115-5817; e-mail: weiner@cnd.bwh.harvard.edu*

Contents

Preface

For many, the most important and interesting things in this world are complex systems—biological (brains, immune systems, ant colonies), social (economic and political systems), and "artificial" (the internet, computer programs for artificial intelligence). We can define a **system** as an interacting collection of entities or **agents**, such as cells or ants. If a system is termed **distributed**, it is implied that the system is composed of many entities (such as the trillions of cells in the immune system), and that activities of the system are accomplished by the combined action of many of these entities. Here we confine our attention to **autonomous systems**, those without a boss.

As usual, definitions are slippery. An army can be regarded as a system, and one can defend calling it distributed. We would not call an army autonomous, for it is directed by a general. Yet if we peer closely at the general, we find that his or her brain is composed of a great number of cells that have no obvious directing core. If such a core is discovered some day, it is likely that this subsystem will turn out to be a distributed autonomous system. The solution of this paradox is the familiar one that concepts appropriate to one level of discourse may not be appropriate to another. If we look at both an army and an ant colony as an aggregate of (biological) cells, then both aggregates might be characterized as distributed autonomous systems. This common characterization is not appropriate, however, if we compare task allocation in an army and an ant colony. The general is an individual soldier who has a dominant influence on how the army deploys and maneuvers in peace and war (though the troops may not accept the general's orders) while no single ant has more than a tiny influence on how the ant colony forages, fights, and cares for its young. Indeed, task allocation is one of the subjects of this book.

This book deals with three biological examples of distributed autonomous systems—respectively, molecular, cellular, and community-wide—namely metabolism, the immune system, and ant colonies. The lion's share of our concern will be devoted to the immune system. Because of the immune system's cardinal biological and medical importance, a great deal of information is available concerning its operation—from submolecular events that might last only milliseconds through various cellular population changes that take place on time scales ranging from minutes to years, to modifications in animal immune systems that evolved over eons. Moreover, although the immune system is extremely complex, the broad outlines of its functions seem fairly clear. Because of its importance, its complexity, and the depth to which it is understood, the immune system is an ideal focus for an effort to better understand distributed autonomous systems.

The elements of the immune system include various "fighting arms" or **effectors**—such as macrophage cells that "swallow" bacteria and cellular debris, chemicals called antibodies of various specificities and types that can block pathogen action or mark pathogens for macrophage attack, and cytotoxic T cells that destroy pathogen-ridden cells. There is also a "signal corps" exemplified by helper T cells and the messenger chemicals called cytokines that they secrete. Cytokines mold cellular actions. A given action is typically affected by several cytokines, and a given cytokine typically affects several actions.

The immune system faces the problem of **effector choice**, choosing the right effectors for the various tasks with which the system is faced. How does the immune system "know" how to selectively mobilize the widely different responses it is observed to use when the body is attacked by variegated species and strains of bacteria, viruses, and worms? How does the system know how intensively to activate a chosen effector class, and how to extinguish that activation when the pathogen menace wanes. Is something optimized?

Another problem is **organization of the signaling network**. What principles lie behind the complexity of this network, principles that presumably lead it to an essential role in forging an appropriate immune response?

Other distributed autonomous biological systems must solve problems that are analogous to those facing the immune system. Consider effector choice. This problem is faced by ants when they build a nest or repel attackers, for various ant types have overlapping special abilities. Metabolic networks must choose suitable pathways to achieve appropriate utilization of shifting food supplies amidst changing demands.

Organizing their signaling network is clearly vital for ant colonies, and the network is certainly a complex one. For example, the "alarm pheromone" is composed of several chemicals, and the same alarm signal induces different responses in different ant types. Complex signaling appears to be an implicit challenge in the metabolic system, too. For example, the activity of a single enzyme may be regulated by a variety of different substances.

Not only biological systems are scrutinized here. This book reflects a major trend in contemporary science and technology to look to biology for inspiration and paradigms for understanding and constructing nonbiological systems. Artificial intelligence seeks ideas from brain research, computer vision considers aping natural vision, and in artificial life much effort is devoted to applying concepts of ecological and evolutionary theory to all manner of complex interactions. Recently there have been the first glimmerings of artificial immunology, the exploitation of concepts from immunology to the design of complex systems. In this spirit, we examine here how biological ideas can help design better tools for searching the web, for foiling computer intruders, and for improving an "agent based" high-level artificial intelligence program for constructing analogies. Keep in mind that there is a two-way street between natural and artificial biology. For example, contemplation of what seems a good idea in a distributed approach to finding and fighting computer viruses may provide interesting conjectures as to heretofore unrecognized methods by which the immune system is organized to repel pathogens.

We turn now to an overview of what the book contains.

PART I: AN OVERVIEW OF IMMUNOLOGY

As we have mentioned, the case is strong that the immune system provides a superb opportunity to plumb the depths of complex systems. The problem is that a complex system is complex, so that it is not easy to obtain enough background to understand the vocabulary and concepts that biologists employ when they discuss immune operation. In chapter 1, Hofmeyr attempts to alleviate this problem with his "gentle introduction to the immune system." The chapter emphasizes the types of issues that concern us here—architecture, algorithms, and principles.

PART II: CASE STUDIES IN IMMUNE COMPLEXITY: EXPERIMENTS

This part reviews experimental findings in certain aspects of immunology. Even minimally comprehensive textbook-level accounts of current knowledge in immunology comprise hundreds of pages, so that the reviews here are necessarily restricted in the topics that they cover and in the depth with which a given topic can be covered. Nonetheless, these chapters can provide information that is not only useful in itself but also can give a good feeling for the complexity of phenomena that must be comprehended in a satisfactory overview of immune system operation.

The actions of the immune system (and of other physiological systems) are mediated by chemical cytokines, which are secreted by immune cells and by other cells. More than one hundred cytokines have been discovered, and the number is increasing. Denny provides an extensive but yet (necessarily) highly selective survey of the field (ch. 2). Among the concepts exemplified are the following. There are three ways that a cytokine acts—**autocrine** (on the same cell that secretes it), **paracrine** (on a nearby cell), or **endocrine** (on a distant cell). Cytokines often are **pleiotropic** (acting in multiple ways on a variety of target cells). Cytokine action is frequently characterized by **redundancy** and **antagonism** or **synergism**.

T-helper (Th) cells somehow integrate information in their environment in such a way as to induce secretions of cytokines that suitably guide immune responses. There are two main classes of Th cells. Th1 cells channel the immune system into actions that, typically, are appropriate for combatting intracellular viruses and bacteria while Th2 cells typically spur means for fighting challenges by extracellular pathogens. A central problem in contemporary immunology is to understand how the choice between Th1 and Th2 is made. In chapter 3 Jankovic and Sher review current opinion concerning this matter.

If animals are fed proteins, their subsequent immune response to these proteins is diminished. Such "oral tolerance" is being clinically tested as a way to combat autoimmune disease. Weiner reviews the current status of this field, and thereby provides a case study of how a variety of immune principles combine to produce phenomenology that is important both physiologically and medically (ch. 4).

PART III: DESIGN PRINCIPLES FOR THE IMMUNE SYSTEM

Several chapters discuss in varying degrees of detail a variety of design principles that are exhibited by the immune system. Major principles discussed by Orosz in chapter 5 include the following.

- **Layering**: "new processes are built on top of older, less-effective processes" —which yields **scaffolding**, wherein early steps in the immune response provide the conditions needed for later steps.
- **Parallel processing**.
- **Dynamic engagement**: the cells of the immune system act only briefly, and then are replaced by other cells.
- **Variable network connectivity**.

Orosz gives the name of **immune ecology** to the study of how the various cells work together, and **immune informatics** to the examination of the role of communication in immune ecology.

In chapter 6 Cohen discusses how a specific immune response flows from **receptor degeneracy** (immune receptors bind many molecules with varying degrees of affinity) and **pleiotropy** (interacting multiple and redundant effects of immune components). In the process of **co-respondence** "a composite picture of the antigen is created by **synergy**: the mutual cooperativity of semiautonomous agents, each perceiving a different molecular world."

In their study of diversity in the immune system, Borghans and De Boer take a different point of view from Cohen regarding receptors. They note in chapter 7 that adaptive immune systems have enormous **receptor diversity** and, hence, each antigen can in all likelihood be recognized very specifically. Similarly there is large diversity in a population of organisms among the major histocompatibility (MHC) molecules, which "present" pieces of antigen to the immune system. Yet individual MHC diversity is limited. What can be the evolutionary reasons for the "diverse diversity" that is observed? A particularly interesting part of Borghans and De Boer's reply to this question stems from the hypothesis that the **memory cells** of the adaptive immune system store appropriate effector mechanisms against the various antigens that they encounter. High receptor diversity reconciles **comprehensive reactivity** with **minimal inappropriate crossreactivity**.

Noest's contribution is a theoretical attempt to see how basic design principles can emerge from reasonable assumptions concerning the function of a complex system (ch. 8). Noest assumes that a dominating task for T cells in the immune system is **signal detection**—perceiving signals, amidst a noisy background, that betray the presence of unknown intruders (pathogens). Given this, application of classical optimal detection theory yields a remarkable variety of properties that are possessed by T cells.

In chapter 9 Segel provides a view of immune informatics that he terms a **diffuse informational network**, which can be profitably viewed both as a **command network** and an **informational network**. He argues that it is helpful to view the immune system as simultaneously pursuing a variety of overlapping and even contradictory goals. He discusses how a variety of sensors can provide a **diffuse feedback**, which can provide an overall improvement in achieving the variety of goals.

Butcher and his colleagues summarize recent experimental results on how immune cells get to the right place at the right time (ch. 10), by a **multistep homing process**, mediated at each step by a variety of receptor-ligand pairs. As is illustrated quantitatively by theoretical considerations, multistep homing gives flexibility and a specificity which exceeds that of the individual steps. The authors point out that the homing process in immunology can serve as a paradigm for understanding many analogous processes in developmental biology.

"Anarchic" is how Sercarz (ch. 11) terms immune organization. He argues that immune organization is profitably viewed as a collection of interacting modular assemblies. Like the mythical Golem, the individual units have limited adaptability. Their variability and complexity make unpredictable the detailed reaction of the immune system to an antigenic challenge.

PART IV: BIOCHEMICAL SYSTEMS

Biochemistry is replete with interlocking groups of chemical reactions. The "agents" here are molecules, and most of the tasks are connected "merely" with housekeeping functions. More understanding of its structure is required before one can puzzle out how the biochemical network is organized to perform the tasks that it faces. In chapter 12 Ross and Vlad present a summary of some recent research that can aid in acquiring such understanding. Particularly relevant is his discussion of the use of genetic algorithms to produce reaction networks that provide **optimal performance** of some preassigned task.

PART V: SOCIAL INSECTS

Social insects provide splendid biological examples of distributed autonomous systems. Bonabeau (ch. 13) briefly reviews some of the remarkable collective feats of these "superorganisms." He goes on to discuss how the following three principles can explain much of the observed behavior.

1. **Self organization**, wherein the interactions among simple individuals can give rise to complex collective behavior, with time and length scales that by far exceed the corresponding quantities for the individual insects.
2. **Differential and variable response thresholds** that a stimulus must exceed before an individual insect will perform a certain task.
3. **Templates**, environmental factors such as temperature, humidity, and light that guide colony activity—which in turn may alter the templates.

In chapter 14 Gordon analyzes the operation of colonies of harvester ants, based on her own experimental and theoretical work. Ants have a number of tasks to perform. For ants, **task allocation** corresponds to effector choice in the immune system: how many ants should be allocated to each task. In contrast to the immune system, the time history of task allocation can be observed quite fully and easily. Gordon finds that ants adjust their behavior to changing conditions both by changing tasks and switching between inactive and active states. Decisions concerning adjustments are modulated by local information, including interactions between ants and environmental cues. Interesting experimental and theoretical advances have been made in understanding task allocation.

PART VI: APPLICATIONS TO COMPUTER SCIENCE

The final section concerns applications of ideas concerning the performance of distributed autonomous networks to computer science.

In chapter 15, Rocha and Bollen discuss how biologically motivated ideas can be applied to distributed information systems such as the Internet and library information retrieval systems. They seek **adaptability, evolvability,** and **creativity**. Specific goals include the ability to recommend information to users about topics that should interest them but concerning which they have not explicitly inquired; the ability to promote dialogue between users and information resources; and the ability to infer new and different categories of key words for different groups of users.

Chapter 16 (Mitchell) shows the operation of an environmental exploration scheme called the **parallel terraced scan** in a high-level artificial intelligence program named Copycat. "In this scheme, many possibilities are explored in parallel, each being allocated resources according to feedback about its current promise, which is determined dynamically as new information is obtained." Construction of Copycat was motivated by the belief that analogy making is a central process in higher cognitive function. Copycat can solve problems such as: if *ABC* becomes *ABD*, what should *IJK* become? How about *XYZ*? The effectors of Copycat are "codelets" whose ultimate task is to build relationships among the strings of letters that form Copycat's input and, more importantly, among descriptions of relationships that might characterize various possible transformations of these strings. It is instructive to compare the parallel terraced scan with the system that the immune system employs in **resource allocation**. For example, both systems use searches with random elements, controlled by a combination of bottom-up and top-down forces. There is a correspondence between the codelets that explore structures and the lymphocytes that patrol the body "looking for trouble."

Viruses not only invade our bodies but also our computers. Immune principles have been applied to combat computer viruses and other malicious intrusions. In chapter 17 Forrest and Hofmeyr describe a number of projects where the concept of **immunology as information processing** can be helpful in better understanding both immunology itself and possible "artificial immune systems" for computer defense. To give just one example, the chapter describes how theories for the way that the immune system detects "nonself" inspired a distributed system for detecting intrusions into local area networks of computers.

ACKNOWLEDGMENTS

The genesis of this book was a workshop with the same title held at the Santa Fe Institute in July 1999. The book is not merely a chronicle of the meeting, but rather is the fruit of thoughts expressed at the meeting and then refined by later interactions among the participants. Not all the meeting participants contributed chapters, and some of the chapters were written by individuals that were not present at the meeting.

There was crucial early support for the meeting from SFI's Theoretical Immunology Program, under the auspices of Joseph P. and Jeanne M. Sullivan Foundation. The bulk of the funds for the meeting was provided by NIH (grant 1 R25 GM60043-01) and NSF (grant DMS-9876803). The authors are grateful to SFI for encouragement and very considerable assistance, both during the meeting and afterward, and to the various funding agencies for making the workshop possible.

DEDICATION

We dedicate this book to the memories of a mother, a daughter, and a granddaughter.

Minna Margolis Segel (1900–2000)
Michal Cohen (1961–1978)
Avital Hanna Dym (1988–2000)

One lived a long and full life; two also deserved to, but didn't.

Lee A. Segel and Irun R. Cohen
Rehovot, Israel

April, 2000

Part I: An Overview of Immunology

An Interpretative Introduction to the Immune System

Steven A. Hofmeyr

1 INTRODUCTION

This chapter is intended as a gentle introduction to the immune system for researchers who do not have much background in immunology. It is not a comprehensive overview, and is certainly not a substitute for a good immunology textbook. The interested reader should consult Janeway and Travers [4], Paul [13], and Piel [14]. The goal of this chapter is to sketch an outline of how the immune system fits together, so that readers may then go and consult detailed research papers, knowing where to look. For this reason, the emphasis here is on interpretation, not details, with an interpretative bias toward viewing the immune system from the perspective of information processing, that is, in terms of the architecture, algorithms, and principles embodied by the immune system.

The basis of the interpretation is the teleological viewpoint that the immune system has evolved for a particular purpose. Fundamentally, such a viewpoint is wrong, but it is useful for expository purposes: it is easier to understand the immune system to a first approximation if the components and mechanisms are viewed with the assumption that they exist to solve a particular problem. It is thus assumed that the "purpose" of the immune system is to protect the body from threats posed by toxic substances and pathogens, and to do so in a way that minimizes harm to the body and ensures its continued functioning.[1] The term **pathogen** embraces a plethora of inimical microorganisms, such as bacteria, parasites, viruses, and fungi, that constantly assault the body. These pathogens are the source of many diseases and ailments; for example, pneumonia is caused by bacteria, AIDS and influenza are caused by viruses, and malaria is caused by parasites. Replicating pathogens can lead to a rapid demise of the host if left unchecked.

There are two aspects to the problem that the immune system faces: the identification or *detection* of pathogens, and the efficient *elimination* of those pathogens while minimizing harm to the body, from both pathogens and the immune system itself. The detection problem is often described as that of distinguishing "self" from "nonself" (which are elements of the body, and pathogens/toxins, respectively). However, many foreign microorganisms are not harmful, and an immune response to eliminate them may damage the body. In these cases it would be healthier not to respond, so it would be more accurate to say that the problem faced by the immune system is that of distinguishing between *harmful* nonself and everything else [7, 8].[2] Once pathogens have been detected, the immune system must eliminate them in some manner. Different pathogens have to be eliminated in different ways, and the components of the immune system that accomplish this are called *effectors*. The elimination problem facing the immune system is that of choosing the right effectors for the particular kind of pathogen to be eliminated.

This chapter is structured as a narrative, introducing details as they become relevant to the story, to avoid overwhelming the reader with terms and technicalities. Whenever new terms are introduced, they appear in boldface, and are explained soon afterward. In addition a glossary of these terms appears in the appendix. The next section (2) gives a high-level description of the main components of the immune system architecture, and in the sections that follow these components are described in more detail. The problem of the detection of specific pathogens is discussed in section 3. Section 4 explains how the immune system produces sufficient cellular diversity to provide protection against a wide variety of pathogens. Section 5 describes how the immune system adapts to specific kinds of pathogens, and section 6 discusses how the immune system "remembers" pathogenic structures to facilitate rapid secondary responses. Section 7 describes how the immune system remains tolerant of the body, i.e., why the immune system does not attack the body. In section 8, the problem of detecting and eliminating pathogens hidden within cells is discussed. Some consequences of this are described in section 9, and in section 10 the problem of effector selection to eliminate pathogens is discussed.

[1]This is a limited view of "purpose"; in general, it may be better to adopt the viewpoint that the purpose of the immune system is to maintain homeostasis, which includes protecting the body from pathogens and toxins that could disrupt that homeostasis.

[2]However, the fact that the immune system does react to "harmless" microorganisms is essential to immunization (see section 5).

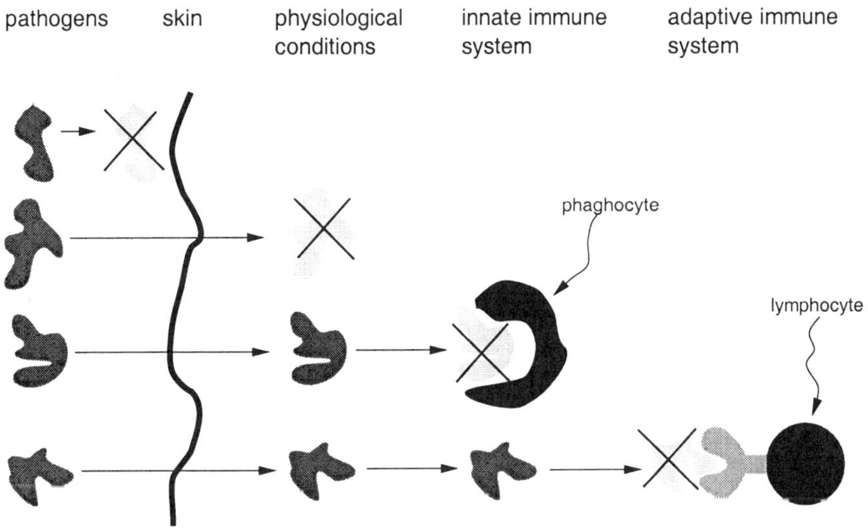

FIGURE 1 Immune system defenses are multilayered. The blobs on the left represent pathogens that could infect the body. The first layer of protection is the skin, which blocks most pathogens. Elimination is indicated by a cross through a faded representation of the pathogen. The second layer of defense is physiological, where conditions such as temperature make the bodily environment more hostile for pathogens. The third layer is the innate immune system, consisting of roaming scavenger cells such as phagocytes which engulf pathogens and debris. The final layer is the adaptive immune system, which consists of cells called lymphocytes that adapt to the structure of pathogens to eliminate them efficiently.

2 THE ARCHITECTURE OF THE IMMUNE SYSTEM

The architecture of the immune system is multilayered, with defenses on several levels (see fig. 1). Most elementary is the skin, which is the first barrier to infection. Another barrier is physiological, where conditions such as pH and temperature provide inappropriate living conditions for foreign organisms. Once pathogens have entered the body, they encountered the **innate** immune system and the acquired or **adaptive** immune system. Both systems consist of a multitude of cells and molecules that interact in a complex manner to detect and eliminate pathogens. Both detection and elimination depend upon chemical bonding: surfaces of immune system cells are covered with various receptors, some of which chemically bind to pathogens and some of which bind to other immune system cells or molecules to enable the complex system of signaling that mediates the immune response.

2.1 THE INNATE IMMUNE SYSTEM

The term "innate" refers to that part of the immune system with which we are born; that is, it does not change or adapt to specific pathogens (unlike the adaptive immune system). The innate immune system provides a rapid first line of defense—to keep early infection in check—giving the adaptive immune system time to build up a more specific response. Innate immunity consists primarily of a chemical response system called **complement**, and the **phagocytic** system, which involves roaming "scavenger" cells, such as **macrophages**. These cells detect and engulf extracellular molecules and materials, clearing the system of both debris and pathogens.[3]

2.1.1 The Complement System. The complement system provides the earliest innate response. When complement molecules that exist in the plasma bind to certain kinds of bacteria, they help eliminate the bacteria through **lysis** or **opsonization**. Lysis is the process whereby complement ruptures the bacterial membrane, which results in destruction of the bacterium. Opsonization refers to the coating of bacteria with complement (or antibodies; see section 5), enabling the bacteria to be detected by macrophages. Self cells have regulatory proteins on their surfaces that prevent complement from binding to them, and so are protected against the effects of complement. The activation of complement and macrophages in the early stages of infection typically happens in the first few hours.

2.1.2 Macrophages. Macrophages are "scavenger" cells found in tissues throughout the body. They play a crucial role in all stages of immune response. In the early stages they have several different functions. For example, they have receptors for certain kinds of bacteria, and for complement; thus, they engulf those bacteria and bacteria opsonized by complement. Additionally, macrophages that are activated by binding secrete molecules called **cytokines**. The release of cytokines activates the next phase of host defense, termed the early induced response.

2.1.3 Cytokines and Natural Killer Cells. Cytokines are molecules that function as a variety of important signals. Cytokines are not only produced by macrophages and other immune system cells, but also by cells which are not a part of the immune system, for example, cells that secrete cytokines when damaged. A major effect of the cytokines is to induce an **inflammatory response**, which is characterized by an increase in local blood flow and permeability between blood and tissues. These changes allow large numbers of circulating immune system cells to be recruited to the site of infection. Another effect of cytokines is inducing the increase in body temperature associated with fever. Fever is thought to be beneficial because the activity of

---

[3] As an aid to remembering these terms, it is useful to recall their Greek origins: endo means "inside," phago means "eat," cyto means "cell." So, for example, endocytic means "inside cells."

pathogens is reduced with an increase in temperature, whereas elevated temperatures increase the intensity of the adaptive immune response. Yet another effect of cytokines is to reinforce the immune response by triggering the liver to produce substances known as **acute phase proteins** (ATP), which bind to bacteria, thus activating macrophages or complement.

Certain cells produce proteins, called **interferons**, when infected by viruses. Interferons are so-called because they inhibit viral replication, but they also have many other functions. For example, they also activate **natural killer** (NK) cells to kill virus-infected host cells. NK-cells bind to carbohydrates on normal host cells, but are normally not activated because healthy cells express molecules that act as inhibitory signals. Some virally infected cells cannot express these inhibitory signals and are killed by activated NK-cells. Activated NK-cells release chemicals that trigger **apoptosis** in the infected cell. Apoptosis is programmed cell death, a normal cellular response that is essential in many bodily functions other than immunity.

2.2 THE ADAPTIVE IMMUNE SYSTEM

The adaptive immune system is so-called because it adapts or "learns" to recognize *specific* kinds of pathogens, and retains a "memory" of them for speeding up future responses. The learning occurs during a **primary response** to a kind of pathogen not encountered before by the immune system. The primary response is slow, often first only becoming apparent several days after the initial infection, and taking up to three weeks to clear an infection. After the primary response clears an infection, the immune system retains a memory of the kind of pathogen that caused the infection. Should the body be infected again by the same kind of pathogen, the immune system does not have to relearn to recognize the pathogens, because it "remembers" their specific appearance, and will mount a much more rapid and efficient **secondary response**. The secondary response is often quick enough so that there are no clinical indications of a reinfection. Immune memory can confer protection up to the lifetime of the organism (measles is a good example).

The adaptive immune system primarily consists of certain types of white blood cells, called **lymphocytes**, which circulate around the body via the blood and lymph systems.[4] Lymphocytes cooperate in the detection of pathogens, and assist in pathogen elimination. However, we can abstractly view lymphocytes as mobile, independent *detectors*. There are trillions of these lymphocytes, forming a system of distributed detection, where there is no centralized control, and little, if any, hierarchical control. Detection and elimination of pathogens is a consequence of trillions of cells—detectors—interacting through simple, localized rules.

[4]Lymphocytes are so called because they exist in the lymph; an alternative term is **leukocyte**, where leuko means "white," a reference to the fact that leukocytes are "white blood cells."

The remainder of this chapter concentrates on the adaptive immune system because it appears to have the most complex and interesting architecture. Where necessary, the interaction between the innate and the adaptive immune systems is mentioned, because they are closely linked in the detection and elimination of pathogens.

3 SPECIFIC RECOGNITION IN THE IMMUNE SYSTEM

A detection or recognition event occurs in the immune system when chemical bonds are established between **receptors** on the surface of an immune cell and **epitopes**, which are locations on the surface of a pathogen or protein fragment (a **peptide**). Both receptors and epitopes have complicated three-dimensional structures that are electrically charged. The more complementary the structure and charge of the receptor and the epitope, the more likely it is that binding will occur. See figure 2.

The strength of the bond between a receptor and an epitope is termed the **affinity**. Receptors are deemed *specific* because they bind tightly only to a few similar epitope structures or patterns. This specificity extends to the lymphocytes themselves: receptor structures may differ between lymphocytes, but on a single lymphocyte, all receptors are identical, making a lymphocyte specific to a particular set of similar epitope structures (this feature is termed **monospecificity**). Pathogens have many different epitopes, reflecting their molecular structures, so many different lymphocytes may be specific to a single kind of pathogen.

A lymphocyte has on the order of 10^5 receptors on its surface, all of which can bind epitopes. Having multiple identical receptors has several beneficial effects. First, it allows the lymphocyte to "estimate" the affinities of its receptors for a given kind of epitope, through frequency-based sampling: as the affinities increase, so the number of receptors binding will increase. The number of receptors that bind can be viewed as an estimate of the affinity between a single receptor and an epitope structure.[5] Second, having multiple receptors allows the lymphocyte to estimate the number of epitopes (and thus infer the number of pathogens) in its immediate neighborhood: the more receptors bound, the more pathogens in the neighborhood. Finally, monospecificity is essential to the immune response, because if lymphocytes were not monospecific, reaction to one kind of pathogen would induce a response to other, unrelated epitopes.

The behavior of lymphocytes is strongly influenced by affinities: a lymphocyte will only be *activated* (this can be termed a "detection event") when

[5]Of course, this is an idealization, because the receptors may bind different epitope structures, so what is being "estimated" is a rather arbitrary mean of the affinities between the receptors and different epitope structures. However, as the affinities increase, the estimation becomes more accurate because the epitope structures must be increasingly similar.

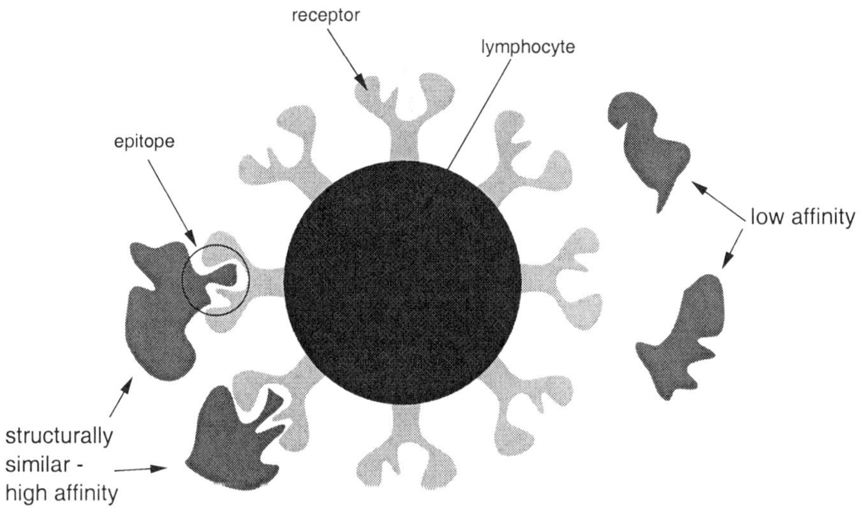

FIGURE 2 Detection is a consequence of binding between complementary chemical structures. The surface of a lymphocyte is covered with receptors. The pathogens on the left have epitope structures that are complementary to the receptor structures and so the receptors have higher affinities for those epitopes than for the epitopes of the pathogens on the right, which are not complementary.

the number of receptors bound exceeds some threshold.[6] Thus, a lymphocyte will only be activated by pathogens if its receptors have sufficiently high affinities for particular epitope structures on the pathogens, and if the pathogens exist in sufficient numbers in the locality of the lymphocyte. Such activation thresholds allow lymphocytes to function as *generalized* detectors: a single lymphocyte detects (is activated by) structurally similar kinds of epitopes. If we consider the space of all epitope structures as a set of patterns, then a lymphocyte detects or "covers" a small subset of these patterns. Hence, there does not have to be a different lymphocyte for every epitope pattern to cover the space of all possible epitope patterns. There is evidence to suggest that certain kinds of lymphocytes (memory cells) have lower activation thresholds than other lymphocytes, and so need to bind fewer receptors to become activated (more of this in section 6).

[6]Lymphocytes require additional signals to be activated. This is termed costimulation, and is discussed in section 7.

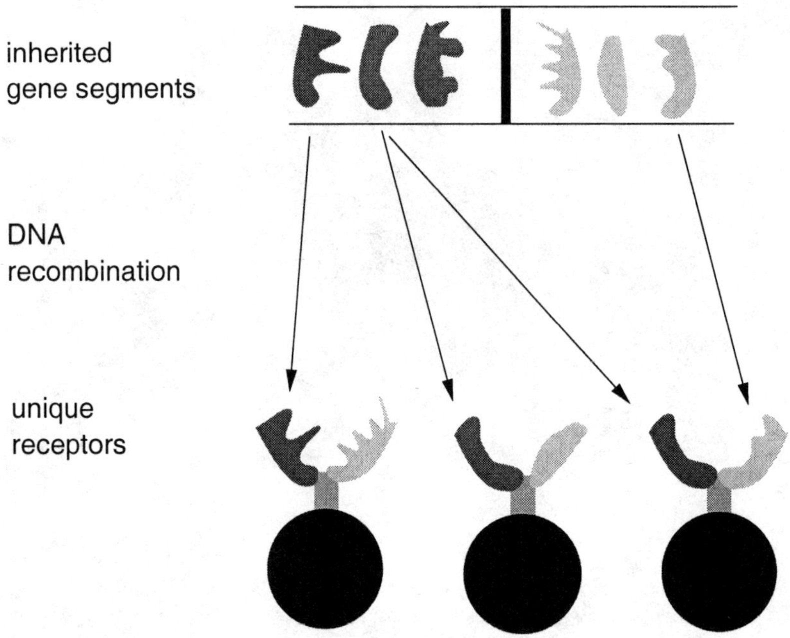

FIGURE 3 Generating receptor diversity. "Random" recombination of gene segments generates a combinatorial number of receptor varieties.

4 GENERATING RECEPTOR DIVERSITY

Because detection is carried out by binding with nonself,[7] the immune system must have a sufficiently diverse **repertoire** of lymphocyte receptors to ensure that at least some lymphocytes bind to any given pathogen. Generating a sufficiently diverse repertoire is a problem, because the human body does not manufacture as many varieties of proteins as there are possible pathogen varieties of epitopes. Inman [3] has estimated that the immune system has available about 10^6 different proteins, and that there are potentially 10^{16} different foreign proteins or patterns to be recognized. One of the main mechanisms for producing the required diversity is a pseudo-random process, in which recombination of DNA results in different lymphocyte genes, and hence different receptors.[8]

Tonegawa [19] has estimated that there are at most 10^8 different varieties of receptors. If we assume that there are 10^{16} different epitope varieties,

[7]Recall from section 1 that the term nonself encompasses all pathogens, toxins, etc., which are foreign to the body.

[8]This is a simplification. For research concerning receptor diversity, see Oprea and Forrest [11].

then there will be insufficient repertoire diversity to bind every single possible pathogen. This problem is exacerbated because replicating pathogens are likely to evolve to evade detection from the existing repertoire. The immune system appears to address this problem by dynamic protection. There is a continual turnover of lymphocytes: each day approximately 10^7 new lymphocytes are generated [12]. Assuming that there are at any given time 10^8 different lymphocytes, and these are turned over at a rate of 10^7 per day, it will take ten days to generate a completely new lymphocyte repertoire. Over time, this turnover of lymphocytes (together with immune memory; see section 6) increases the protection offered by the immune system.

5 ADAPTATION

The immune system needs to be able to detect and eliminate pathogens as quickly as possible, because pathogens can replicate exponentially. The more specific a lymphocyte is to a particular variety of pathogen epitope, presumably the more efficient it will be at detecting and eliminating that kind of pathogen. Thus, the immune system incorporates mechanisms that enable lymphocytes to "learn" or adapt to specific kinds of epitopes, and to "remember" these adaptations for speeding up future responses. Both of these principles are implemented by a class of lymphocytes called **B cells**.[9]

When a B cell is activated, it migrates to a **lymph node**. The lymph nodes are glands in which the adaptive response develops. There are hundreds of lymph nodes distributed throughout the body. In the lymph node, the B cell produces many short-lived (on the order of a few days) clones through cell division. B-cell cloning is subject to a form of mutation termed **somatic hypermutation** (because the mutation rates are nine orders of magnitude higher than ordinary cell mutation rates). These high mutation rates increase the chance that the clones will have different receptor structures from the parent, and hence different epitope affinities. The new B-cell clones have the opportunity to bind to pathogenic epitopes captured within the lymph nodes. If they do not bind, they will die after a short time. If they succeed in binding, they will leave the lymph node and differentiate into **plasma** or **memory** B cells (see fig. 4). Plasma B cells secrete a soluble form of their receptors, called **antibodies**, which play a key role in immunological defense.[10] Antibodies that bind to pathogen epitopes have two beneficial effects: first, they opsonize pathogens, and secondly, they **neutralize** pathogens; i.e., antibodies

[9]The "B" in B cell refers to the fact that B cells mature only in the bone marrow.

[10]Immune responses are often measured in terms of antibody production. Anything that causes the production of antibodies is known as **antigen**, a term compounded from **anti**body-**gen**erating, but the term antigen has come to mean anything that evokes an immune response. The B-cell response is called the **humoral response** because the antibodies form a fluid or "humor." This is in contrast to the alternative **cellular response** mediated by T cells; see section 8.

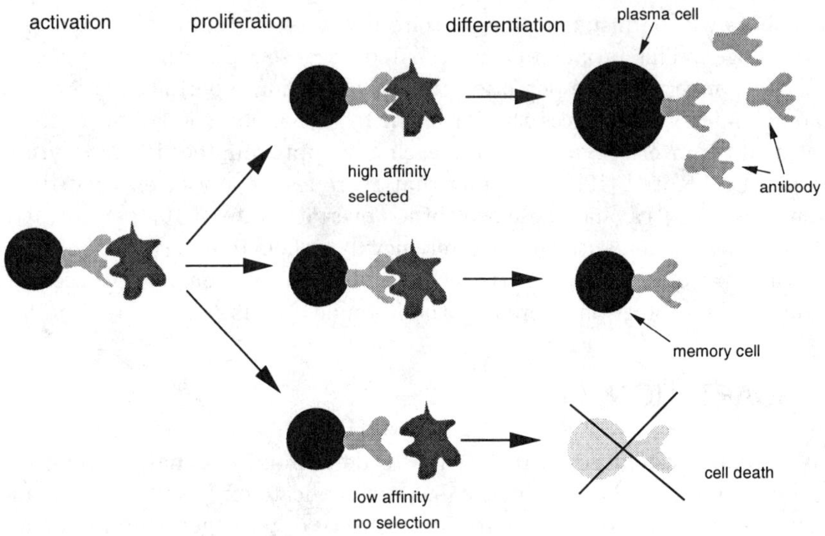

FIGURE 4 Affinity Maturation. Activated B-cells proliferate, producing mutated clones, which are subject to selection via epitope affinities. On the left is an activated B cell. It proliferates, producing clones with mutated receptors. Clones with the highest affinity for the pathogenic epitopes survive and differentiate to become plasma or memory cells.

block binding between pathogens and self cells. The role of memory cells is described in section 6.

This cycle of activation-proliferation-differentiation is repeated, resulting in increasing selection of high-affinity B cells, because the higher the affinity of the clones for the presented epitopes, the more likely it is that those clones will survive. This process, called **affinity maturation**, is essentially a Darwinian process of variation and selection: clones compete for available pathogens, with the highest affinity clones being the "fittest," and hence replicating the most (see fig. 5). This primary response may take several weeks to eliminate the pathogens.

6 IMMUNOLOGICAL MEMORY

A successful immune response results in the proliferation of memory B cells that have higher than average affinities for the pathogen epitopes that caused the response. Retention of the information encoded in these memory B cells constitutes the "memory" of the immune system: if the same pathogens are encountered in the future, the preadapted subpopulation of B cells can provide

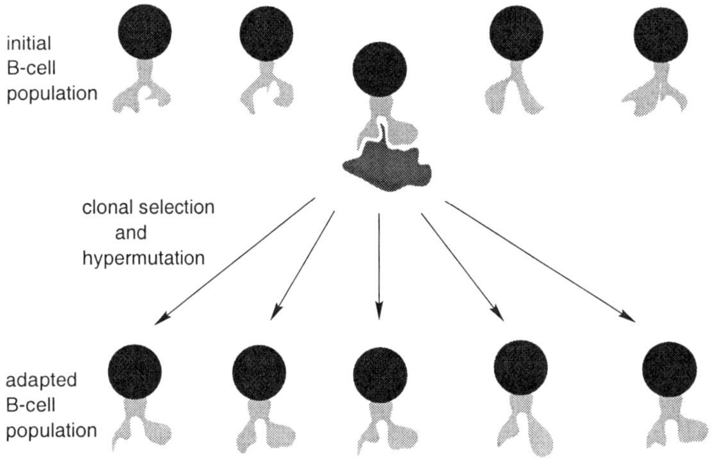

FIGURE 5 Affinity maturation is a Darwinian process of variation and selection. The variation is provided by somatic hypermutation, and the selection is provided by competition for pathogen epitopes.

a secondary response that is more rapid than the original primary response (see fig. 6).

Our understanding of immune memory is problematic because B cells typically live only a few days, and once an infection is eliminated, we do not know what stops the adapted subpopulation of B cells from dying out. There are currently two dominant theories. According to one of the theories, the adapted memory cells are long-lived, surviving for up to the lifetime of the organism [5]. The other theory postulates that the adapted B cells are constantly restimulated by traces of nonself proteins that are retained in the body for years [1].

A secondary response (via memory cells) is not only triggered by reintroduction of the same pathogens, but also by infection with new pathogens that are similar to previously seen pathogens; in computer science terms, immune memory is *associative* [17]. This feature underlies the concept of **immunisation**, where exposure to benign forms of a pathogen engenders a primary response and consequent memory of the pathogen enables the immune system to mount a more rapid secondary response to similar but virulent forms of the same pathogen (see fig. 7).

7 TOLERANCE OF SELF

The picture described thus far has a fatal flaw: receptors that are randomly generated and subject to random changes from somatic hypermutation could

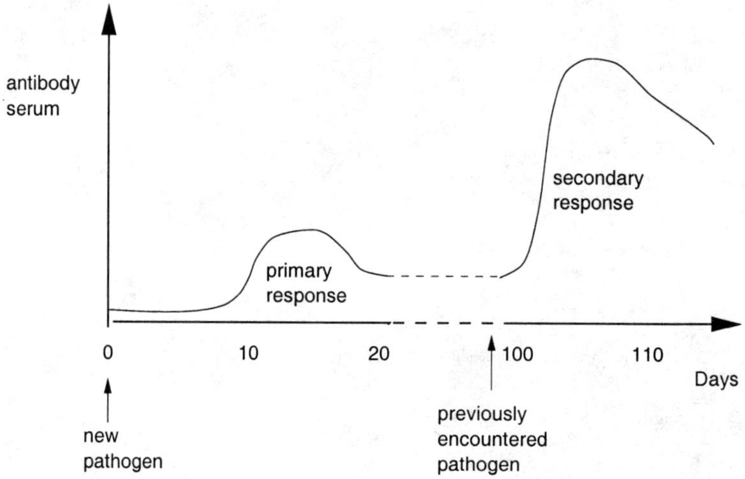

FIGURE 6 Responses in immune memory. Primary responses to new pathogen epitopes take on order of weeks; memory of previously seen pathogens epitopes allows the immune system to mount much faster secondary responses (on the order of days). The y-axis (**antibody**) is a measure of the strength of the immune system response.

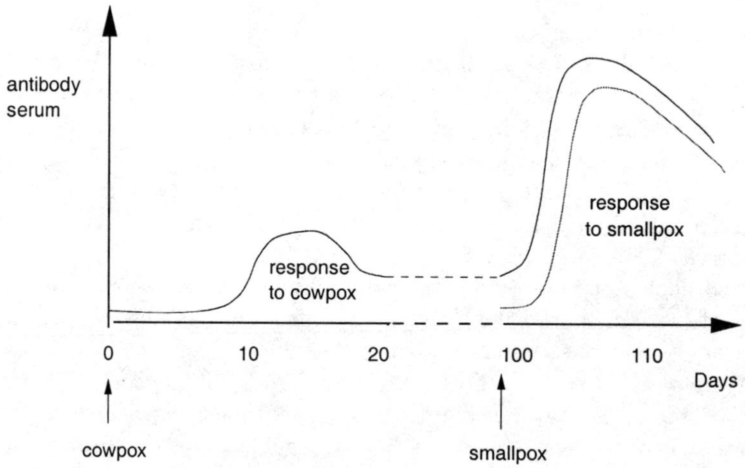

FIGURE 7 Associative memory underlies the concept of immunization. At time zero, the cowpox pathogen is introduced. Although harmless, it is recognized as foreign, so the immune system mounts a primary response to it, clears the infection, and retains a memory of the cowpox. Smallpox is so similar to cowpox, that the memory population generated by the cowpox reacts to the smallpox, eliminating the smallpox in a more efficient secondary response.

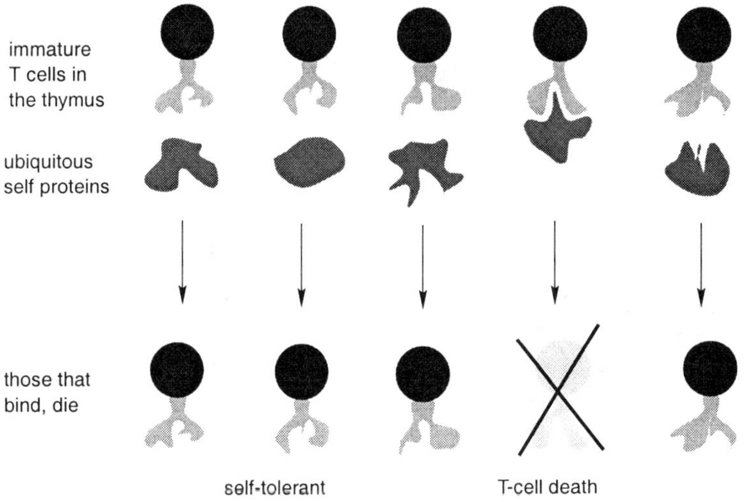

FIGURE 8 T cells undergo central tolerization via clonal deletion in the thymus.

bind to self and initiate **autoimmunity**. Autoimmunity occurs when the immune system attacks the body. Autoimmunity is rare[11]; generally the immune system is **tolerant** of self; that is, it does not attack self.

Tolerance is among the responsibilities of another class of lymphocytes, the **T-helper** cells (Th cells), so called because they mature in the thymus, and "help" the B cells. Most self epitopes are expressed in the **thymus** (an organ located behind the breast bone) so during maturation Th cells are exposed to most self epitopes. If an immature Th cell is activated by binding self, it will be censored (i.e., it dies by programmed cell death) in a process called **clonal deletion** or **negative selection** (see fig. 8). Th cells that survive the maturation process and leave the thymus will be tolerant of most self epitopes. This is called **central tolerance**, because the immature Th cells are tolerized in a single location (the thymus).

B cells are tolerized in the bone marrow, but this is not sufficient to prevent the development of **autoreactive** B cells (those that bind to self epitopes). During affinity maturation, B cells hypermutate, which can result in previously tolerant B cells producing autoreactive clones. Because affinity maturation occurs in many distributed locations (the lymph nodes), a form of peripheral (or distributed) tolerization is required. Th cells provide this through a mechanism known as **costimulation**. To be activated, a B cell must receive costimulation in the form of two disparate signals: **signal I** occurs when the number of pathogens binding to receptors exceeds the affinity

[11]At the most, five percent of adults in Europe and North America suffer from autoimmune disease [18].

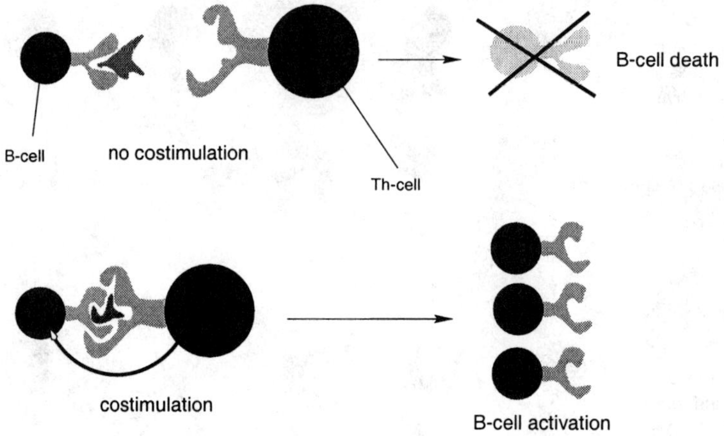

FIGURE 9 Costimulation from Th cells implements distributed or peripheral tolerization for B cells.

threshold (as described in section 3), and **signal II** is provided by Th cells. If a B cell receives signal I in the absence of signal II, it dies.

To provide signal II to a B cell, a Th cell must "verify" the epitopes detected by the B cell. The way in which it performs this verification is complex. In a process known as **antigen processing**, B cells engulf pathogenic peptides and present these peptides on the surface of the B cell, using molecules of the **major histocompatibility complex** (MHC). These MHC molecules show the Th cells what is inside the B cell; that is, what the B cell has detected. **T-cell receptors** differ from B-cell receptors in that they bind to MHC/peptide complexes. If a Th cell binds to an MHC/peptide complex presented on the surface of a B cell, it will provide signal II to that B cell, and the B cell will be activated. Because Th cells undergo central tolerization in the thymus, most mature Th cells are self-tolerant, and so will not costimulate B cells that recognize self. The Th cell "verifies" that the detection carried out by the B cell is correct, and not autoreactive (see fig. 9).

Unfortunately, the picture is not as simple as this. Some peripheral self proteins are never presented in the thymus (the exact fraction is unknown), and so Th cells emerging from the thymus may still be autoreactive. Self-tolerance in Th cells is also assured through costimulation: once again, signal I is provided by exceeding the affinity threshold, but signal II is provided by cells of the innate immune system. These innate system cells are thought to give out signal II in the presence of tissue damage.[12]

[12]This is a simplification. For a more detailed exposition of possible tolerization and costimulation mechanisms, see Marrack et al. [6] and Matzinger [7, 8].

An autoreactive T cell could survive in regions of tissue damage, but as soon as it leaves the region of tissue damage, it will receive signal I in the absence of signal II and die. This can be termed *frequency-based* tolerization because an autoreactive Th cell should encounter self in the absence of tissue damage with higher frequency than self in the presence of tissue damage, assuming healthy self is generally more frequent than nonself. If these frequencies change, then frequency tolerization will lead to a loss of immune function, which has been observed when overwhelming initial viral doses result in nonself becoming frequent [10]. The utility of frequency tolerization is emphasized by the fact that the loss of the thymus does not result in devastating autoimmunity, which is what we would expect if Th cells were only tolerized centrally.

The specialization of the different lymphocytes gives the immune system the ability to provide a faster adaptive response that is not self-reactive. Th cells have the "responsibility" for self-tolerance, thus freeing B cells to hypermutate and adapt to a specific pathogen. Although Th cells must be able to recognize the peptides presented by the B cells, both classes of lymphocytes are necessary. Th cells are general, nonspecific detectors, and so are not efficient at detecting specific pathogens. B cells, by contrast, adapt to become more specific, and thus more effective at detecting particular pathogens. It has been estimated that B cells detect specific pathogens 10 to 10,000 times more efficiently than Th cells [7].

8 DETECTION OF INTRACELLULAR PATHOGENS

The immune system is vastly more complex than portrayed so far. Another important facet is the problem of *intracellular* pathogens. Intracellular pathogens are organisms such as viruses and certain bacteria which live inside host cells. Such pathogens are not "visible" to B cells; all that the B cell potentially binds to is the outside of the host cell, which has only self epitopes. What the immune system needs is some way to "look inside" host cells to see if they are infected.

The MHC molecules described in section 7 provide the solution. Almost all cells in the body have MHC molecules, which function as transporters, carrying fragments of proteins (peptides) from within the cell to the cell surface, where they are presented to the immune system in the form of **MHC/peptide complexes**. If a cell is infected with a virus, MHC carries viral peptides to the surface and presents them to the immune system.

MHC molecules are divided into two **classes**: class I MHC and class II MHC. The immune system differs in its response to peptides presented with class I MHC and class II MHC. Class II MHC occurs only in cells of the immune system, such as macrophages and B cells. As discussed before, Th cells bind to class II MHC/peptide complexes, and when activated they stimulate an immune response in the presenting cell, for example, macrophages are

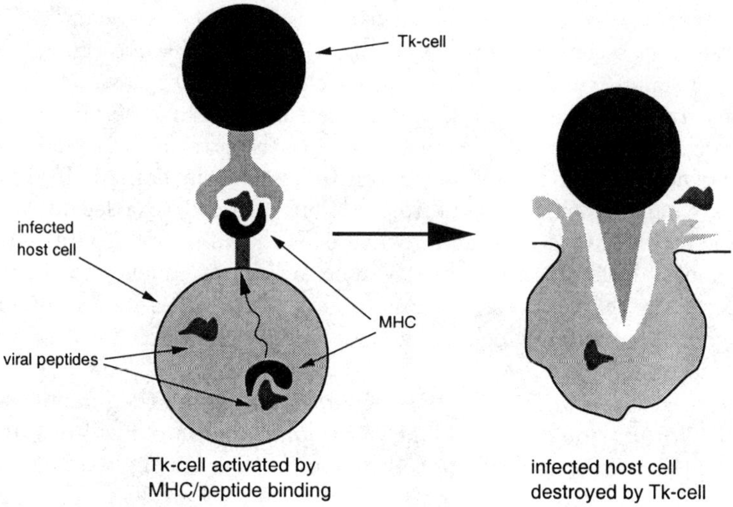

FIGURE 10 Tk cells eliminate intracellular pathogens. Tk cells activated by binding to a MHC/peptide complexes on a self cell and costimulated by the innate immune system will destroy the self cell that presented the MHC/peptide complex.

stimulated to destroy whatever is in their vesicles, and B cells are stimulated to proliferate and differentiate.

Class I MHC/peptide complexes are recognized by another class of lymphocytes, called **cytotoxic** or **killer** T cells (Tk cells). Both Tk cells and Th cells are types of T cells. All T cells mature in the thymus, are tolerized via clonal deletion, and do not hypermutate when cloning. Tk cells, like all T cells, are only activated by binding to MHC/peptide complexes with costimulation from the innate immune system. If a Tk cell is activated, it will kill the infected host cell (hence the name). It does this through apoptosis, triggering the host cell into programmed death; or physically, by punching holes in the cell wall; or chemically, by the secretion of toxic chemicals (see fig. 10).

9 MAJOR HISTOCOMPATIBILITY COMPLEXT AND DIVERSITY

It is essential for host defense that MHC forms MHC/peptide complexes with as many foreign peptides as possible, so that those foreign peptides are recognized by Th cells. Because of the nature of molecular bonding, a single type of MHC can form complexes with multiple, but not all pathogenic peptides. Hence, there is selective pressure on pathogens to evolve so that their characteristic peptides cannot be bound by MHC, because then they will be effectively hidden from the immune system. Therefore, it appears to be essential that the body have as many varieties of MHC as possible. However, as the diversity of MHC types increases, there is a resulting increase in the chance that immature Th cells will bind to complexes of MHC and self, which means that more Th cells will be eliminated during negative selection. Eliminated Th cells are a waste of resources, so evolution should favor lower rates of Th-cell elimination during negative selection. Hence the number of MHC types is constrained from below by the requirement for diversity to detect pathogens, and from above by resource limitations imposed by negative selection. Mathematical models of this tradeoff indicate that the number of MHC types present in the human body (about four to eight) is close to optimal [9].

MHC types do not change over the life of an organism and are determined by genes that are the most polymorphic in the body. Hence, MHC is representative of genetic immunological diversity within a population. This diversity is crucial in improving the robustness of a population to a particular type of pathogen. For example, there are some viruses, such as the Epstein-Barr virus, which have evolved dominant peptides that cannot be bound by particular MHC types, leaving individuals who have those MHC types vulnerable to the disease [4]. The genetic diversity conferred by MHC is so important that it has been proposed that the main reason for the continuance of sexual reproduction is to confer maximally varied MHC types upon offspring [2]. There are some studies with mice that support this theory. These studies indicate that mice use smell to choose mates whose MHC differs the most from theirs [15].

10 EFFECTOR SELECTION AND ITS ROLE IN PATHOGEN ELIMINATION

The immune system has a variety of *effector functions* because different pathogens must be eliminated in different ways. For example, intracellular pathogens such as viruses are eliminated via Tk cells, whereas extracellular bacteria are eliminated by macrophages or complement, etc. After pathogens have been detected, the immune system must select the appropriate effectors so that the pathogens are efficiently eliminated. Selection of effectors is determined by chemical signals in the form of cytokines, but it is not clear how selection actually works. Mathematical models indicate ways in which selection could occur, if cytokines reflect the local state of the system (i.e., the damage suffered from pathogens, the damage suffered from the immune system, etc.) [16].

Both B and T cells play a role in effector selection. After proliferation, Th cells differentiate into two kinds: **Th1** and **Th2** cells. Only Th1 cells can activate B cells. Th2 cells (which are known as **inflammatory** T cells, on the other hand, do not interact with B cells, but instead are responsible for the

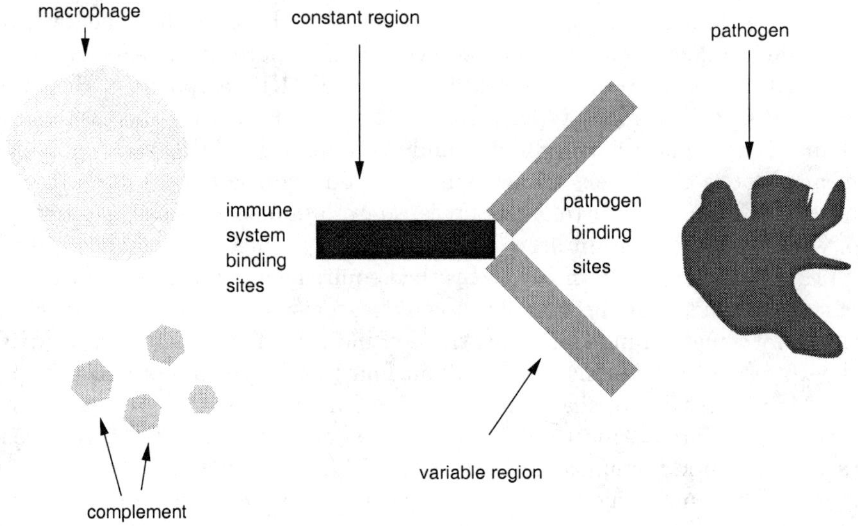

FIGURE 11 Variable and constant antibody regions. An antibody is y-shaped, with the arms of the y being the variable regions and the tail being the constant region. The variable regions provide specific binding of pathogen epitopes; the constant region defines the isotype that binds to other immune system components, such as macrophages and complement.

activation of macrophages. When macrophages are infected with bacteria in their vesicles, they must be stimulated to destroy those bacteria; this is the role of the Th2 cells. To a first approximation, Th1 cells are implicated in the response against extracellular pathogens, whereas Th2 cells are implicated in the response against intracellular pathogens.

If Th cells differentiate into the incorrect effectors for the pathogen threat, the consequences can be disastrous. This is clearly illustrated in the case of leprosy, a disease caused by the leprosy bacterium, which inhabits macrophage vesicles. In most cases of leprosy, Th cells differentiate into Th2 cells, and stimulate the macrophages to destroy the bacteria, but in some cases, for reasons not understood, the Th cells differentiate into Th1 cells, having little effect on bacteria which are isolated from the effects of B cells. The consequence of this misguided response is that the bacteria proliferate in the macrophages, resulting in gross tissue damage which eventually leads to death.

B cells play a role in effector selection via the antibodies they secrete. Antibodies have a y-shaped structure (see fig. 11), with three different regions. The arms of the y are termed the **variable** regions, and the tail of the y is the **constant** region. The variable regions are randomly generated (as described in section 4) so that they bind to specific pathogen epitopes. The constant region, on the other hand, is not randomly generated (hence the name), but comes in a few structural varieties, called **isotypes.**[13] The constant region is the part of the antibody to which other immune system cells (such as macrophages) bind. Depending on the isotype of the constant region, different responses will be triggered upon binding, so it is this part of the antibody that determines effector function.

A single B cell can clone multiple B cells, each with a different isotype, even while the receptor variable regions remain the same. This is known as **isotype switching**, and enables the immune system to choose between various effector functions via chemical binding. For example, the isotype determines whether the primary immune response is neutralisation, opsonization, sensitization for killing by NK cells (see section 2.1), or activation of the complement system.

[13] There are many different isotypes, for example, IgA, IgG, and IgM.

APPENDIX: GLOSSARY

Acute phase proteins: proteins produced largely by the liver whose appearance in the circulation marks acuteacute inflammation.

Adaptive immune system: that part of the immune system that uses antigen receptors and adapts or "learns" to recognize specific antigens, and retains a memory of those antigens to speed up future responses.

Affinity: the cumulative strength of noncovalent chemical bonds between a receptor and a ligand.

Affinity maturation: Darwinian process of variation and selection that occurs to B-cell receptors in lymph nodes and spleen, leading to the evolution of B-cell populations better adapted to recognize specific epitopes.

Antibody: a soluble form of B-cell receptors secreted by plasma B cells.

Antigen: originally, any substance that causes the production of antibodies; now, more generally, anything that is recognized by antibodies or by the antigen receptors of lymphocytes

Antigen processing: the process whereby antigen-presenting cells (such as B cells and macrophages) engulf antigens, digest them, and present fragments of their proteins in MHC/peptide complexes.

Apoptosis: programmed cell death.

Autoimmunity: a state in which the immune system recognizes the body's own antigens.

Autoreactive lymphocytes: autoimmune lymphocytes.

B cells: lymphocytes that mature in the bone marrow, and secrete antibody.

Cellular response: a historical term, now used mostly to refer to the part of the immune response that is connected with T cells.

Central tolerance: process of tolerizing immature T cells in the thymus.

Class I MHC: MHC, which occurs in almost all cells, that presents (in an "MHC/peptide complex") peptides generated in the cell's interior.

Class II MHC: MHC, which only occurs in certain cells of the immune system or when particular cells are activated, that presents fragments of proteins that are bound to the cell surface or that have been ingested.

Clonal deletion: process whereby immature T cells that recognized an antigen in the thymus die by apoptosis.

Complement: molecules that participate in the immune response by binding to bound antibodies or certain microbial products.

Constant antibody region: the tail of the antibody, which includes the Fc receptor, that binds to other antibodies or to immune effectors.

Costimulation: the mechanism whereby two or more signals participate in the process of activating a lymphocyte.

Cytokines: signal molecules that transmit information between cells; "hormones" of the immune system.

Cytotoxic T cells: T cells that are activated to kill target cells.

Epitopes: the locations on an antigen to which antigen receptors bind.

Humoral response: a historical term for immune responses involving mainly B cells and their antibodies.

Hypermutation: accelerated mutation; typically occurs during affinity maturation.

Immunization: the process of inducing an immune response.

Inflammation: a complex of processes initiated in response to injury and leading to healing. Among the characteristics of inflammation are an increase in local blood flow and in permeability between blood and tissues, resulting in the recruiting of large numbers of immune cells to the site of the response and the secretion of cytokines.

Innate immune system: that part of the immune system that depends on germline encoded receptors: it does not change or adapt to specific pathogens.

Immunogen: a molecule that can elicit an adaptive immune response.

Interferons: types of cytokines originally discovered because of their anti-viral action.

Isotypes: the several structural varieties of the constant regions of antibodies. Istotypes are called IgM, IgG, etc.

Isotype switching: the ability of a B-cell clone to generate daughter B cells with different isotypes.

Killer T cells: another name for cytotoxic T cells.

Leukocytes: white blood cells, including lymphocytes, neutrophils, eosinphils, basophils, monocytes, and macrophages.

Lymph nodes: small organs, distributed throughout the body, in which an adaptive immune response can develop.

Lymphocytes: T cells or B cells.

Lysis: the rupture of cell membranes.

Macrophages: scavenger cells that engulf pathogens, process antigens and also provide many signals to the immune system.

Memory cells: lymphocytes that have been activated in the past and retain a memory of previous antigens, engendering a secondary response to future contact with the antigen.

Monospecifity: the characteristic of lymphocytes that all receptors on a single lymphocyte are identical.

Natural killer cells: cells of the innate immune system that kill tumor cells and intracellular pathogens.

Negative selection: same as clonal deletion.

Neutralization: the process whereby antibodies binding to pathogens prevent binding between the pathogen and its receptors on host cells.

Opsonization: the coating of cells with complement or antibodies, leading to phagocytosis.

Pathogens: microorganisms such as bacteria, parasites, viruses and fungi that invade the body and cause illness.

Peptides: protein fragments.

Phagocytic system: the part of the immune system involving scavenger cells that detect extracellular molecules and material, clearing the body of both debris and pathogens.

Plasma B cells: activated B cells that secrete antibodies.

Positive selection: the stimulation and maturation in the thymus of T cells with sufficient affinity to antigens presented by self MCH; T cells that do not undergo positive selection in the thymus die.

Primary response:	the immune response to antigens that the immune system has never before encountered.
Receptors:	molecular structures on the surface of cells that bind to complementary molecules (called ligands).
Repertoire:	the diversity of lymphocyte receptors present in the immune system.
Secondary response:	the memory-based immune response to antigens that the body has previously encountered.
Signal I:	a necessary signal for activating a lymphocyte; typically provided by the binding of antigen receptors to an antigen epitope.
Signal II:	a second signal required for activating a lymphocyte; typically provided by other cells and molecules of the immune system. (There may be more than one "signal II".)
Tolerant lymphocytes:	immune cells that do not respond to an immunogen.
T-helper cell	a type of lymphocyte that matures in the thymus and "helps" other lymphocytes by providing a second signal for costimulation.
Thymus:	an organ located behind the breast-bone where T cells mature.
Th1 cells:	T-helper cells that secrete a characteristic set of cytokines (such as interleukin-2 and interferon gamma) and tend to encourage a cellular response.
Th2 cells:	T-helper cells that secrete a characteristic set of cytokines (such as interleukin-4 and interleukin-10) and tend to encourage a humoral response.
Variable antibody regions:	the arms of an antibody that are somatically generated and have evolved to bind to specific antigen epitopes.

REFERENCES

[1] Gray, D. *Sem. Immunol.* **4** (1992): 29–34.

[2] Hamilton, W. D., R. Axelrod, and R. Tanese. "Sexual Reproduction as an Adaptation to Resist Parasites." *Proc. Natl. Acad. Sci. USA* **87** (1990): 3566–3573.

[3] Inman, J. K. "The Antibody Combining Region: Speculations on the Hypothesis of General Multispecificity." *Theor. Immunol.* (1978).

[4] Janeway, C. A., and P. Travers. *Immunobiology: The Immune System in Health and Disease*, 3d ed. London: Current Biology Ltd., 1996.

[5] MacKay, C. R. "Immunological Memory." *Adv. Immunol.* **53** (1993): 217–265.

[6] Marrack, P., and J. W. Kappler. "How the Immune System Recognizes the Body." *Sci. Am.* **269(3)** (1993): 48–54.

[7] Matzinger, P. "Tolerance, Danger, and the Extended Family." *Ann. Rev. Immunol.* **12** (1994): 991–1045.

[8] Matzinger, P. "An Innate Sense of Danger." *Sem. Immunol.* **10** (1998): 399–415.

[9] Mitchison, A. "Will We Survive?" *Sci. Am.* **269(3)** (1993): 102–108.

[10] Moskophidis, D., F. Lechner, H. Pircher, and R. M. Zinkernagel. "Virus Persistence in Acutely Infected Immunocompetent Mice by Exhuastion of Antiviral Cytotoxic Effector T Cells." *Nature* **362** (1993): 758–761.

[11] Oprea, M., and S. Forrest. "How the Immune System Generates Diversity: Pathogen Space Coverage with Random and Evolved Antibody Libraries." In *GECCO 99, Real-World Applications Track*. San Francisco, CA: Morgan Kaufmann, 1999.

[12] Osmond, D. G. "The Turn-Over of B-Cell Populations." *Immunol. Today* **14(1)** (1993): 34–37.

[13] Paul, W. E. *Fundamental Immunology*, 2d ed. New York: Raven Press Ltd., 1989.

[14] Piel, J. "Life, Death, and the Immune System." Special Issue. *Sci. Am.* **269(3)** (1993): 20–102.

[15] Potts, W. K., C. J. Manning, and E. K. Wakeland. "Mating Patterns in Semi-natural Populations of Mice Influenced by MHC Genotype." *Nature* **352** (1991): 619–621.

[16] Segel, L. A. "The Immune System as a Prototype of Autonomous Decentralized Systems." In *Proc. IEEE Conf. Sys., Man & Cyber.* Orlando, FL: IEEE Computer Society Press, 1997.

[17] Smith, D., S. Forrest, and A. S. Perelson. "Immunological Memory is Associative." In *Workshop Notes, Workshop 4: Immunity Based Systems, Intnl. Conf. on Multiagent Systems*, 62–70. Menlo Park, CA: AAAI Press, 1996.

[18] Steinman, L. "Autoimmune Disease." *Sci. Am.* **269(3)** (1993): 75–83.

[19] Tonegawa, S. "Somatic Generation of Antibody Diversity." *Nature* **302** (1983): 575–581.

Part II: Case Studies in Immune Complexity: Experiments

Cytokines: A Common Signaling System for Cell Growth, Inflammation, Immunity, and Differentiation

Thomas N. Denny

1 INTRODUCTION

Cytokines represent a large group of polypeptides having a diverse motif of activity. The cytokine field of research has evolved from the 1950s study of interferons, the 1960s study of lymphokines, and the more recent study of growth factors and their influence on hematopoietic and nonhematopoietic cells. Cytokines are probably the most important biologically active group of molecules to be identified since the discovery of endocrine hormones and their respective pathways. Most cytokine molecules are small- to medium-sized proteins (about 20–25kDa) or glycoproteins that mediate potent biological affects on many cell types. These molecules were first identified by their role in initiating or maintaining a inflammatory response, development, and maintenance of immune responses and supporting hematopoiesis. However, it is now clear that cytokines are involved in many, if not all, physiological functions, being the common signaling system for cell growth, inflammation, immunity, differentiation, and tissue repair processes.

The discovery of new cytokine molecules is occurring at an amazing rate with more than one hundred cytokines having currently been identified. Many of these agents possess pleiotropic functions, thus making it more difficult to link specific functional tasks to a specific cytokine. They are produced by immunocompetent cells and other cell types (e.g., epithelial, fibroblast, endothelial, etc.) and they may also affect these cells. Cytokines normally possess a transient short range of activity, they are usually active at picomolar concentrations and interact with specific receptors, which may be induced upon cell activation. Typical cytokine production is regulated by various inducing stimuli at the level of transcription or translation, resulting in autocrine or paracrine cell influence. Although the range of actions displayed by individual cytokines can be broad and diverse, at least some action(s) of each cytokine is (are) targeted at hematopoietic cells. Phenotypically, cytokine actions lead to an increase or decrease in the rate of proliferation, a change in cell differentiation state, and/or a change in the expression of some different function.

This chapter will summarize the important characteristics of cytokines related to their production and effects on host cells or pathological processes such as inflammation, immunity, and cell differentiation. Since the number of cytokine molecules in existence is extensive and beyond the scope of this chapter, discussion will be limited to key cytokines linked to the processes noted above. However, at the end of this chapter in table 5, a comprehensive summary of cytokines and their actions is provided for the reader.

2 CYTOKINE NOMENCLATURE

Cytokine nomenclature is less than ideal though a look at history will help to understand how the present system has evolved. Early on, the secreted products of activated lymphocytes were termed "lymphokines" while monocyte products were called "monokines." The term lymphokines was originally denoted to describe products of lymphocytes exposed to specific antigen [57]. However, less discriminate use of this term eventually led to its use to describe secreted proteins from a variety of cell sources, affecting the growth or functions of many cell types. In 1974, to counter the wrong impression that such proteins were produced by lymphocytes alone, Cohen et al. proposed the term "cytokines" [45]. Following a period of hesitation, the term "cytokines" has become the generally accepted name for this group of proteins. Later, in 1979 an international workshop proposed the term "interleukin" to denote a system of individual proteins that act as communication signals between different populations of leukocytes [1].

Thus, many cytokines are now called interleukins though others still remain known by their older names such as interferon (IFN) IFNα/β or IFN-γ and tumor necrosis factor (TNF). Though, for some cytokine enthusiasts, these older names may be easier to remember, they suggest only one function of these pleiotropic agents. Usually, that one function is the function that was first recognized for the agent.

3 CHARACTERISTICS OF CYTOKINES

Cytokine production is regulated by various inducing stimuli (fig. 1), usually at the level of transcription or translation. A cytokine produces its actions by binding to specific high-affinity cell surface receptors in close proximity to where it is produced. Cytokine production following cell activation is usually short lived and the process over within a few days. A particular cytokine may exhibit *autocrine* action, binding to receptors on the membrane of the same cell that secreted it; it may exhibit *paracrine* action, binding to receptors on a

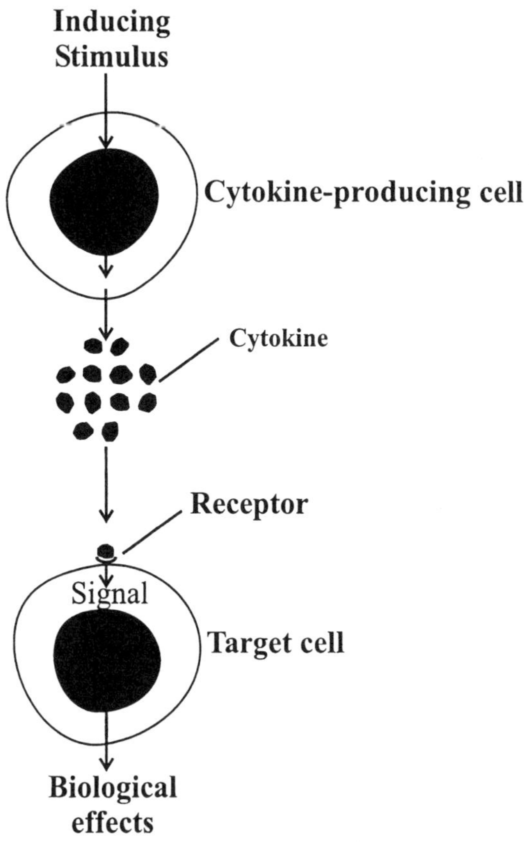

FIGURE 1 Overview of cytokine induction. An inducing stimulus such as viral, bacterial, or fungal antigen and/or a cytokine (e.g., macrophage production of IL-12 to induce T cells to produce cytokines) stimulates a cytokine producing cells to release cytokine. This cytokine is then available to act on a target cell.

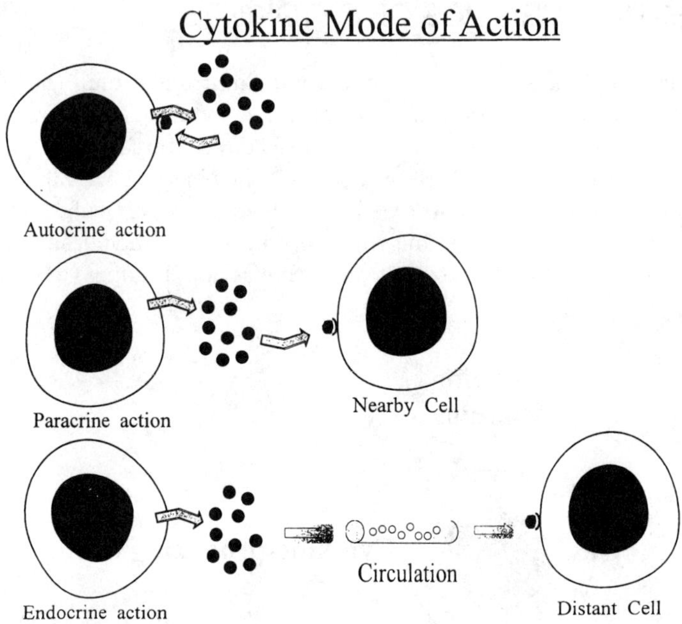

FIGURE 2 Three types of cytokine action: (a) autocrine action where cytokine produced by a cell acts on the cell itself (e.g., macrophage production of TNFα); (b) paracrine action where cytokine produced by a cell acts on cells in close proximity (e.g., macrophage production of IL-12 that induces T cells to release other cytokines); and (c) endrocrine action where a cell produces a cytokine that travels through the circulation for action on a distantly located cell (e.g., macrophage production of IL-1 travels to the CNS to participate in production of fever.)

target cell in close proximity to the producer cell; and in a few rare examples (e.g., TNF, IL-1, or IL-6 in septic shock) it may exhibit *endocrine* action, binding to target cells located in distant parts of the body (fig. 2).

Most cytokine actions can be attributed to an altered pattern of gene expression in the target cells. Cytokines act on multiple target cells (fig. 3) with *pleiotrophy.* They possess *redundancy* (fig. 4(a)) in that different cytokines have similar actions and they can act with *synergism* (fig. 4(b)) or *antagonism* (fig. 4(c)) since exposure of cells to two or more cytokines at a time may lead to qualitatively different responses.

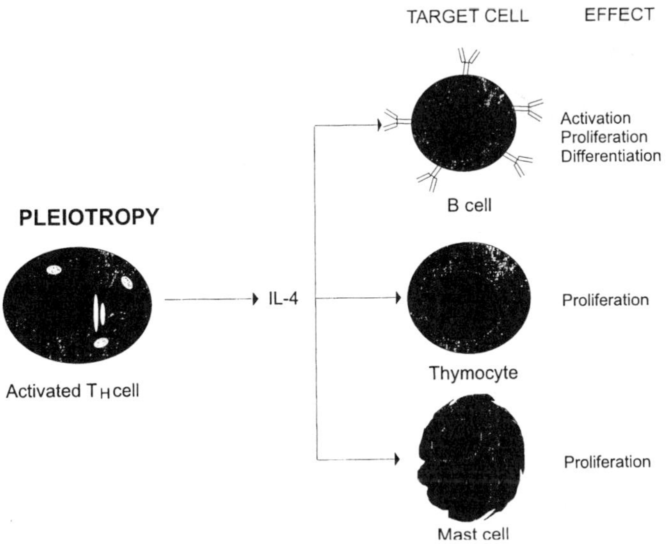

FIGURE 3 A specific cytokine may act on multiple target cells showing a pleiotrophic action. For example, IL-4 produced by a T_H cell may stimulate a B cell to undergo activation, proliferation, and differentiation. IL-4 may also stimulate a thymocyte or mast cell to proliferate.

TABLE 1 Characteristic features of cytokines.

- Most cytokines are simple polypeptides or glycoproteins with a molecular weight of 25 κDa or less (some cytokins form higher molecular weight oligomers and one cytokine (IL-12) is a heterodimer).
- Constitutive production of cytokines is usually low or absent; production is regulated by various inducing stimuli at the level of transcription or translation.
- Cytokine production is transient and the action radius is usually short (typical action is autocrine or paracrine, not endocrine).
- Cytokines produce their actions by binding to specific high-affinity cell surface receptors (K_d in the range 10^{-9} to 10^{-12} M).
- Most cytokine actions can be attributed to an altered pattern of gene expression in the target cells. Phenotypically, cytokine actions lead to an increase (or decrease) in the rate of cell proliferation, a change in cell differentiation state and/or a change in the texpression of some differentiated functions.
- Although the range of actions displayed by individual cytokines can be broad and diverse, at least some action(s) of each cytokine is (are) targeted at hematopoietic cells.

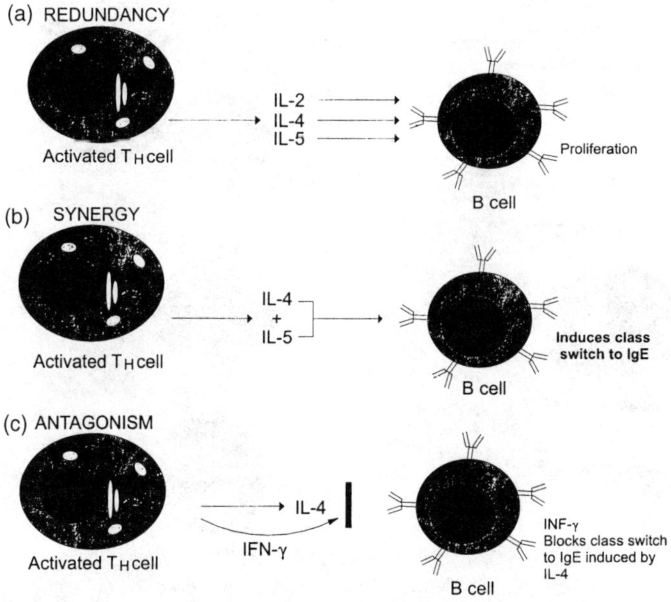

FIGURE 4 Cytokines may function with redundancy (e.g., T_H cell IL-2, IL-4, and IL-5 all stimulate B-cell proliferation); synergy (T_H cell IL-4 and IL-5 cooperate to induce Ig class switch) or in a antagonistic manner (IL-4 and IFN-γ are produced by the T_H cell through IFN-γ can block the effects of IL-4).

A cytokine may increase or decrease the production of another cytokine and it may participate in receptor transmodulation such that the expression of receptors for another cytokine or growth factor are up- or downregulated. Cytokines may also participate in receptor transsignaling by increasing or decreasing signaling of receptors for another cytokine or growth factor.

The primary characteristics of cytokines are shown in table 1. From this summary it is apparent that these properties are shared by two additional groups of protein mediators, namely growth factors and hormones. Though the dividing lines between cytokines and growth factors are subtle, one clear difference is that the production of growth factors such as transforming growth factor-β (TGF-β) and platelet-derived growth factor (PDGF) tends to be constitutive and not as tightly regulated as are the cytokines. In addition, unlike cytokines, the major actions of growth factors are targeted at non-hematopoietic cells.

The distinction between classic polypeptide hormones and cytokines is another area that is not easily defined (table 2). For example, one of the major distinguishing features of classic hormones (e.g., insulin) is that they

TABLE 2 Distinguishing features between polypeptide hormones and cytokines.

Hormones		Cytokines	
Characteristic Features	**Exceptions**	**Characteristic Features**	**Exceptions**
Secreted by one type of specialized cells.		Made by more than one type of cell.	IL-2, IL-3, IL-4, IL-5, and TNF-α are made only by lymphoid cells.
Each hormone is unique in its actions.		Structurally dissimilar cytokines have an overlapping spectrum of actions ("redundancy").	
Restricted target cell specificity and a limited spectrum of actions.	Insulin	Multiple target cells and multiple actions ("pleiotropy").	
Act at a distant site (endocrine mode of action).		Usually have short action radius (autocrine or paracrine mode of action).	TNF, IL-1, or IL-6 in septic shock.

are produced by specialized cells (e.g., β cells of the pancreas). Other examples exist such as the production of growth hormone by the anterior pituitary and parathormone by the parathyroid. In comparison, cytokines are usually produced by less specialized cells and the same cytokine can be produced by one or more phenotypically unrelated cells (e.g, production of IL-1 by mononocytes—macrophages, mesangial cells, NK cells, B cells, T cells, neutrophils, endothelial cells, smooth muscle cells, fibroblasts, astrocytes, and microglial cells). However, as with most biological systems, exceptions to this rule do occur. Most notably, cytokines IL-2, IL-3, IL-4, IL-5, and lymphotoxin (TNF-β) are produced primarily only by T cells. Thus, it appears that the most characteristic features of cytokines, those that distinguish them from hormones, are the redundancy and pleiotropy of cytokine actions. More specifically, many examples exist where dissimilar cytokines (e.g., TNF and IL-1) can be remarkably equivalent in their actions and where individual cytokines tend to exert a multitude of actions on different cells and tissues.

Having shown that some differences clearly exist, it is equally important to recognize that cytokines, growth factors, and polypeptide hormones all function as extracellular signaling molecules displaying virtually similar mechanisms of actions. Data supporting this conclusion can be based in the findings that receptors for several cytokines and hormones (e.g., IL-2, IL-3, IL-4, IL-5, IL-6, IL-7, granulocyte-macrophage colony stimulating factor (GM-CSF), granulocyte colony stimulating factor (G-CSF), erythropoietin, prolactin, and growth hormone) show several common structural features [25, 26, 49, 50, 76, 106, 168]. Further, similar molecular pathways transmit signals from growth factor, polypeptide, hormone, or cytokine receptors to the nucleus and several components in the signal transduction pathways are shared by cytokines, growth factors, and polypeptide hormones.

4 CYTOKINES AND THEIR COMMUNICATION NETWORK

An extensive effort to develop the molecular understanding or explanation of the processes involved in regulating immune responses have occurred in recent years. Such processes include cytokine receptor—cytokine signaling and their signal transduction pathways. For example, discoveries of Janus kinases (JAKs) and signal transducers and activators of transcription (STATs) help to explain signaling via cytokine receptors. The JAK/STAT pathway offers a trail or route of signaling from the membrane to gene regulation. Cytokine binding to such receptors activates receptor-associated tyrosine kinases of the JAK family, so-called because they have two symmetrical kinaselike domains, and thus resemble the mythical two-headed Roman god Janus. These kinases then phosphorlyate cytosolic proteins called STATs. Phoshorylation of STAT proteins leads to their homo- and heterodimerization; STAT dimers can then translocate to the nucleus, where they activate various genes. The proteins encoded by these genes contribute to the growth and differentiation of particular subsets of lymphocytes. Though JAKs do not offer an explanation for the specificity of cytokine signaling, cytokine receptors and STATs do provide for specificity.

In this pathway, gene transcription is activated very soon after the cytokine binds to its receptor, and specificity of signaling in response to different cytokines is achieved by using different combinations of JAKs and STATs. This signaling pathway is used by most of the cytokines that are released by T cells in response to antigen. Though cytokines are not themselves antigen-specific, their directed release in antigen-specific cell-cell interactions and their selective action is on the cell that triggers their production.

Cytokine receptors lack intrinsic catalytic activity (e.g., enzymatic activity such as tyrosine and serine/threonine kinases), unlike other growth factor receptors such as the transforming growth factor β/activatin family of receptors. However, as we have mentioned, cytokine receptors are associated with the structurally unique family of molecules termed JAKs. A specific cytokine can exert its effects through its specific receptor expressed on the membrane of a responsive target cell. Since these receptors are expressed by many cells, the cytokines can affect a diverse array of cells.

Sequence analysis of cloned receptor proteins have revealed some common structural features among different cytokine receptors. All cytokine receptors have an extracellular domain, a single membrane spanning domain, and a cytoplasmic domain. Conserved amino acid sequence motifs have been identified in the extracellular domain of many cytokine receptors. These motifs can be used to define the cytokine-receptor family.

An additional feature found in many cytokine receptors is the presence of two polypeptide chains: a cytokine-specific subunit and a signal-transducing β subunit, which often is not specific for the cytokine. The β subunit is required for high-affinity binding of a cytokine as well as for transduction of an activating signal across the membrane. The transducing β subunits of all the cytokine receptors studied to date have been shown to induce tyrosine phosphorylation, although none has tyrosine kinase activity. This finding suggests that the transmembrane or cytoplasmic domain of the transducing subunit is closely associated with an intracellular protein kinase.

Some cytokine receptors have been shown to possess a shared common signal-transducing subunit, a phenomenon that explains the redundancy and antagonism exhibited among some cytokines. For example, IL-3, IL-5, and GM-CSF each bind to a unique low-affinity cytokine-specific receptor consisting of an α subunit only. All three low-affinity α subunits can associate noncovalently with a common β subunit designated KH97. This subunit increases the affinity of the subunit for cytokine and functions to transduce the signal across the membrane (fig. 5). It is noteworthy that IL-3, IL-5, and GM-CSF all exhibit considerable redundancy.

Since the receptors for IL-3, IL-5, and GM-CSF share a common signal-transducing β subunit, each of these cytokines would be expected to induce a similar activation signal, accounting for the redundancy among their biological effects. In fact, all three cytokines induce the same patterns of protein phosphorylation and phosphorylate the protein kinase *Raf.* Additionally, IL-3 and GM-CSF exhibit antagonism. Furthermore, IL-3 binding has been shown to be inhibited by GM-CSF and, conversely, the binding of GM-CSF has been shown to be inhibited by IL-3. Since the signal-transducing β subunit is shared among these two cytokines, their antagonism is due to competition for a limited number of β subunits by the cytokine-specific α subunits. The examples of receptor function outlined above are not restricted to the cytokines mentioned as similar situations have been described for other cytokines such as IL-6.

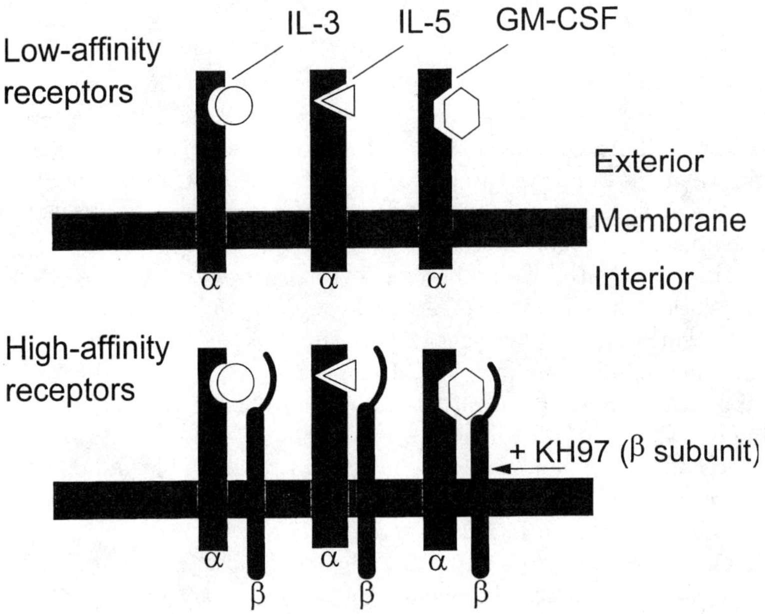

FIGURE 5 Comparison of low-affinity and high-affinity receptors for IL-3, IL-5, and GM-CSF. Cytokine α subunits display low-affinity binding and do not transduce a signal following activation. When the common β subunit associates with the α subunit a high-affinity dimer is formed. This dimer receptor unit efficiently transduces a signal across the cell membrane.

5 ROLE OF CYTOKINES IN THE INFLAMMATORY RESPONSE

Inflammation is a physiologic response to a variety of stimuli such as microbial infections (e.g., pathogenic bacteria/viruses fungi); physical agents (e.g., trauma, thermal, or radiant energy); chemicals (e.g., silicosis); and immunologic reactions (e.g., delayed type hypersensitivity or allergic state). Inflammation is fundamentally a protective response that serves to destroy, dilute, or wall off the offending agent. This process also sets in motion a series of events that as far as possible, heal, repair, and reconstitute damaged tissue. The inflammatory response can occur as an acute or chronic process. Acute inflammation is generally accompanied by a systemic response, known as the acute-phase response, which is characterized by a rapid alteration in the levels of several plasma proteins. In some diseases, persistent immune activation can lead to chronic inflammation resulting in pathologic consequences.

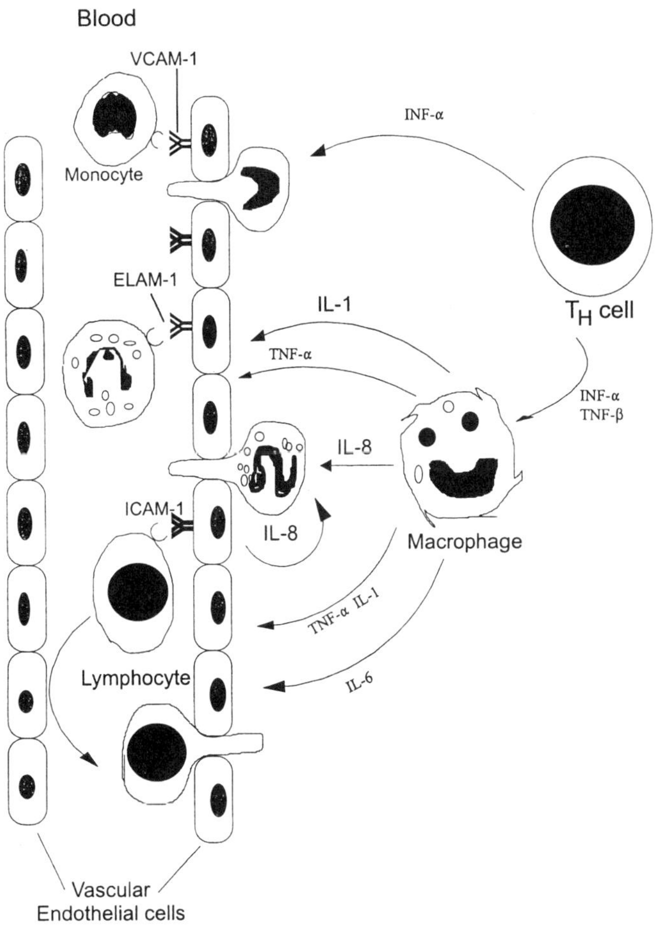

FIGURE 6 Schematic of a localized acute inflammatory response that is mediated mostly by TNF-α, IL-1, and IL-6. Vascular endothelial/leukocyte cell adherence increases and leukocytes undergo extravasation into tissue spaces. IL-1 and TNF-α will induce increased expression of leukocyte adhesion molecules (e.g., ICAM-1, VCAM-1, ELAM-1) and increased macrophage production of IL-8. The production of IL-8 serves to chemotactically attract neutrophils and increase their adherence to endothelial cells. Macrophages are recruited to the area by increased production of TFN-α, TNF-β, and IFN-γ enhance the phagocytic activity of macrophages and neutrophils and their subsequent release of lytic enzymes.

6 ACUTE INFLAMMATORY RESPONSE

The acute inflammatory reaction is initiated following activation of tissue macrophages and their release of TNF-α, IL-1, and IL-6 (fig. 6). IL-1 is a prototypic multifunctional cytokine that exists in two forms, IL-1α and IL-1β, and affects nearly every cell. Its properties are compared to other proinflammatory cytokines, TNF-α and IL-6, in table 3. Subsequent release of these cytokines will induce many of the localized and systemic changes observed in the acute response (table 4). This response includes the induction of fever, increased synthesis of hormones such as ACTH and hydrocortisone, increased production of leukocytes, and production of a large number of hepatocyte-derived acute-phase proteins including C-reactive protein (CRP) and serum amyloid A (SAA).

TNF-α, IL-1, and IL-6 act locally on fibroblasts and endothelial cells, inducing coagulation and an increase in vascular permeability. Both TNF-α and IL-1 induce increased expression of adhesion molecules on vascular endothelial cells. TNF-α stimulates expression of ELAM-1, an endothelial leukocyte adhesion molecule that selectively binds to neutrophils. IL-1 can also induce increased expression of ICAM-1 and VCAM-1, the intercellular adhesion molecules for lymphocytes and monocytes. IL-1 and TNF-α also act on macrophages and endothelial cells, inducing production of IL-8. IL-8 is a potent chemotactic factor and contributes to the recruitment of neutrophils by increasing their adhesion to vascular endothelial cells. IFN-γ also has chemotactic properties for macrophages, attracting them to a site where antigenic material has been localized.

The IL-1, TNF-α, and IL-6 triad are also responsible for many of the systemic acute-phase effects that occur during acute inflammatory response. Each of these cytokines acts on the hypothalmus to induce a fever response. Within 12–24 h of an acute-phase inflammatory response, increased levels of IL-1, TNF-α , and IL-6 (in addition to Oncostatin M and leukemia inhibitory factor) induce production of acute phase proteins by hepatocytes. TNF-α also acts on vascular endothelial cells and macrophages to induce secretion of colony-stimulating factors (M-CSF, G-CSF, and GM-CSF). These colony stimulating factors stimulate hematopoiesis, resulting in transient increases in the numbers of leukocytes that are required to defend the host from an infection.

Cells of the monocyte-macrophage lineage are the main cellular source for IL-1, although most cell types have the potential to secrete this cytokine [39]. In the absence of an in vitro or in vivo stimulation, the IL-1 genes are not expressed. Diverse inducers, including bacterial products (e.g., LPS), complement components and cytokines (TNF, IFN-γ, GM-CSF, and IL-1 itself), cause transcription though this does not always result in translation. For example, adhesion causes accumulation of IL-1 mRNA, which requires a triggering stimulus (small amounts of LPS) for translation into protein.

The synthesis of IL-1 is inhibited by endogenous agents, especially prostaglandins and glucocorticosteriods. In monocytes or monocytic cell lines grown in vitro, IL-1 production is inhibited by the addition of PGE_2. This type of inhibition is also considered a negative feedback mechanism since, when cells are stimulated (with LPS or phorbol esters) to produce IL-1, they also produce PGE_2, which will down regulate or limit IL-1 production [107, 112]. Conversely, the same authors have shown that the addition of prostaglandin inhibitors upregulate IL-1 production.

Glucocorticosteriods inhibit the synthesis of IL-1, both at the transcriptional and the translational level as they do the synthesis for most of the proinflammatory cytokines such as TNF-α, IL-2, IL-6, IL-8, and MCP-1 [107, 182]. Stimuli which induce IL-1 can also induce T cells to produce IL-1Ra (receptor antagonist protein), which may counterbalance the effects of IL-1. For the most part, signals that induce IL-1 also cause production of IL-1Ra, which may counteract the action of agonist molecules. However, expression of IL-1 and IL-1Ra can be discordant. Immune complexes and glycans preferentially trigger production of IL-1Ra vs. IL-1 [6, 7, 155, 159, 170]. In addition, IL-4, IL-13 and IL-10 inhibit IL-1 expression though they amplify IL-1Ra production [31, 135, 159].

Many cell types can be stimulated in vitro to express IL-6. Various cytokines and other agents have been shown to induce IL-6 production. Glucocorticoids, for example, are potent inhibitors of IL-6 production. IL-6 is not stored as a preformed molecule in the cell to any significant amount. When a stimulus triggers a cell, IL-6 mRNA levels increase and the cytokine is rapidly produced (e.g., minutes) and secreted. Monocytes and macrophages appear to be the first to release IL-6 upon a inflammatory signal. The monocyte/macrophage inflammatory products IL-1 and TNF are strong inducers of IL-6 release from stromal cell populations such as fibroblasts, endothelial cells, and kertinocytes though they are not the exclusive producers of this cytokine. However, they do represent a significant number of cells that potentially contribute to the rapid rise in the local synthesis of IL-6 and its eventual systemic distribution.

While many cells can express IL-6 upon in vitro stimulation or during different disease states under different in vivo conditions, it is not clear which specific subset of cells are expressing IL-6. The use of in situ hybridization studies have not been productive in identifying IL-6-expressing tissue in various disease states. IL-6 is produced by monocyte/macrophages in response to stimuli such as LPS [73, 126], IL-1 [12, 13], TNF [12, 126], IFN-γ, and GM-CSF [141], fibrin fragments D and E [165], protease complexes [114], and certain pharmacologic agents [185], as well as some metal cations [176]. IL-6 production in monocytes is inhibited by IL-4 [53], IL-10 [56, 97], and glucocorticosteroid [4, 206, 213].

TABLE 3 Comparison of major properties of IL-1, TNF-a and IL-6 acute inflammatory response cytokines.

IL-1α, IL-β, and IL-1RA	
Protein:	IL-1α: Precursor—271 amino acid, intracellular proform, no signal sequence, biologically active Mature—159 amino acids, extracellular mature form IL-1β: Precursor—269 amino acids, intracellular proform, no signal sequence, biologically inactive Mature—153 amino acid, extracellular mature form, 12 β pleated sheets IL-1RA: Secreted form—signal sequence Intracellular form—no signal sequence
Producers:	Both IL-1α and IL-1β produced by many nucleated cell types, high levels by macrophages, also by kerainocytes, endothelial cells, and some T and B cells IL-1RA produced by macrophages, PMN, keratinocytes, and epithelial cells
Targets:	Many cell types respond to IL-1α or IL-1β in a variety of ways: Immune—T and B cells: Increased proliferation Inflammatory – fibroblasts: PGE_2, proliferation – macrophages: PGE_2, monokines, maintain cytotoxicity – endothelium: PAI, PGE_2, tissue factor, ICAM-1 – participate in septic shock – CNS: fever, sleep, anorexia, CRF – bone and cartilage: resorption
Receptors:	Affinity = 10^{-10} M, 50–5000 sites/cell Type 1-1g superfamily, binds IL-1α, IL-1β, or IL-1RA Type II-Ig superfamily, binds IL-1β > IL-1α, human binds IL-1RA, mouse does not, may not signal, occurs in soluble form
TNFα	
Protein:	Propeptide—233 amino acids, intracellular, membrane form, no signal sequence Mature—157 amino acids, myristilated, homotrimer
Producers:	Macrophages, many other cell types including T cells

TABLE 3 continued

TNFα (continued)	
Targets:	Many cells types react in a variety of ways; T cells: IL-2R expression, IFN-γ production B cells: proliferation and Ig production Macrophages: activation, monokines, PGE_2 tissue factor Endothelial cells: ICAM-1, PGE_2, PAI, tissue factor Some tumor cells: lysis Adipocytes: suppresses lipoprotein lipase Neutrophils: activation endothelial binding Bone and cartlage: resorption Fibroblasts: PGE_2, proliferation, cytokines CNS: fever Participates in septic shock, acute GVHD
Receptors:	Affinity = 10^{-10} M to 10^{-11} M, 1000–10,000 sites/cell Two types (both bind lymphotoxin, both occur in soluble form, both signal by aggregation): 55 κD—426 amin acids, 221 intracellular – mediates antiviral, proliferation, gene induction 75 κD—439 amino acids, 174 intracellular – mediates T-cell proliferation
IL-6	
Protein:	Propeptide–212 amino acids with signal sequence Mature–190 amino acids, variably glycosylated and myristilated
Producers:	Macrophages, fibroblasts, T cells, endothelial cells
Targets:	Many cell types react in a variety of ways: Some plasmacytomas and hybridomas: growth factor B lymphoblasts: cofactor for Ig secretion thymocytes and T cells: comitogenic Hepatocytes: acute-phase reactants Osteoclasts (bone resorption) Autocrine for myeloma and Kaposi's sarcome
Receptors:	Affinity = 10^{-11} M, two chains: gp80-binds IL-6 gp130-transduces signal (also used by several other cytokines)

TABLE 4 Redundant and Pleiotropic Effects of IL-1, TNF-α, and IL-6.

Effect	IL-1	TNF-α	IL-6
Endogenous pyrogen fever	+	+	+
Synthesis of acute-phase proteins by liver	+	+	+
Increased vascular permeability	+	+	+
Increased adhesion molecules on vascular endothelium	+	+	−
Fibroblast proliferation	+	+	−
Platelet production	+	−	+
Induction of IL-8	+	+	−
Induction of IL-6	+	+	−
T-cell activation	+	+	+
B-cell activation	+	+	+
Proliferation of Kaposi's sarcoma	+	−	+
Increased immunoglobulin syntheses	−	−	+

Fibroblasts from different tissues can produce IL-6 following stimulation by IL-1α, IL-1β [88, 89, 157, 174], TNF-α, TNF-β [127, 164], PDGF [109], LPS [91], oncostatin M [26, 173], PGE [221], and viruses [174], though not IL-4 [53].

Other cell types have been shown to produce IL-6 following stimulation, such as endothelial cells with IL-1/TNF [103, 145] or oncostatin M [26, 123]; epithelial cells with IL-1/TNF [121]; keratinocytes with IL-1 [13]; bone marrow stroma with IL-1 [40] in which IL-6 is inhibited by estradiol [101]; astrocytes and microglia with IL-1/TNF [67, 168]; mesangial cells [94]; T and B cells [92, 94, 93]; and mast cells [115].

It was originally thought that TNF-α production was limited to monocytes and macrophages. However, current view suggests that following stimulation, TNF-α can be produced by many cell types at least in vitro. However, TNF-β (lymphotoxin) is exclusively made by T-lymphocytes. Intracellular TNF-α mRNA levels may not be linked to the amount of protein synthesized and secreted. Treatment with LPS increases gene expression about 3-fold, intracellular TNF-α mRNA 100-fold, while TNF-α protein production is increased 1000-fold [18]. In addition to LPS, which represents the main stimulus, viral, fungal and parasital antigens, enterotoxin, mycobacterial cord factor, C5a anaphylatoxin, immune complexes, IL-1 and IL-2, all stimulate production of TNF-α. In addition, TNF-α can also stimulate production of itself in an autocrine manner.

Synthetic metallo-proteinase inhibitors can specifically inhibit TNF-α processing and secretion at a posttranslational step in in vitro and in vivo settings [77, 128, 131]. TNF-β is produced by CD4 and CD8 cells following antigenic stimulation in the context of class II and class I restriction, respectively [147]. TNF-β can be induced by IL-2 or IL-2 plus IFN-γ, or some viruses such as vesicular stomatitis virus or herpes simplex-2 [147].

Antagonists of TNF production include glucocorticoids and prostaglandin E_2 (PGE_2). These agents inhibit synthesis both at the transcriptional and posttranscriptional level [19, 177]. Transforming growth factor (TGF-β), IL-4, and IL-10 are all important antagonists of TNF [63, 66, 90]. Pharmacologic agents such as pentoxifylline and thalidomide selectively inhibit TNF production by monocytes and macrophages without affecting IL-1 and IL-6 production [172, 206]. The agent cyclosporin A inhibits TNF production macrophages at the translational level and in T and B cells at the transcriptional level [82].

7 CHRONIC INFLAMMATORY RESPONSE

This process develops in response to prolonged persistence of an antigen, it typically follows the acute inflammatory response. These antigens can come from microorganisms or self-antigens in autoimmune diseases where T cells are continually activated. Two cytokines in particular, IFN-γ and TNF-α produced by T_H1 cells and macrophages, respectively, play a central role in the development of chronic inflammation. IFN-γ is a pleiotrophic cytokine with one of its most striking effects being the activation of macrophages that results in increased expression of class II MHC molecules, increased cytokine production and increased microbicidal activity. In a chronic inflammatory response, the accumulation of large numbers of activated macrophages is responsible for much of the tissue damage. These cells release several hydrolytic enzymes and reactive oxygen and nitrogen intermediates that also damage the surrounding tissue. Activated macrophages secrete TNF-α and this agent contributes to much of the tissue wasting that is characteristic of chronic inflammation.

8 CELLULAR GROWTH, DIFFERENTIATION, AND IMMUNITY

Cell growth can be altered by increased or decreased apoptosis, changing rates of proliferation and changes in rates of cellular differentiation. Immunity is traditionally defined as a state of resistance or protection from a pathogenic microorganism or protection from the effect of toxic substances such as snake or insect venom. While to some degree the individual processes that make up immunity, they are inexorably linked to each other as is the group of cytokines that participate in generating the signals that influence the effector cells. Many cytokines such as IL-2, IL-3, IL-4, IL-5, IL-8, IL-10, IL-11, IL-12, and IFN-γ play a major or supporting role in these functions and, thus, it is beyond the scope of this chapter to be inclusive. Therefore, attention and focus here will be limited to the key cytokines IL-2, IL-4, IL10, IL-12, and IFN-γ and their link to cell growth, differentiation, and immunity.

Interleukin-2 (IL-2) was first identified in 1975 as a growth-promoting activity for bone-marrow-derived T lymphocytes [132]. Since then, the spectrum of recognized biologic activity for IL-2 has expanded to include direct effects on the growth and differentiation not only of T cells but also B cells, natural killer (NK) cells, lymphokine-activated killer (LAK) cells, monocytes, macrophages, and oligodendrocytes. The biological effects of IL-2 are mediated through specific receptors present on these target cells. The functional high-affinity IL-2 receptor (IL-2R) is composed of three distinct membrane-associated subunits: a 55-kDa α chain (IL-2R , Tac, p55, CD25), a 70-75-kDa β chain (IL-2Rα, p70/75, CD122), and a 64-kDa γ chain (IL2Rγ, γc, p64).

It appears that IL-2 is produced exclusively by T lymphocytes with both $CD4^+$ and $CD8^+$ cells participating in the production. Both T_H1 and T_H2 subpopulations of T-helper cells can produce IL-2, though it is possible that T_H1 cells may produce the cytokine at higher concentrations than T_H2 cells. T cells require stimulation with antigen or a mitogen (e.g., PHA) if they are going to produce and secrete detectable levels of IL-2. This process involves activation through the T-cell receptor, and when the cytokine is secreted, it will activate cells to secrete additional cytokines or other biologically active substances that can help mediate secondary effects.

IL-2 stimulates dramatic proliferation of activated T lymphocytes. It acts on all immune subsets of T cells and promotes progression through the G_1 phase of the cell cycle, resulting in growth of cells and increase in cell numbers [180]. It can also cause proliferation of "resting" T cells though these cells require high amounts of IL-2 since they lack the IL-2R α chain. IL-2 is also capable of stimulating cytolytic activity of subsets of T lymphocytes, it enhances T-cell motility and it induces secretion of other cytokines such as IFN-γ, IL-4, and TNF. It is, therefore, a T-cell differentiation factor [60, 95].

The proliferative effects of IL-2 on thymocytes have been demonstrated [222] and the cytokine may have a role in thymic development [30, 195]. The effects of the cytokine generally enhance and potentiate immune responses, thereby determining the strength of an immune response. If IL-2 is lacking, antigen-specific anergy will occur. Activated T cells will show increased responses to IL-2 if IL-4 is also present though nonactivated T cells seem refractory to this stimulus [125, 130].

IL-2 stimulates the proliferation of antigen or anti-IgM-activated B lymphocytes [75, 199]. It also promotes the induction of immunoglobulin secretion [157] and J chain synthesis [20] by B cells. As shown with T cells, IL-2 will enhance immune effects mediated by activated B cells [129].

Proliferation of large granular lymphocytes is also achieved when IL-2 is secreted. The cytokine will enhance natural killer (NK) cell activity and induces activity of the cells referred to as the "lymphokine-activated killers" [140]. IL-2 stimulates the cytolytic activity of these cells and will induce them to secrete additional cytokines such as IFN-γ [144, 198].

IL-2 enhances the cytolytic activity of monocytes and some reports suggest that it promotes proliferation and differentiation of these cells [8]. The cytokine will enhance macrophage antibody-dependent tumoricidal activity [158]. Other cells of the monocytic lineage (e.g., oligodendrocytes) show enhanced growth and proliferation in response to IL-2 [16, 175]. The effects of IL-2 on nonhemopoietic cells are not entirely clear, though IL-2 receptors have been detected on a number of different cell types that include fibroblasts [152], squamous cell carcinoma lines [208], and rat epithelial cells [42]. It is likely that IL-2 can enhance the proliferative capacity of some of these cells.

Interleukin-4 (IL-4) is produced primarily by subsets of T cells, mast cells and basophils. In humans, $CD4^+$ CD45RA memory cells produce most of the IL-4 [52]; this is in contrast to the small number of $CD8^+$ T-cell clones that have been shown to secrete detectable levels of IL-4 [117, 178, 218]. Induction of IL-4 is accomplished in these cell types via stimulation of the T-cell receptor with antigen, lectin, anti-CD3 or anti-CD2 and phorbol esters in the presence of calcium ionophore [218]. Mast cells can be induced to produce IL-4 via crosslinkage of high affinity Fc receptors [25, 153]. Eosinophils will produce IL-4 in the presence of anti-IgE antibodies [143].

IL-4 is striking in that it can promote activity of B cells, T cells, monocytes/macrophages, LAK and NK cells, neutrophils, eosinophils and mast cells. IL-4 will not induce resting B cells to proliferate [50]. However, it does induce phenotypic changes and the expression of sIgM (IgM on the cell surface), CD23, soluble 23, CD40 and a slight increase of MHC class II molecules [46, 83, 142, 179, 201, 218]. IL-4 activates the cells but its signal is not strong enough to induce resting B cells to enter G1 phase of the cell cycle [43]. However, IL-4 does exhibit some chemotactic effect for B cells [110].

When IL-4 is present, it will act as a costimulator of B-cell proliferation in cooperation with anti-IgM [96]. IL-4 can also induce proliferation in cooperation with phorbol ester and calicium ionophore, immobilized anti-CD40 and the super antigen *Staphylococcus aureus* (SAC) [51, 83, 201]. When B cells are stimulated by antigen, they will also proliferate if IL-4 is present (120). IL-4 in itself will not induce B-cell immunoglobulin secretion. Activation with additional simulators such as SAC or phorbol esters are required for immunoglobulin synthesis and with such agents IgG, IgM, and low levels of IgA can be detected [10]. IL-4 is capable of inducing B cells to produce IgE though only when $CD4^+$ T cells are present [148, 149]. The same conditions are required for IL-4-assisted IgG_4 production (122). IL-4 alone is capable of inducing switching of IgM to IgG_1, and/or IgE [28, 175].

Activated $CD4^+$ and $CD8^+$ T cells proliferate in response to IL-4 and show an increased response to IL-2 [130]. However, resting T cells treated with IL-2 are inhibited from proliferating in the presence of IL-4 [125]. IL-4 can regulate the induction of cell-mediated cytolytic activity by T cells and is inhibited by IL-12 p40 [2]. IL-4 has been shown to block the generation of antigen nonspecific T-cell cytotoxicity [183] and to induce proliferation of postnatal thymocytes, resulting in the growth of $CD3^+$ thymocytes and the differentiation of pro-T cells into mature T cells [11]. Furthermore, IL-4 has been shown to promote T-cell chemotaxis in association with other cytokines [189].

IL-4 can influence monocytes and their surface antigen expression as well as increasing their expression of the class II MHC antigen, CD13, CD23, CD18, CD11b, and CD11c [119, 191, 203, 204]. IL-4 downregulates the expression of CD14, CD64, CD32, and CD16 [192, 193]. IL-4 will inhibit the production of a wide spectrum of cytokines from stimulated monocytes (44, 90,59,184) and induce the production of IL-1 receptor antagonist, G-CSF, GM-CSF [61, 210]. IL-4 will cancel monocytic activation activity induced by IFN-γ [192, 193].

Studies in vitro have shown that IL-4 alone is unable to induce LAK cell activity in mononuclear cells and strongly inhibits IL-2-induced production of LAK cells [9, 151, 183]. IL-4 has also been shown to inhibit IL-2-induced expression of CD69 and the production of IFN-γ by NK cells [151]. IL-4 will upregulate the expression of CD23 on eosinophils and decreases the amount of Fc receptors present. IL-4 can stimulate mast cells and astrocytes though, in contrast, it will inhibit osteoblast growth [162, 161]. Finally IL-4 has been shown to induce expression of adhesion molecules on endothelial cells [146].

Interleukin-10 (IL-10) is produced by a variety of cell types. LPS-activated monocytes [54], SAC activated B cells [205], B cells immediately following infection with EBV [27], EBV-transformed cell lines [15, 205], Burkitt lymphoma [15, 205], HIV-related lymphoma [58], and activated T-cell clones [219]. Production of IL-10 by memory T cells is about 5-10 times grater than that found in naive T cells. IL-10 production is enhanced by TNF-α where IFN-γ, IL-4, IL-13, and IL-10 itself have been shown to inhibit IL-10 in response to LPS-activated monocytes and thus has autoregulatory properties [41, 54, 55, 202, 207].

IL-10 decreases the typical adherence to plastic properties that monocytes display. Monocyte production of IL-1, IL-6, IL-8, IL-10 (itself), IL-12p35, IL-12p40, IFN-α, GM-CSF, M-CSF, MIP-1, and TNF-α [39, 47, 54, 86, 158]. However, IL-10 does not inhibit the production of all monokines and has a enhancing effect on the expression of IL-1Ra [31, 54, 99]. IL-10 has been shown to downregulate the spontaneous expression of MHC II antigens, ICAM-1 (CD54), and B70 (CD86) in addition to their expression on monocytes following stimulation by IFN-γ or IL-4 [54, 111, 211].

In contrast to the above, IL-10 upregulates the expression of FcγRI (CD64) on monocytes to levels that are typically seen via IFN-γ induction [194]. This increased surface receptor expression also correlated to enhanced antibody-dependent cell-mediated cytotoxicity. IL-10 has also been shown to inhibit to migration inhibitory factor (MIF), IFN-γ, or LPS-inducible production of nitric oxide (NO) in murine macrophages [22]. The inhibition of NO has also shown enhanced survival of *Toxoplasma gondii, Leishmania donovani, Trypanosoma cruzi, Listeria monocytogenes, Mycobacterium bovis*, and *Candida albicans* [34, 65, 68, 74, 215, 216].

Upregulation of IL-10 strongly curbs antigen presentation by monocytes/macrophages, which results in a loss of proliferative activity and cytokine production by the T cells [54, 63]. Increased IL-10 also contributes to the downregulation of MHC class II antigens, of costimulatory molecules such as CD54, CD80, and CD86, of monokine production including IL-12 and the antigen-presenting/accessory cell function [54, 111, 134]. While IL-10 inhibits antigen-specific responses toward protein antigens, it also has been shown to have a similar action on alloantigens [14, 32].

On T cells, IL-10 will inhibit the production of IL-2 and TNF-α following antigen-presenting cell independent activation, though production of IFN-γ, IL-4 and IL-5 is not affected [55]. Chemotactic effect of IL-8 on $CD4^+$ T cells is inhibited by IL-10. However, IL-10 does exhibit chemotactic activity on $CD8^+$ cells [104]. In addition, IL-10 has been shown to inhibit apoptosis of human T cells and T-cell clones when growth factors are lacking or following infectious mononucleosis [187, 188].

IL-10 has significant B-cell activity including functioning as a costimulator for the proliferation of human B-cell precursors and B cells following stimulation by anti-IgM monoclonal antibodies, SAC or cross-linking their CD40 antigen [169, 171]. IL-10 has been shown to induce B-cell differentiation and it may act as a switch factor for the production of IgA in association with TGF-β or IgG_1 and IgG_3 [24, 51].

IL-10 has been shown to inhibit the enhanced production of IFN-γ and TNF-α by IL-2 activated NK cells in the presence of monocytes. However, the cytotoxicity of human NK cells was not affected by IL-10 [47, 98]. IL-10 does not effect hematopoiesis though, in combination with other cytokines, it enhances the growth of mast cells, megakaryocytes, and certain stem cells from mouse bone marrow [160, 197].

Interleukin-12 (IL-12) is a potent immunomodulator of NK and cytoloytic T cells, being able to synergize with IL-2 to potentiate the secretion of IFN-γ [35, 70, 79, 108, 214]. The primary cellular source of IL-12 production is the monocyte/macrophage. Data also exists suggesting that B cells can also produce IL-2, though their production capacity is much lower [48]. IL-12 can be induced by LPS, killed *Mycobacterium tuberculosis* and SAC, though the latter is the most potent simulator of IL-12 [48]. Stimulation of mononuclear cells with phorbol esters does not induce IL-12 [48]. Furthermore, IL-12 levels do not appear to be influenced in response to IL-1α, IL-1β, IL-2, IL-4, IL-6, IFN-γ, IFN-β, TNF-α, TNF-β, or GM-CSF [48].

IL-12 was first shown to activate spontaneously cytotoxic human NK/lymphokine-activated killer (NK/LAK) cells to become cytolytic [108]. Overnight *in vitro* activation of $CD56^+$ NK cells with recombinant IL-12 resulted in enhanced killing of NK-sensitive and NK-resistant tumor target cells [37, 166] antibody-coated tumor target cells [118, 166], and virus-infected fibroblasts or T cells [37, 38]. IL-12 effects are independent of IL-12, IFN-α, IFN-β, IFN-γ, and TNF-α [38, 166].

IL-12 has been linked to upregulation of cell surface expression of adhesion/activation molecules and cytokine receptors on cytolytic cells, including CD2, CD11a, CD54, CD56, CD69, CD71, HLA-DR, TNF (75 kDa) receptor and receptors for IL-2α and IL-2β subunits [80, 138, 139, 156, 166]. IL-12 has also been shown to assist the induction of CTL responses to weak immunogenic allogenic melanoma tumor cells [72, 214].

IL-12 has been shown to enhance the proliferative responses of T cells in short-term in vitro assays. However, in contrast with IL-2 and IL-7, IL-12 stimulates minimal proliferation of resting peripheral blood mononuclear cells though it can stimulate proliferation of cells that have been previously activated by various agents [17, 72, 87, 108, 150, 212]. The maximum proliferation induced by IL-12 on PHA-activated lymphocytes is approximately one-half that induced by IL-2, though similar to what is achieved with IL-4 or IL-7 [72]. In addition, suboptimal amounts of IL-2 with IL-12 have shown additive effects on the proliferation of the activated lymphocytes [5, 72, 150]. However, at higher amounts of IL-2, IL-12 has been shown to inhibit or reduce IL-12-induced proliferation depending on the specific cell type [150].

When IL-12 is added to cultures of resting or activated peripheral blood cells, it results in a dose-dependent induction of IFN-γ from both T cells and NK cells [35, 36, 108, 150, 212, 215, 216]. IL-12 has been shown to induce the secretion of low amounts of TNF-α from alloactivated or resting NK cells [138] though significantly more TNF-α is secreted by IL-2-induced compared to IL-12-induced NK cells [138]. NK cells also produce low levels of GM-CSF and IL-8 in response to IL-12 [140].

Effects of IL-12 on T_H1 vs. T_H2 cell development has been studied. IL-12 has been shown to be a key cytokine in determining which pathway the cells follow [124, 167, 198]. $CD4^+$ T-cell lines activated with *Dermatophagoides pteronysinus* group 1 antigen generally show a T_H2-type phenotype, producing little IFN-γ though significant IL-4. However, when the same cell lines were cultured in the presence of IL-12, a T_H1-type cytokine pattern emerged with the cells producing IFN-γ as the predominant cytokine [124].

IL-12 has been shown to markedly inhibit IgE production following IL-4 stimulation, but it does not appear to effect pokeweed mitogen-stimulated production of IgG, IgM, or IgA [105]. IL-12 does not inhibit IgE responses by B cells that have already switched to membrane IgE expression [133]. Finally, IL-12 may influence T-cell development in the thymus, as addition of IL-12 to fetal thymic organ cultures may result in a reduction in the total cell number along with changes in the distribution of immunophenotypically distinct subpopulations of cells [81].

Interferon gamma (IFN-γ) and some of its properties were first recognized in the early 1960s [84, 85, 209].The substance was discovered as a acid-labile viral inhibitorylike substance present in the cerebrospinal fluid of patients with infectious and noninfectious diseases. For many years after its discovery this substance was referred to as "acid-labile interferon," or "type II" interferons opposed to the acid-stable "type I" interferons and now classified as IFN-α and IFN-β [49].

In contrast to the production of type I interferons, which can take place in any cell, the synthesis of IFN-γ is limited to T cells and NK cells. T cells can be stimulated to produce IFN-γ in response to mitogens or antibodies in a clonally restricted antigen-specific manner. Human CD8+ cells will produce IFN-γ following a viral infection or vaccination and subsequent in vitro stimulation with the same antigen [33, 217]. IFN-γ has been shown to preferentially inhibit the proliferation of T_H2 cells.

The processing of antigen into short peptide fragments is a critical step for immune recognition. This process includes cytosolic degradation of the antigen with portions of it being bound to MHC class I molecules, the complex is recognized by $CD8^+$ T cells. Peptides from proteins degraded in endosomal cellular vesicles bind to MHC class II molecules and the complex then migrates to the cell surface. This peptide-MHC class II complex is then recognized by $CD4^+$ T-helper cells [78]. IFN-γ stimulates this critical step of immune recognition by enhancing the expression of MHC class II molecules on macrophages and T cells. Though this response is quick it does require several hours for IFN-γ to activate the genes encoding the expression of MHC class II molecules. This is in contrast to the early activation genes that become activated via the signal tranduction pathway (e.g., JAK/STAT) within minutes of the challenge [3, 21, 116].

In macrophages, IFN-γ induces activation of hydrogen peroxide release which contributes to the macrophages intracellular killing of parasites or other pathogens [64, 137, 136]. Other interferons (e.g., IFN-α and IFN-β) antagonize this process. Interestingly, this is one of only a few examples where different interferons do not act synergistically [69]. IFN-γ has also been shown to activate macrophage tumoricidal capacity, which is part of the host's natural antitumor resistance. IFN-γ also collaborates with TNF-α and other lymphokines to activate macrophage function [200].

IFN-γ has also been shown to induce indoleamine 2,3 dioxygenase (IDO) activity. IDO is an enzyme of tryptophan catabolism that is responsible for the conversion of tryptophan to kynurenine. There is evidence that IDO plays a role in killing intracellular parasites such as *Toxoplasma gondii* or *Chlamydia trachomatis* and *Chlamydia psitacci*, probably via tryptophan starvation [74, 190, 196].

NK cells are part of the innate immune response and possess a naturally occurring cytolytic effector cell function that is not restricted to the MHC complex. As first-line responders against tumor cells and infectious pathogens, NK cells are producers of IFN-γ and their cytolytic activity is also stimulated by IFN-γ as well as IFN-α and IFN-β.

TABLE 5 Cytokines

Cytokine (alternative names)	Producer cells	Actions	Effect of cytokine or receptor knock-out (where known)
FAMILY: Hematopoietins (four-helix bundles)			
Epo (erythropoietin)	Kidney cells, hepatocytes	Stimulates erythroid progenitors	Epo or EpoR: embryonic lethal
IL-2 (T-cell growth factor)	T cells	T-cell proliferation	IL-2 deregulated T-cell proliferation
			IL-2Rα incomplete T-cell development, IL-2Rβ: increased T-cell autoimmunity, IL2γ_c severe combined immunodeficiency
IL-3 (multicolony CSF)	T cells, thymic epithelial cells	Synergistic action in early hematopoiesis	IL-3 impaired eosinophil development; Bone marrow unresponsive to IL-5, GM-CSF
IL-4 (BCGF-1, BSF-1)	T cells, mast cells	B-cell activation, IgE switch suppresses T_H1 cells	IL-4 decreased, IgE synthesis
IL-5 (BCGF-2)	T cells, mast cells	Eosinophil growth, differentiation	IL-5: decreased IgE, IgG1 synthesis (in mice); decreased levels of IL-9, IL-10, and eosinophils
IL-6 (IFN-β_2, BSF-2, BCDF)	T cells, macrophages, endothelial cells	T- and B-cell growth and differentiation, acute-phase protein production, fever	IL-6 decreased acute-phase reaction, reduced IgA production
IL-7	Non-T cells	Growth of pre-B cells and pre-T cells	IL-7: Early thymic and lymophocyte expansion severely impaired

TABLE 5 Cytokines (continued)

Cytokine (alternative names)	Producer cells	Actions	Effect of cytokine or receptor knock-out (where known)
FAMILY: Hematopoietins (four-helix bundles) [continued]			
IL-9	T cells	Mast cell enhancing activity	
IL-11	Stromal fibroblasts	Synergistic action with IL-3 and IL-4 in hematopoiesis	
IL-13 (P600)	T cells	B-cell growth and differentiation, inhibits macrophage inflammatory cytokine production and T_H1 cells	IL-13 defective regulation of isotype-specific responses
G-CSF	Fibroblasts and monocytes	Stimulates neutrophil development and differentiation	Defective myelopoiesis, neutropenia
IL-15 (T-cell growth factor)	T cells	IL-2-like, stimulates growth of intestinal epithelium, T cells, and NK cells	
GM-CSF (granulocyte macrophage colony stimulating factor)	Macrophages, T cells	Stimulates growth and differentiation of myelomonocytic lineage	GM-CSF, GM-CSFR: pulmonary alveolar proteinosis
OSM (OM, oncostatin M)	T cells, macrophages	Stimulates Kaposis's sarcoma cells, inhibits melanoma growth	
LIF (leukemia inhibitory factor)	Bone marrow stroma, fibroblasts	Maintains embryonic stem cells, like IL-6, IL-11, OSM	LIFR die at or soon after birth, decreased hematopoietic stem cells

TABLE 5 Cytokines (continued)

Cytokine (alternative names)	Producer cells	Actions	Effect of cytokine or receptor knock-out (where known)
FAMILY: Interferons			
IFN-γ	T cells, natural killer cells	Macrophage activation, increased expression MHC molecules and antigen-processing components, Ig class switching	IFN-γ, IFN-γR: Decreased resistance to bacterial infection, especially mycobacteria and certain viruses
IFN-α	Leukocytes	Antiviral, increased MHC class 1 expression	IFN-α, impaired antiviral defences
IFN-β	Fibroblasts	Antiviral increased MHC class 1 expression	
FAMILY: Immunoglobulin superfamily			
B7.1 (CD80)	Antigen-presenting cells	Costimulation of T-cell responses	CD28: decreased T cells responses
B7.2 (B70, CO86)	Antigen-presenting cells	Costimulation of T-cell responses	B7.2: decreased co-stimulator response to alloantigen. CTLA-4: Massive lymphoproliferation, early death.
FAMILY: TNF family			
TNF-α (cachectin)	Macrophages, NK cells, T cells	Local inflammation, endothelial activation	TNF-αR: resistance to septic stock, susceptibility to *Listeria*
TNF-β (lymphotoxin, LT, LT-α)	T cells, B cells	Killing, endothelial activation	TNF-β: absent lymph nodes decreased antibody, increased IgM
Lt-β	T cells, B cells	Lymph node development	Defective development of peripheral lymph nodes, Peyer's patches and spleen

TABLE 5 Cytokines (continued)

Cytokine (alternative names)	Producer cells	Actions	Effect of cytokine or receptor knock-out (where known)
FAMILY: TNF family (continued)			
CD40 ligand (DC40L)	T cells, mast cells	B-cell activation, class switching	CD40L: poor antibody response, no class switching, diminished T-cell priming (hyper IgM syndrome)
Fas ligand (FasL)	T cells, stroma?	Apoptosis, Ca^{2+}-independent cytotoxicity	Fas, FasL: mutant forms lead to lymphoproliferation, and autoimmunity
CD27 ligand (CD27L)	T cells	Stimulates T cells proliferation	
CD30 ligand (CD30L)	T cells	Stimulates T and B cell proliferation	CD30: Increased thymic size, alloreactivity
4-1BBL	T cells	Costimulates T and B cells cells	
FAMILY: Unassigned			
TGF-β	Chondrocytes, monocytes, T cells	Inhibits cell growth, anti-inflammatory	TGFβ: lethal inflammation
IL-1α	Macrophages, epithelial cells	Fever, T-cell activation, macrophage activation	IL-1RI: decreased IL-6 production
IL-1β	Macrophages, epthelial cells	Fever, T-cell activation, macrophage activation	IL-1β: impaired acute phase response
IL-1 RA	Monocytes, macrophages, neutrophils hepatocytes	Binds to but doesn't trigger IL-1 receptor, acts as a natural antagonist of IL-1 function	IL-1RA: reduced body mass, increased sensitivity to endotoxins (septic shock)

TABLE 5 Cytokines (continued)

Cytokine (alternative names)	Producer cells	Actions	Effect of cytokine or receptor knock-out (where known)
FAMILY: Unassigned (continued)			
L-10 (cytokine synthesis inhibitor F)	T cells, macrophages, EBV-transformed B cells	Potent suppressant of macrophage functions	IL-10 or CRF2-4: reduced growth, anemia, chronic enterocolitis
IL-12 (NK cell stimulatory factor)	B cells, macrophages	Activates NK cells, induces CD4 T-cell differentiation to T_H1-like cells	IL-12 impaired in IFN-γ production and in T_H1 responses
MIF	T cells, pituitary cells	Inhibits macrophages migration, stimulates macrophage activation	
IL-16	T cells, mast cells, eosinophils	Chemoattractant for CD4 T cells, monocytes and eosinophils, anti-apoptotic for IL-2-stimulated T cells	
IL-17 (mCTLA-8)	CD4 memory cells	Induce cytokine production by epithelia, endothelia, and fibroblasts	
IL-18 (IGIF, interferon-γ inducing factor)	Activated macrophages and Kupffer cells	Induces IFN-γ production by T cells and NK cells, favors T_H1 induction	Defective NK activity and T_H1 responses

IFN-γ serves as a regulatory agent assisting the host in the determination of Ig isotype response in that it stimulates the expression of immunoglobulin of the IgG2a isotype while inhibiting production of IgG3, IgG1, IgG2b, and IgE [23, 181].

Nitric-oxide synthase (NOS) is an enzyme that is responsible for the conversion of 1-arginine to 1-citrulline, yielding the gaseous molecule NO. The enzyme exists in a constitutively expressed form (cNOS) and an inducible form (iNOS). The indicible form is produced in macrophages in response to challenge by lipopolysaccharide and IFN-γ. This inducible response generates very high levels of NO, which are part of the host defense mechanism to antiviral, antimicrobial, antiparasitic and antitumor agents mediated by IFN-γ. Induction of NO in macrophages by IFN-γ is considered by many as one of the major mechanisms by which IFN-γ inhibits replication of viruses such as ectromelia, vaccina, and herpes simplex virus [29, 104]. This provides yet another example of how cytokines link host defense mechanisms that include innate and adaptive processes.

9 SUMMARY

The material presented here has been arranged to demonstrate the complex cellular interactions involving cells of the immune, inflammatory, and hematopoietic system and their relationship to cyotkines as a common signaling system. However, it is impossible to do justice to the subject matter in the limited space of this chapter. Therefore, an extensive list of references have been cited to help guide the reader when more information or details are desired.

The important biological activities of pleiotropy, redundancy, synergy, and antagonism which contribute to the complexity of cytokine networks are fundamental principles of the subject matter. These areas have been discussed at various levels along with molecular features of cytokines and their JAK/STAT communication network. Thus, it is important to note that cytokines can serve as agents that provide a signaling network between cells and within a cell via a transmembrane receptor. Thus, in many means, they are analogous to a local area computer network interacting with other individual personal computers via the internet.

With hundreds of cytokinelike molecules having been identified, it was necessary to restrict discussion here to a few key cytokines and their most important properties. However, we have summarized many other cytokines, the cells that produce them, and their actions in table 5.

Finally, we hope that the material presented here stimulates others to pursue the study of these molecules and their systems, so that the "unknowns" will become known, and that such knowledge will eventually prevent or eliminate suffering from the plethora of diseases that challenge us each day.

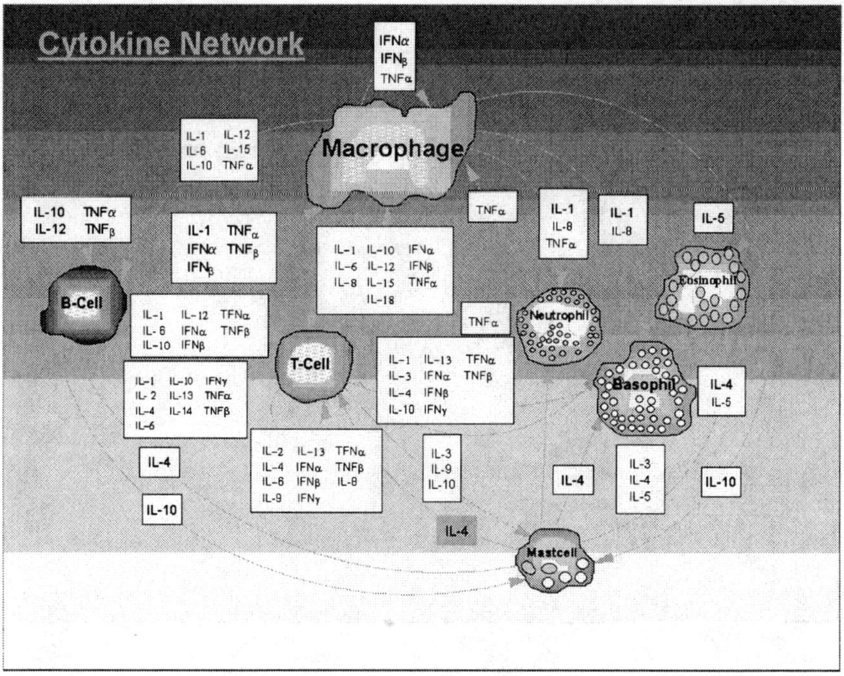

FIGURE 7 Schematic of overall cytokine network associated with host immune responses. Arrows indicate cells producing the cytokine and the type of action (e.g., autocrine or paracrine). The complexity of this network is clearly visible as are the multiple target cells that exist for many of these molecules. Printed by permission. Copyright © MedSystems.

10 ACKNOWLEDGMENTS

The skillful assistance of Carmen Laracuente for her help preparing this manuscript and Anthony Scolpino for his contribution creating the figures and illustrations are greatly appreciated. Special thanks to the Santa Fe Institute for hosting the workshop and Lee Segel and Irun Cohen for making the event happen.

REFERENCES

[1] Aarden, L. A., T. K. Brunner, J.-C. Cerottini, et al. "Revised Nomenclature for Antigen-Nonspecific T Cell Proliferation and Helper Factors." *J. Immunol.* **123** (1979): 2928–2929.

[2] Abdi, K., and S. H. Herrmann. "CTL Generation in the Presence of IL-4 is Inhibited by Free p40: Evidence for Early and Late IL-12 Function." *J. Immunol.* **159(7)** (1997): 3148–3155.
[3] Amaldi, I., W. Reith, C. Berte, and B. Mach. "Induction of HLA Class II Gene by IFN-γ is Transcriptional and Requires a Trans-acting Protein." *J. Immunol.* **142** (1989): 999–1004.
[4] Amano, Y., S. Lee, and A. Allison. "Inhibition of Glucocorticoids of the Formation of Interleukin-1A, Interleukin-1B, and Interleukin-6: Mediation by Decreased mRNA Stability." *Mol. Pharmacol.* **43** (1992): 176–182.
[5] Andrews, J. V. R., D. D. Schoof, M. M. Bertagnolli, G. E. Peoples, P. S. Goedegebuure, and T. J. Eberlein. "Immunomodulatory Effects of Interleukin-12 on Human Tumor-Infiltrating Lymphocytes." *J. Immunother.* **14** (1993): 1–10.
[6] Arend, W. P. "Interleukin 1 Receptor Antagonist. A New Member of the Interleukin 1 Family." *J. Clin. Invest.* **88** (1991): 1445.
[7] Arend, W. P. "Interleukin-1 Receptor Antagonist." *Adv. Immunol.* **54** (1993): 167.
[8] Baccarini, M., R. Schwinzer, and M. L. Lohmann Matthes. "Effect of Human Recombinant IL-2 on Murine Macrophage Precursors. Involvement of a Receptor Distinct from the p55 (Tac) Protein." *J. Immunol.* **142** (1989): 118–125
[9] Banchereau, J. "Human Interleukin-4 and Its Receptor." In *Hematological Growth Factors in Clinical Applications*, edited by R. Mertelsmann and F. Herrman, 433–469. New York: Marcel Dekker, 1999.
[10] Banchereau, J., F. Brière, J. P. Galizzi, P. Miossec, and F. Rousset. "Human Interleukin-4." *J. Lipid Mediators Cell Signal* **9** (1994): 43–53.
[11] Bàrcena, A., M.-L. Toribio, L. Pezzi, and A.-C. Martinez. "A Role for Interleukin-4 in the Differentiation of Mature T Cell Receptor γ/δ^+ Cells from Human Intrathymic T Cell Precursors." *J. Exp. Med.* **172** (1990): 439–464.
[12] Bauer, J., M. Birmelin, G. H. Northoff, W. Northemann, T. A. Tran-Thi, H. Ueberberg, K. Decker, and P. C. Heinrich. "Induction of Rat Alpha-2-Macroglobulin in vivo and in Hepatocyte Primary Cultures: Synergistic Action of Glucocorticoids and a Kupffer Cell Derived Factor." *FEBS Lett.* **177** (1984): 89–94.
[13] Bauer, J., U. Ganter, T. Geiger, U. Jacobshagen, T. Hirano, T. Matsuda, T. Kishimoto, T. Andus, G. Acs, and W. Gerok. "Regulation of Interleukin-6 Expression in Cultured Human Blood Monocytes and Monocyte-Derived Macrophages." *Blood* **72** (1988): 1134–1140.
[14] Bejarano, M. T., R. de Waal Malefyt, J. S. Abrams, M. Bigler, R. Bacchetta, J. E. de Vries, and M. G. Roncarolo. "Interleukin 10 Inhibits Allogeneic Proliferative and Cytotoxic T Cell Responses Generated in Primary Mixed Lymphocyte Cultures." *Int. Immunol.* **4** (1992): 1389–1497.

[15] Benjamin, D., C. D. Park, and V. Sharma. "Human B Cell Interleukin 10." *Leuk. Lymphoma* **12** (1994): 205–210.

[16] Benveniste, E. N., and J. E. Merrill. "Stimulation of Oligodendroglial Proliferation and Maturation of Interleukin-2." *Nature* **321** (1986): 610–613.

[17] Bertagnoli, M. M., B.-Y. Lin, D. Young, and S. H. Hermann. "IL-12 Augments Antigen-Dependent Proliferation of Activated T Lymphocytes." *J. Immunol.* **149** (1992): 3778–3783.

[18] Beutler, B., and A. Cerami. "Tumor Necrosis, Cachexia, Shock, and Inflammation: A Common Mediator." *Ann. Rev. Biochem.* **57** (1988): 505–518.

[19] Beutler, B., J. Han, V. Kruys, and B. P. Giroir. "Coordinate Regulation of TNF Biosynthesis at the Levels of Transcription and Translation. Patterns of TNF Expression in vivo." In *Tumor Necrosis Factors. The Molecules and Their Emerging Role in Medicine*, edited by B. Beutler, 561–574. New York: Raven Press, 1992.

[20] Blackman, M. A., M. A. Tiggs, M. E. Minie, and M. E. Koshland. "A Model System for Peptide Hormone Action in Differentiation: Interleukin 2 Induces a B Lymphoma to Transcribe the J Chain Gene." *Cell* **47** (1986): 609–617.

[21] Blanar, M. A., E. C. Boettger, and R. A. Flavell. "Transcriptional Activation of *HLA-DR*α by Interferon Requires Trans-acting Protein." *Proc. Natl. Acad. Sci. USA* **85** (1988): 4672–4676.

[22] Bogdan, C., Y. Vodovotz, and C. Nathan. "Macrophage Deactivation by Interleukin 10." *J. Exp. Med.* **174** (1991): 1549–1555.

[23] Bossie, A., and E. S. Vitetta. "IFN-γ Enhances Secretion of IgG2α-Committed LPS-Stimulated Murine B Cells: Implication for the Role of IFN-γ in Class Switching." *Cell. Immunol.* **135** (1991): 95–104.

[24] Brière, F., D. C. Servet, J. M. Bridon, R. J. Saint, and J. Banchereau. "Human Interleukin 10 Induces Naive Surface Immunoglobulin D^+ (sIgD^+) B Cells to Secrete IgG1 and IgG3." *J. Exp. Med.* **179** (1994): 757–762.

[25] Brown, M. A., J. H. Pierce, C. J. Watson, J. Falco, J. N. Ihle, and W. E. Paul. "B Cell Stimulatory Factor-1/Interleukin-4 mRNA is Expressed by Normal and Transformed Mast Cells." *Cell* **50** (1987): 809–818.

[26] Brown, T. J., J. M. Rowe, J. Lui, and M. Shoyab. "Regulation of Interleukin-6 Expression by Oncostatin M." *J. Immunol.* **147** (1991): 2175–2180.

[27] Burdin, N., C. Peronne, J. Banchereau, and F. Rousset. "Epstein-Barr Virus Transformation Induces B Lymphocytes to Produce Human Interleukin 10." *J. Exp. Med.* **177** (1993): 295–304.

[28] Callard, R. E. "Immunoregulation by Interleukin-4 in Man." *Br. J. Haematol.* **78** (1991): 293–299.

[29] Campbell, I. L., A. Samini, and C. S. Chiang. "Expression of the Inducible Nitric Oxide Synthase. Correlation with Neuropathology and

Clinical Features in Mice with Lymphocytic Choriomeningitis." *J. Immunol.* **153** (1994): 3622–3629.
[30] Carding, S. R., E. J. Jenkinson, R. Kingston, A. C. Hayday, K. Bottomly, and J. J. Owen. "Developmental Control of Lymphokine Gene Expression in Fetal Thymocytes During T-Cell Ontogeny." *Proc. Natl. Acad. Sci. USA* **86** (1989): 3342–3345.
[31] Cassatella, M. A., L. Meda, S. Gasperini, F. Calzetti, and S. Bonora. "Interleukin 10 (IL-10) Upregulates IL-1 Receptor Antagonist Production from Lipopolysaccharide-Stimulated Human Polymorphonuclear Leukocytes by Delaying mRNA Degradation." *J. Exp. Med.* **179** (1994): 1695.
[32] Caux, C., C. Massacrier, B. Vandervliet, C. Barthelemy, Y. J. Liu, and J. Banchereau. "Related Art Interleukin-10 Inhibits Alloreaction Induced by Human Dendritic Cells." *Int. Immunol. 6(8)* (1994): 1177–1185.
[33] Celis, E., R. M. Miller, T. J. Wiktor, B. Dietzschold, and H. Koprowski. "Isolation and Characterization of Human T Cell Lines and Clones Reactive to Rabies Virus: Antigen Specificity and Production of Interferon-Gamma." *J. Immunol.* **136** (1986): 692–697.
[34] Cenci, E., L. Romani, A. Menacci, R. Spaccapelo, E. Schiaffella, P. Puccetti, and F. Bistoni. "Interleukin-4 and Interleukin-10 Inhibit Nitric Oxide-Dependent Macrophage Killing of Candida Albicans." *Eur. J. Immunol.* **23** (1993): 10340–1038.
[35] Chan, S. H., B. Perussia, J. W. Gupta, M. Kobavashi, M. Pospisil, H. A. Young, S. F. Wolf, D. Young, S. C. Clark, and G. Trinchieri. "Induction of Interferon Production by Natural Killer Cells Stimulatory Factor: Characterization of the Responder Cells and Synergy with Other Inducers." *J. Exp. Med.* **173** (1991): 869–879.
[36] Chan, S. H., M. Kobayashi, D. Santoli, B. Perussia, and G. Trinchieri. "Mechanisms of IFN- Induction by Natural Killer Cell Stimulatory Factor (NKSF/IL-12). Role of Transcription and mRNA Stability in the Synergistic Interaction between NKSF and IL-2." *J. Immunol.* **148** (1992): 92–98.
[37] Chehimi, J., S. E. Starr, I. Frank, M. Rengaraiu, S. J. Jackson, C. Llames, M. Kobayashi, B. Perussia, D. Young, and E. Nickbarg. "Natural Killer (NK) Cell Stimulatory Factor Increases the Cytotoxic Activity of NK Cells from Both Healthy Donors and Human Immunodeficioency Virus-Infected Patients." *J. Exp. Med.* **175** (1992): 789–796.
[38] Chehimi, J., N. M. Valiante, A. D'Andrea, M. Rengaraju, Z. Rosando, M. Kobayashi, B. Perussia, S. F. Wolf, S. E. Starr, and G. Trinchieri. "Enhancing Effect of Natural Killer Cell Stimulatory Factor (NKSF/Interleukin-12) on Cell-Mediated Cytotoxicity Against Tumor-Derived and Virus-Infected Cells." *Eur. J. Immunol.* **23** (1993): 1826–1830.
[39] Chin, J., and M. J. Kostura. "Dissociation of IL-1 Beta Synthesis and Secretion in Human Blood Monocytes Stimulated with Bacterial Cell Wall Products." *J. Immunol.* **151** (1993): 5574–5585.

[40] Chiu, C-P., C. Moulds, R. L. Coffman, D. Rennik, and F. Lee. "Multiple Biological Activities are Expressed by a Mouse Interleukin 6 cDNA Clone Isolated from Bone Marrow Stromal Cells." *Proc. Natl. Acad. Sci. USA* **85** (1988): 7099–7103.

[41] Chomarat, P., M. C. Rissoan, J. Banchereau, and P. Miossec. "Interferon Gamma Inhibits Interleukin 10 Production by Monocytes." *J. Exp. Med.* **177** (1993): 523–527.

[42] Ciacci, C., Y. R. Mahida, A. Dignass, M. Koiozumi, and D. K. Podolsky. "Functional Interleukin-2 Receptors on Intestinal Epithelial Cells." *J. Clin. Invest.* **92** (0993): 527–532.

[43] Clark, E. A., G. L. Shu, Lüscher, K. E. Draves, J. Banchereau, J. A. Ledbetter, and M. A. Valentine. "Activation of Human B Cells. Comparison of the Signal Transduced by Interleukin-4 (IL-4) to Four Different Competence Signals." *J. Immunol.* **143** (1989): 3873–3879.

[44] Cluitmans, F. H. M., B. H. J. Esendam, J. E. Landegent, R. Willemze, and J. H. F. Falkenburg. "IL-4 Down-Regulates IL-2-, IL-3-, and GM-CSF-Induced Cytokine Gene Expression in Peripheral Blood Monocytes." *Ann. Hematol.* **68** (1994): 293–298.

[45] Cohen, S., P. E. Bigazzi, and T. Yoshida. "Commentary. Similarities of T-Cell Function in Cell-Mediated Immunity and Antibody Production." *Cell. Immunol.* **12** (1974): 150–159.

[46] Conrad, D. H., T. J. Waldschmidt, W. T. Lee, M. Rao, A. D. Keegan, R. J. Noelle, R. G. Lynch, and M. R. Kehry. "Effect of B Cell Stimulatory Factor-1 (Interleukin-4) on Fc, and Fc Receptor Expression on Murine B Cells and B Cell Lines." *J. Immunol.* **139** (1987): 2290–2296.

[47] D'Andrea, A., A. M. Aste, N. M. Valiante, X. Ma, M. Kubin, and G. Trinchieri. "Interleukin 10 (IL-10) Inhibits Human Lymphocyte Interferon Gamma-Production by Suppressing Natural Killer Cell Stimulatory Factor/IL-12 Synthesis in Accessory Cells." *J. Exp. Med.* **178** (1993): 1041–1048.

[48] D'Andrea, A., M. Rengaraju, N. M.,Valiante, J. Chehimi, M. Kubin, M. Aste, S. H. Chan, M. Kobayashi, D. Young, and E. Nickbarg. "Production of Natural Killer Cell Stimulatory Factor (Interleukin-12) by Peripheral Blood Mononuclear Cells." *J. Exp. Med.* **176** (1992): 1387–1398.

[49] De Maeyer, E., and J. De Maeyer-Guignard. *Interferons and Other Regulatory Cytokines.* New York: Wiley-Interscience, 1988.

[50] Defrance, T., A. C. Fluckiger, J. F. Rossi, J. P. Magaud, J. J. Sotto, and J. Banchereau. "Antiproliferative Effects of Interleukin-4 on Freshly Isolated Non-Hodgkin Malignant B-Lymphoma Cells. *Blood* **79** (1992): 990–1002.

[51] Defrance, T., B. Vanbervliet, F. Briere, I. Durand, F. Rousset, and J. Banchereau. "Interleukin 10 and Transforming Growth Factor Beta Cooperate to Induce Anti-CD40-Activated Naive Human B Cells to Secrete Immunoglobulin A." *J. Exp. Med.* **175** (1992): 671–682.

[52] DeKruyff, R. H., Y. Fang, H. Secrist, and D. T. Umetsu. "IL-4 Synthesis by in vivo-Primed Memory CD4$^+$ T Cells: II. Presence of IL-4 is Not Required for IL-4 Synthesis in Primed CD4+ T Cells." *J. Clin. Immunol.* **15 (2)** (1995): 105–115.
[53] Donnelly, R. P., L. J. Crofford, S. L. Freman, J. Buras, E. Remmers, R. L. Wilder, and M. J. Fenton. "Tissue-Specific Regulation of IL-6 Production by IL-4." *J. Immunol.* **151** (1993): 5603–5612.
[54] deWaal, M. R., J. Abrams, B. Bennett, C. G. Figdor, and J. E. de Vries. "Interleukin 10 (IL-10) Inhibits Cytokine Synthesis by Human Monocytes: An Autoregulatory Role of IL-10 Produced by Monocytes." *J. Exp. Med.* **174** (1991): 1209–1220.
[55] deWaal, M. R., C. G. Figdor, R. Huijbens, S. Mohan-Peterson, B. Bennett, J. Culpepper, W. Dang, G. Zurawski, and J. E. de Vries. "Effects of IL-13 on Phenotype, Cytokine Production, and Cytotoxic Function of Human Monocytes. Comparison with IL-4 and Modulation by IFN-Gamma or IL-10." *J. Immunol.* **151** (1993): 6370–6381.
[56] deWaal, M. R., H. Yasel, M. G. Roncaralo, H. Spits, and J. E. de Vries. "Interleukin-10." *Curr. Opin. Immunol.* **4** (1992): 314–320.
[57] Dumonde, D. C., R. A. Wolstencroft, G. S. Panayi, M. Matthew, J. Morley, and W. T. Howson. "'Lymphokines': Non-Antibody Mediators of Cellular Immunity Generated by Lymphocyte Activation." *Nature* **224** (1969): 38–42.
[58] Emilie, D., P. Galanaud, M. Raphael, and I. Joab. "Interleukin 10 and Acquired Immunodeficiency Syndrome Lymphomas." *Blood* **81** (1993): 1106–1107.
[59] Essner, R., K. Rhoades, W. H. McBride, D. L. Morton, and J. S. Economou. "IL-4 Down-Regulates IL-1 and TNF Gene Expression in Human Monocytes." *J. Immunol.* **142** (1990): 3857–3861.
[60] Farrar, J. J., W. R. Benjamin, M. L. Hilfiker, M. Howard, W. L. Farrar, and J. Fuller-Farrar. "The Biochemistry, Biology, and Role of Interleukin 2 in the Induction of Cytotoxic T Cell and Antibody Forming B Cell Responses." *Immunol. Rev.* **63** (1982): 129–166.
[61] Fenton, M. J., J. A. Buras, and R. P. Donnelly. "IL-4 Reciprocally Regulates IL-1 and IL-1 Receptor Antagonist Expression in Human Monocytes." *J. Immunol.* **149** (1992): 1283–1288.
[62] Fiorentino, D. F., A. Zlotnik, P. Viera, T. R. Mosmann, M. Howard, K. W. Moore, and A. O'Garra. "IL-10 Acts on the Antigen-Presenting Cell to Inhibit Cytokine Production by Th1 Cells." *J. Immunol.* **146** (1991): 3444–3451.
[63] Fiorentino, D. F., A. Zlotnik, T. R. Mosmann, M. Howard, and A. O'Garra. "IL-10 Inhibits Cytokine Production by Activated Macrophages." *J. Immunol.* **147** (1991): 3815–3822.
[64] Flesch, I. E., and H. E. Kaufmann. "Mechanisms Involved in Mycobacterial Growth Inhibition by Gamma Interferon-Activated Bone Marrow

Macrophages: Role of Reactive Nitrogen Intermediates." *Infect. Immun.* **59** (1991): 3213–3218.

[65] Flesch I. E. A., J. H. Hess, L. P. Oswald, and S. H. E. Kaufmann. "Growth Inhibition of Mycobacterium Bevis by IFN-γ Stimulated Macrophages: Regulation by Endogenous Tumor Necrosis Factor-α and by IL-10." *Int. Immunol.* **6** (1994): 693–700.

[66] Flynn, R. M., and M. A. Palladino. "TNF and TGF-β: The Opposite Sides of the Avenue?" In *Tumor Necrosis Factors. The Molecules and Their Emerging Role in Medicine*, edited by B. Beutler, 131–144. New York: Raven Press, 1992.

[67] Frei, K., U. V. Malipiero, T. P. Leist, R. M. Zinkermagel, M. E. Schwab, and A. Fontana. "On the Cellular Source and Function of Interleukin-6 Produced in the Central Nervous System in Viral Diseases." *Eur. J. Immunol.* **19** (1989): 689–694.

[68] Frei, K., D. Nadal, H. W. Pfister, and A. Fontana. "Listeria Meningitis: Identification of a Cerebrospinal Fluid Inhibitor of Macrophage Listericidal Function as Interleukin 10." *J. Exp. Med.* **178** (1993): 1255–1261.

[69] Garotta, G., K. W. Talmadge, J. R. L. Pink, B. Dewald, and M. Aggiolini. "Functional Antagonism between Type I and Type II Interferon on Human Macrophages." *Biochem. Biophys. Res. Commun.* **140** (1986): 948–954.

[70] Gately, M. K., and R. Chizzonite. "Measurement of Human and Mouse Interleukin 12." In *Current Protocols in Immunology*, edited by J. E. Coligan, A. M. Kruisbeck, D. H. Margulies, E. M. Shevach, and W. Strober, vol.1, 6.16.1–6.16-8. New York: Wiley & Sons, 1988.

[71] Gately, M. K., T. D. Anderson, and T. J. HaYes. "Role of Asialo-GM1-Positive Lymphoid Cells in Mediating the Toxic Effects of Recombinant IL-2 in Mice." *J. Immunol.* **141** (1988): 189–200.

[72] Gately, M. K., B. B. Desai, A. G. Wolitzky, P. M. Quinn, C. M. Dwyer, F. J. Podlaski, P. C. Familletti, F. Sinigaglia, R. Chizonnite, and U. Gubler. "Regulation of Human Lymphocyte Proliferation by a Heterodimeric Cytokine, IL-12 (Cytotoxic Lymphocyte Maturation Factor)." *J. Immunol.* **147** (1991): 874–882.

[73] Gauldie, J., C. Richards, D. Harnish, P. Lansdorp, and H. Baumann. "Interferon-Beta2/B-Cell Stimulatory Factor Type 2 Shares Identify with Monocyte Hepatocyte-Stimulating Factor and Regulates the Major Acute Phase Protein Response in Liver Cells." *Proc. Natl. Acad. Sci. USA* **84** (1987): 7251–7255.

[74] Gazinelli, R. T., L. P. Oswald, S. Hieny, S. L. James, and A. Sher. "The Microbicidal Activity of Interferon-Gamma-Treated Macrophages Against *Trypanooma crusi* Involves an L-Arginine-Dependent, Nitrogen Oxide-Mediated Mechanism Inhibitable by Interleukin-10 and Transforming Growth Factor-Beta." *Eur. J. Immunol.* **22** (1992): 2501–2506.

[75] Gearing, A., R. Thorpe, C. Bird, and M. Spitz. "Human B Cell Proliferation is Stimulated by Interleukin 2." *Immunol. Lett.* **9** (1985): 105–108.

[76] Gearing, D. P., J. A. King, N. M. Gough, and N. A. Nicola. "Expression Cloning of a Receptor for Human Granulocyte-Macrophage Colony-Stimulating Factor." *EMBO J.* **8** (1989): 3667–3676.

[77] Gearing, A. J. H., P. Beckett, M. Christodoulou, M. Churchill, J. Clements, A. H. Davidson, A. H. Drummond, W. A. Galloway, R. Gilbert, and J. L. Gordon. "Processing of Tumour Necrosis Factor—Precursor by Metalloproteinases." *Nature* **370** (1994): 555–557.

[78] Germain, R. N. "MHC-Dependent Antigen Processing and Peptide Presentation: Providing Ligands for T Lymphocyte Activation." *Cell* **76** (1994): 287–299.

[79] Germann, T., M. K. Gately, D. S. Schoenhaut, M. Lohoff, F. Mattner, S. Fischer, S. C. Jin, E. Schmitt, and E. Rude. "Interleukin-12/T Cell Stimulating Factor, a Cytokine with Multiple Effects on T Helper Type 1 (T_H1), but Not on T_H2 Cells." *Eur. J. Immunol.* **23** (1993): 1762–1770.

[80] Gerosa, F., M. Tommasi, C. Benati, G. Gandini, M. Libonati, G. Tridente, G. Carra, and G. Trinchieri. "Differential Effects of Tyrosine Kinase Inhibition in CD69 Antigen Expression and Lytic Activity Induced by rIL-2, IL-12, and rIFN-α in Human NK Cells." *Cell. Immunol.* **150** (1993): 382–390.

[81] Godfrey, D. I., J. Kennedy, M. K. Gately, J. Hakimi, B. R. Hubbard, and A. Zlotnik. "IL-12 Influences Intrathymic T Cell Development." *J. Immunol.* **152** (1994): 2729–2735.

[82] Goldfeld, A. E., E. K. Flemington, V. A. Boussiotis, C. M. Theodos, R. G. Titus, J. L. Strominger, and S. H. Speck. "Transcription of the Tumor Necrosis Factor Gene is Rapidly Induced by Anti-Immunoglobulin and Blocked by Cyclosporin A and FK506 in Human B Cells." *Proc. Natl. Acad. Sci. USA* **89** (1992): 12198–12201.

[83] Gordon, J., M. H. Misslum, G. R. Guy, and J. A. Ledbetter. "Resting B Lymphocytes Can Be Triggered Directly through the CD240 (Bp50) Antigen: A Comparison with IL-4 Mediated Signaling." *J. Immunol.* **140** (1988): 1425–1430.

[84] Green, J. A., S. R. Cooperband, and S. Kibrick. "Immune Specific Induction of Interferon Production in Cultures of Human Blood Lymphocytes." *Science* **164** (1969): 1415–1417.

[85] Gresser, I., and K. Naficy. "Recovery of an Interferon-Like Substance from Cerebrospinal Fluid." *Proc. Soc. Exp. Biol. Med.* **117** (1964): 285–289.

[86] Gruber, M. F., C. C. Williams, and T. L. Gerrard. "Macrophage-Colony-Stimulating Factor Expression by Anti-CD45 Stimulated Human Monocytes is Transcriptional Upregulated by IL-1 Beta and Inhibited by IL-4 and IL-10." *J. Immunol.* **152** (1994): 1354–1361.

[87] Gubler, U., A. O. Chua, D. S. Schoenhaut, C. M. Dwyer, W. McComas, R. Motyka, N. Nabavi, A. G. Wolitzky, P. M. Quinn, and P. C. Familletti. "Coexpression of Two Distinct Genes is Required to Gen-

erate Secreted, Bioactive Cytotoxic Lymphocyte Maturation Factor." *Proc. Natl. Acad. Sci. USA* **88** (1991): 4143–4147.

[88] Guerne, P. A., B. L. Zuraw, J. H. Vaughan, D. A. Carson, and M. Lotz. "Synovium as a Source of Interleukin 6 in vitro." *J. Clin. Invest.* **83** (1989): 585–592.

[89] Haegeman, G., J. Content, G. Volckaert, R. Derynck, J. Tavernier, and W. Fiers. "Structural Analysis of the Sequence Coding for an Inducible 26-κ Da Protein in Human Fibroblasts." *Eur. J. Biochem.* **59** (1986): 625–632.

[90] Hart, P. H., G. F. Vitti, D. R. Burgess, G. A. Whitty, D. S. Piccoli, and J. A. Hamilton. "Potential Anti-Inflammatory Effects of Interleukin 4: Suppression of Human Monocyte Tumor Necrosis Factor α, Interleukin 1, and Prostaglandin E_2." *Proc. Natl. Acad. Sci. USA* **86** (1989): 3803–3807.

[91] Helfgott, D. C., L. T. May, Z. Sthoeger, I. Tamm, and P. B. Seghal. "Bacterial Lipopolysaccharide (Endotoxin) Entrances Expression and Secretion of B2 Interferon by Human Fibroblasts." *J. Exp. Med.* **166** (1987): 1300.

[92] Hirano, T., S. Akira, T. Taga, and T. Kashimoto. "Biological and Clinical Aspects of Interleukin 6." *Immunol. Today* **11** (1990): 443–449.

[93] Horii, Y., A. Muraguchi, S. Suematsu, T. Matsuda, K. Yoshizaki, T. Hirano, and T. Kishimoto. "Regulation of BSF-2/IL-6 Production by Human Mononuclear Cells: Macrophage-Dependent Synthesis of BSF-2/IL-6 T-Cells." *J. Immunol.* **141** (1988): 1529–1535.

[94] Horii, Y., A. Muraguchi, M. Iwano, T. Matsuda, T. Hirayama, H. Yamada, Y. Fujii, K. Dohi, H. Ishikawa, and Y. Ohmoto. "Involvement of IL-6 in Mesangial Proliferative Glomerulonephritis." *J. Immunol.* **143** (1989): 3949–3955.

[95] Howard, M., L. Matis, T. R. Malek, E. Shevach, W. Kell, D. Cohen, K. Nakanishi, and W. E. Paul. "Interleukin 2 Induces Antigen-Reactive T Cell Lines to Secrete BCGF-I." *J. Exp. Med.* **158** (1983): 2024–2039.

[96] Howard, M., J. Farrar, M. Hilfiker, B. Johnson, K. Takatsu, T. Hamaoka, and W. E. Paul. "Identification of a T-Cell Derived B-Cell Growth Factor Distinct from Interleukin-2." *J. Exp. Med.* **155** (1982): 914–923.

[97] Howard, M., and A. O'Garra. "Biological Properties of Interleukin 10." *Immunol. Today* **13** (1992): 198–200.

[98] Hsu, D. H., K. W. Moore, and H. Spits. "Differential Effects of IL-4 and IL-10 on IL-2-Induced IFN-Gamma Synthesis and Lymphokine-Activated Killer Activity." *Int. Immunol.* **4** (1992): 563–569.

[99] Jenkins, J. K., M. Malyak, and W. P. Arend. "The Effect of Interleukin-10 on Interleukin-1 Receptor Antagonist and Interleukin-1 Beta Production in Human Monocytes and Neutrophils." *Lymphokine Cytokine Res.* **13** (1994): 84–89.

[100] Jewett, A., and B. Bonavida. "Activation of the Human Immature Natural Killer Cell Subset by IL-12 and Its Regulation by Endogenous TNF-α and IFN-γ Secretion." *Cell. Immunol.* **154** (1994): 273–286.

[101] Jilka, R. L., G. Hangoc, G. Girasole, G. Passeri, D. C. Williams, J. S. Abrams, B. Boyce, H. Broxmeyer, and S. C. Manolagas. "Increased Osteoclast Development after Estrogen Loss: Mediation by Interleukin-6." *Science* **257** (1992): 88–91.

[102] Jinquan, T., C. G. Larsen, B. Gesser, K. Matsushima, and P. K. Thestrup. "Human IL-10 is a Chemoattractant for CD8$^+$ T Lymphocytes and an Inhibitor of IL-8-Induced CD4$^+$ T Lymphocyte Migration." *J. Immunol.* **151** (1993): 4545–4551.

[103] Jirik, F. R., T. J. Podor, T. Hirano, T. Kishimoto, D. J. Loskutoff, D. A. Carson, and M. Lotz. "Bacterial Lipopolysaccharides and Inflammatory Mediators Augment IL-6 Secretion by Human Endothelial Cells." *J. Immunol.* **142** (1989): 144–147.

[104] Karupiah, G., Q. W. Xie, R. M. L. Buller, C. Nathan, C. Duarte, and J. D. MacMicking. "Inhibition of Viral Replication by Interferon-γ-Induced Nitric Oxide Synthase." *Science* **261** (1993): 1445–1448.

[105] Kiniwa, M., M. Gately, U. Gubler, R. Chizzonite, C. Fargeas, and G. Delespesse. "Recombinant Interleukin-12 Suppresses the Synthesis of IgE by Interleukin-4 Stimulated Human Lymphocytes." *J. Clin. Invest.* **90** (1992): 262–266.

[106] Kishimoto, T., T. Taga, and S. Akira. "Cytokine Signal Transduction." *Cell* **76** (1994): 253–262.

[107] Knudsen, P. J., C. A. Dinarello, and T. B. Strom. "Glucocorticoids Inhibit Transcriptional and Posttranscriptional Expression of Interleukin 1 in U937 Cells." *J. Immunol.* **139** (1987): 4129.

[108] Kobayashi, M., L. Fitz, M. Ryan, R. M. Hewick, S. C. Clark, S. Chan, R. Loudon, F. Sherman, B. Perussia, and G. Trinchieri. "Identification and Purification of Natural Killer Cell Stimulatory Factor (NKSF), a Cytokine with Multiple Biological Effects on Human Lymphocytes." *J. Exp. Med.* **170** (1989): 827–845.

[109] Kohase, M., L. T. May, I. Tamm, J. Vilcek, and P. B. Sehgal. "A Cytokine Network in Human Diploid Fibroblasts: Interactions of β-Interferons, Tumor Necrosis Factor, Platelet-Derived Growth Factor, and Interleukin-1." *Mol. Cell Biol.* **7** (1987): 273–280.

[110] Komai-Koma, M., F. Y. Liew, and P. C. Wilkinson. "Interactions between IL-4, Anti-CD40, and Anti-Immunoglobulin as Activators of locomotion of Human B Cells." *J. Immunol.* **155(3)** (1995): 1110–1116.

[111] Kubin, M., M. Kamoun, and G. Trinchieri. "Interleukin-12 Synergizes with B7/CD28 Interaction in Inducing Efficient Proliferation and Cytokine Production of Human T Cells." *J. Exp. Med.* **180** (1994): 212–222.

[112] Kunkel, S. L., S. W. Chensue, and S. H. Phan. "Prostaglandins as Endogenous Mediators of Interleukin 1 Production." *J. Immunol.* **136** (1986): 186.

[113] Kupper, T. S., K. Min, P. Sehgal, H. Mizutani, N. Birchall, A. Ray, and L. May. "Production of IL-6 by Keratinocytes." *Ann. NY Acad. Sci.* **557** (1989): 454–464.

[114] Kurdowska, A., and J. Travis. "Acute Phase Protein Stimulation by Alpha-Antichymotrypsin-Cathepsin G Complexes." *J. Biol. Chem.* **265** (1990): 21023–21026.

[115] Leal-Berumen, L., P. Conlon, and J. S. Marshall. "IL-6 Production by Rat Peritoneal Mast Cells is Not Necessarily Preceded by Histamine Release and Can Be Induced by Bacterial Lipopolysaccharide." *J. Immunol.* **152** (1994): 5468–5476.

[116] Lew, D. J., T. Decker, I. Strehlow, and J. E. Darnell. "Overlaping Elements in the Guanylate-Binding Protein Gene Promoter Mediate Transcriptional Induction by Alpha and Gamma Interferons." *Mol. Cell Biol.* **11** (1991): 182–191.

[117] Lewis, D. B., K. S. Prickett, A. Larsen, K. Grabstein, M. Weaver, and C. B. Wilson. "Restricted Production of Interleukin-4 by Activated Human T Cells." *Proc. Natl. Acad. Sci. USA* **85** (1988): 9743–9747.

[118] Lieberman, M. D., R. K. Sigal, N. N. Williams, II, and J. M. Daly. "Natural Killer Cell Stimulatory Factor (NKSF) Augments Natural Killer Cell and Antibody-Dependent Tumoricidal Response Against Colon Carcinoma Cell Lines." *J. Surg. Resh.* **50** (1991): 410–415.

[119] Littman, B. H., F. F. Dastvan, P. L. Carlson, and K. M. Sanders. "Regulation of Monocyte/Macrophage C2 Production and HLA-DR Expression by IL-4 (BSF-1) and IFN-γ." *J. Immunol.* **142** (1989): 520–525.

[120] Llorente, L., F. Mitjavila, M.-C. Crevon, and P. Galanaud. "Dual Effects of Interleukin-4 on Antigen-Activated Human B Cells: Induction of Proliferation and Inhibition of Interleukin-2-Dependent Differentiation." *Eur. J. Immunol.* **20** (1990): 1887–1892.

[121] Luger, T. A., T. Schwarz, J. Krutmann, R. Kirnbauer, P. Neuner, A. Kock, A. Urbanski, W. Borth, and E. Schauer. "Interleukin-6 is Produced by Epidermal Cells and Plays an Important Role in the Activation of Human T-Lymphocytes and Natural Killer Cells." *Ann. NY Acad. Sci.* **557** (1989): 405–414.

[122] Lundgren, M., U. Persson, P. Larsson, C. Magnusson, C. I. Smith, L. Hammarstrom, and E. Severinson. "Interleukin-4 Induces Synthesis of IgE and IgG4 in Human B Cells." *Eur. J. Immunol.* **19** (1989): 1311–1315.

[123] Malejczyk, J., M. Malejczyk, A. Urbanski, and T. Luger. "Production of Natural Killer Cell Activity Augmenting Factor (Interleukin-6) by Human Epiphyseal Chondrocytes." *Arthritis Rheum.* **35** (1992): 706–713.

[124] Manetti, R., P. Parronchi, M. G. Giudizi, M. P. Piccinni, E. Maggi, G. Trinchieri, and S. Romagnani. "Natural Killer Cell Stimulatory Factor (Interleukin 12 [IL-12]) Induces T Helper Type 1 (Th1)-Specific Immune Responses and Inhibits the Development of IL-4-Producing Th Cells." *J. Exp. Med.* **177** (1993): 1199–1204.

[125] Martinez, O. M., R. S. Gibbons, M. R. Garovoy, and F. R. Aronson. "IL-4 Inhibits IL-2 Receptor Expression and IL-2-Dependent Proliferation of Human T Cells." *J. Immunol.* **144** (1990): 2211–2215.

[126] May, L. T., J. Ghrayeb, U. Santhanam, S. B. Tatter, Z. Sthoeger, D. C. Helfgott, N. Chiorazzi, G. Grieniner, and P. B. Sehgal. "Synthesis and Secretion of Multiple Forms of 2-Interferon/B-Cell Differentiation Factor 2/Hepatocyte-Stimulating Factor for Human Fibroblasts and Monocytes." *J. Biol. Chem.* **263** (1988): 7760–7766.

[127] May, L. T., U. Santhanam, S. B. Tatter, J. Ghrayeb, and P. B. Sehgal. "Multiple Forms of Human Interleukin-6: Phophoglycoproteins Secreted by Many Different Tissues." *Ann. NY Acad. Sci.* **557** (1989): 114–121.

[128] McGeehan, G. M., J. D. Becherer, R. C. Bast, Jr, C. M. Boyer, B. Champion, K. M. Connolly, J. F. Conway, P. Furdon, S. Karp, and S. Kidao. "Regulation of Tumour Necrosis Factor-Processing by a Metalloproteinase Inhibitor." *Nature* **370** (1994): 558–561.

[129] Mingari, M. C., F. Gerosa, G. Carra, R. S. Accolla, A. Moretta, R. H. Zubler, T. A. Waldmann, and L. Moretta. "Human Interleukin-2 Promotes Proliferation of Activated B Cells via Surface Receptors Similar to Those of Activated T Cells." *Nature* **312** (1984): 641–643.

[130] Mitchell, L. C., L. S. Davis and P. E. Lipsky. "Promotion of Human T Lymphocyte Proliferation by IL-4." *J. Immunol.* **142** (1989): 1548–1557.

[131] Mohler, K. M., P. R. Sleath, J. N. Fitzner, D. P. Cerretti, M. Alderson, S. S. Kerwar, D. C. Torrance, C. Otten-Evans, T. Greenstreet, and K. Weerawarna. "Protection Against a Lethal Dose of Endotoxin by an Inhibitor of Tumour Necrosis Factor Processing." *Nature* **370** (1994): 218–220.

[132] Morgan, D. A., F. W. Ruscetti, and R. Gallo. "Selective in vitro Growth of T Lymphocytes from Normal Human Bone Marrows." *Science* **193** (1976): 1007–1008.

[133] Morris, S. C., K. B. Madden, J. J. Adamovicz, W. C. Gause, B. R. Hubbard, M. K. Gately, and F. D. Finkelman. "Effects of IL-12 on in vivo Cytokine Gene Expression and Ig Isotype Selection." *J. Immunol.* **152** (1994): 1047–1056.

[134] Murphy, E. E., G. Terres, S. E. Macatonia, C. S. Hsieh, J. Mattson, L. Lanier, M. Wysocka, G. Trinchieri, K. Murphy, and A. O'Garra. "B7 and Interleukin-12 Cooperate for Proliferation and IFN- Production by Mouse Th1 Clones that are Unresponsive to B7 Costimulation." *J. Exp. Med.* **180** (1994): 223–231.

[135] Muzio, M., F. Re, M. Sironi, et al. "Interleukin-13 Induces the Production of Interleukin-1 Receptor Antagonist (IL-1ra) and the Expression of the mRNA for the Intracellular (Keratinocyte) Form of IL-1ra in Human Myclomonocytic Cells." *Blood* **83** (1994): 1738.

[136] Nathan, C. F., and S. Tsunawaki. "Secretion of Toxic Oxigen Products by Macrophages: Regulatory Cytokines and Their Effects on the Oxidase." In *Biochemistry of Macrophages*, edited by D. Evered, J. Nugent and M. O'Connor, vol. 188, 211–230. Ciba Foundation Symposium. London: Pittman, 1986.

[137] Nathan, C. F., C. R. Horowitz, J. De La Harpe, S. Vadhan-Raj, S. A. Sherwin, H. F. Oettgen, and S. E. Krown. "Administration of Recombinant Interferon to Cancer Patients Enhances Monocyte Secretion of Hydrogen Peroxide." *Proc. Natl. Acad. Sci. USA* **82** (1985): 8686–8690.

[138] Naume, B., M. Gately, and T. Espevik. "A Comparative Study of IL-12-(Cytotoxic Lymphocyte Maturation Factor), IL-2-, and IL-7-Induced Effects on Immunomagnetically Purified CD56+ NK Cells." *J. Immunol.* **148** (1992): 2429–2436.

[139] Naume, B., M. K. Gately, B. B. Desai, A. Sundan, and T. Espevik. "Synergistic Effects of Interleukin 4 and Interleukin 12 on NK Cell Proliferation." *Cytokine* **5** (1993): 38–46.

[140] Naume, B., A.-C. Johnsen, T. Espevik, and A. Sundan. "Gene Expression and Secretion of Cytokines and Cytokine Receptors from Highly Purified CD56+ Natural Killer Cells Stimulated with Interleukin-2, Interleukin-7 and Interleukin-12." *Eur. J. Immunol.* **23** (1993): 1831–1838.

[141] Navarro, S., N. Debili, J. F. Bernaudin, W. Vainchenker, and J. Doly. "Regulation of the Expression of IL-6 in Human Monocytes." *J. Immunol.* **142** (1989): 4339–4345.

[142] Noelle, R., P. H. Krammer, J. Ohara, J. W. Uhr, and E. S. Vitetta. "Increased Expression of Ia Antigens on Resting B Cells: An Additional Role for B-Cell Growth Factor." *Proc. Natl. Acad. Sci. USA* **81** (1984): 6149–6154.

[143] Nonaka, M., R. Nonaka, K. Woolley, E. Adelroth, K. Miura, Y. Okhawara, M. Glibetic, K. Nakano, P. O'Byrne, and J. Dolovich. "Distinct Immunohistochemical Localization of IL-4 in Human Inflamed Airway Tissues. IL-4 is Localized to Eosinophils in vivo and is Released by Peripheral Blood Eosinophils." *J. Immunol.* **155(b)** (1995): 3234–3244.

[144] Ortaldo, J. R., A. T. Mason, J. P. Gerard, L. E. Henderson, W. Farrar, R. F. Hopkins, III, R. B. Herberman, and H. Rabin. "Effects of Natural and Recombinant IL 2 Regulation on IFN Gamma Production and Natural Killer Activity: Lack of Involvement of the Tac Antigen for These Immunoregulatory Effects." *J. Immunol.* **133** (1984): 779–783.

[145] Pador, T. J., F. R. Jirik, D. J. Loskutoff, D. A. Carson, and M. Lotz. "Human Endothelial Cells Produce IL-6: Lack of Responses to Exogenous IL-6." *Ann. NY Acad. Sci.* **557** (1989): 374–385.

[146] Palmer-Crocker, R. L., and J. S. Pober. "IL-4 Induction of VCAM-1 on Endothelial Cells Involves Activation of a Protein Tyrosine Kinase." *J. Immunol.* **154** (1995): 2838–2845.

[147] Paul, N. L., and N. H. Ruddle. "Lymphotoxin." *Ann. Rev. Immunol.* **6** (1988): 407–438.

[148] Pène, J., F. Rousset, F. Bière, I. Chretien, J. Y. Bonnefoy, H. Spits, T. Yokota, N. Arai, K. Arai, and J. Banchereau. "IgE Production by Normal Human Lymphocytes is Induced by Interleukin-4 and Suppressed by Interferons and Prostaglandin E2." *Proc. Natl. Acad. Sci. USA* **85** (1988): 6880–6884.

[149] Pène, J., I. Crétien, F. Rousset, F. Briére, J. Y. Bonnefoy, and J. E. de Vries. "Modulation of IL-4 Induced Human IgE Production in vitro by IFN-γ and IL-5: The Role of Soluble CD23 (s-CD23)." *Eur. J. Immunol.* **142** (1988): 1558–1564.

[150] Perussia, B., S. H. Chan, A. D'Andrea, K. Tsuji, D. Santoli, M. Pospisil, D. Young, S. F. Wolf, and G. Trinchieri. "Natural Killer (NK) Cell Stimulatory Factor or IL-12 has Differential Effects on the Proliferation of TCR-$\alpha\beta^+$, TCR-$\gamma\delta^*$ T Lymphocytes, and NK Cells." *J. Immunol.* **149** (1992): 3495–3502.

[151] Phillips, J. H., A. Nagler, H. Spits, and L. L. Lanier. "Immunomodulating Effects of IL-4 on Human Natural Killer Cells." In *IL-4: Structure and Function*, edited by H. Spits, 169–185. Ann Arbor, MI: CRC Press, 1992.

[152] Plaisance, S., E. Rubinstein, A. Alileche, Y. Sahraoui, P. Krief, Y. Augery-Bourget, C. Jasmin, H. Suarez, and B. Azzarone. "Expression of the Interleukin-2 Receptor on Human Fibroblasts and Its Biological Significance." *Int. Immunol.* **4** (1992): 739–746.

[153] Plaut, M., J. H. Pierce, C. J. Watson, J. Hanley-Hyde, R. P. Nordan, and W. E. Paul. "Mast Cell Lines Produce Lymphokines in Response to Cross Linkage of Fc RI or to Calcium Ionophores." *Nature* **339** (1989): 64–67.

[154] Poupart, P., P. Vandenabeele, S. Cayphas, J. Van Snick, G. Haegeman, V. Kruys, W. Fiers, and J. Content. "B Cell Growth Modulating and Differentiating Activity of Recombinant Human 26-kd Protein (BSF-2), HulFN-beta2, HPGF." *EMBO J.* **6** (1987): 1219–1224.

[155] Poutsiaka, D. D., M. Mengozzi, E. Vanier, B. Sinha, and C. A. Dinarello. "Cross-Linking of the Beta-Glucan Receptor on Human Monocytes Results in Interleukin-1 Receptor Antagonist but Not Interleukin-1 Production." *Blood* **82** (1993): 3695.

[156] Rabinowich, H., R. B. Herberman, and T. L. Whiteside. "Differential Effects of IL12 and IL2 on Expression and Function of Cellular Adhesion

Molecules on Purified Human Natural Killer Cells." *Cell. Immunol.* **152** (1993): 481–498.

[157] Ralph, P., G. Jeong, K. Welte, R. Mertelsmann, H. Rabin, L. E. Henderson, L. M. Souza, T. C. Boone, and R. J. Robb. "Stimulation of Immunoglobulin Secretion in Human B Lymphocytes as a Direct Effect of High Concentrations of IL 2." *J. Immunol.* **133** (1984): 2442–2445.

[158] Ralph, P., I. Nakoinz, A. Sampson-Johannes, S. Fong, D. Lowe, H. Y. Min, and L. Lin. "IL-10, T Lymophocyte Inhibitor of Human Blood Cell Production of IL-1 and Tumor Necrosis Factor." *J. Immunol.* **148** (1992): 808–814.

[159] Re, F., M. Mengozzi, M. Muzio, C. A. Dinarello, A. Mantovani, and F. Colotta. "Expression of Interleukin-1 Receptor Antagonist (IL-1ra) by Human Circulating Polymorphonuclear Cells." *Eur. J. Immunol.* **23** (1993): 570.

[160] Resnick, D., B. Hunte, W. Dang, S. L. Thompson, and S. Hudak. "Interleukin-10 Promotes the Growth of Megakaryocyte, Mast Cell, and Multilineage Colonies: Analysis with Committed Progenitors and $Thy1^{1+}Sca^{1+}$ Stem Cells." *Exp. Hematol.* **22** (1994): 136–141.

[161] Riancho, J. A., M. T. Zarrabeitia, J. M. Olmos, J. A. Amado, and J. Gonzalez-Macias. "Effects of Interleukin-4 on Human Osteoblast-like Cells." *Bone Miner* **21** (1993): 53–61.

[162] Riancho, J. A., J. Gonzalez-Marcias, J. A. Amado, J. M. Olmos, and J. L. Fernandez-Luna. "Interleukin-4 as a Bone Regulatory Factor: Effects on Murine Osteoblast-like Cells." *J. Endocrinol. Invest.* **18(3)** (1995): 174–179.

[163] Richards, C. D., and A. Agro. "Interaction between Oncostatin-M, Interleukin-1 and Prostaglandin E2 in Induction of IL-6 Expression in Human Fibroblasts." *Cytokine* **6** (1994): 40–47.

[164] Richards, C. D., and J. Saklavala. "Molecular Cloning and Sequence of Porcine Interleukin-6 cDNA and Expression of mRNA in Synovial Fibroblasts in vitro." *Cytokine* **3** (1991): 269–276.

[165] Ritchie, D. G., and G. M. Fuller. "Hepatocyte-Stimulating Factor: A Monocyte-Derived Acute-Phase Regulatory Protein." *Ann. NY Acad. Sci.* **408** (1983): 490.

[166] Robertson, M. J., R. J. Soiffer, S. F. Wolf, T. J. Manley, C. Donahue, D. Young, S. H. Herrmann, and J. Ritz. "Responses of Human Natural Killer (NK) Cells to NK Cell Stimulatory Factor (NKSF): Cytolytic Activity and Proliferation of NK Cells are Differentially Regulated by NKSF." *J. Exp. Med.* **175** (1992): 779–788.

[167] Romagnani, S. "Induction of T_H1 and T_H2 Responses: A Key Role for the 'Natural Immune Response?'" *Immunol. Today* **13** (1992): 379–381.

[168] Rose-John, S., and P. C. Heinrich. "Soluble Receptors for Cytokines and Growth Factors: Generation and Biological Function." *Biochem. J.* **300** (1994): 281–290.

[169] Rousset, F., E. Garcia, T. Defrance, C. Peronne, N. Vezzio, D. H. Hsu, R. Kastelein, K. W. Moore, and J. Banchereau. "Interleukin 10 is a Potent Growth and Differentiation Factor for Activated Human B Lymphocytes." *Proc. Nat. Acad. Sci. USA* **89** (1992): 1890–1893.

[170] Roux, L. P., C. Modoux, and J. M. Dayer. "Production of Interleukin-1 (IL-1) and a Specific IL-1 Inhibitor during Human Monocyte-Macrophage Differentiation: Influence of GM-CSF." *Cytokine* **1** (1989): 45.

[171] Sacland, S., V. Duvert, I. Moreau, and J. Bancherau. "Human B Cell Precursors Proliferate and Express CD23 after CD40 Ligation." *J. Exp. Med.* **178** (1993): 113–120.

[172] Sampaio, E. P., E. N. Sarno, R. Galilly, Z. A. Cohn, and G. Kaplan. "Thalidomide Selectively Inhibits Tumor Necrosis Factor Production by Stimulated Human Monocytes." *J. Exp. Med.* **173** (1991): 699–703.

[173] Saneto, R. P., A. Altman, R. L. Knobler, H. M. Johnson, and J. deVellis. "Interleukin 2 Mediates the Inhibition of Oligodendrocyte Progenitor Cell Proliferation in vitro." *Proc. Natl. Acad. Sci. USA* **83** (1986): 9221–9225.

[174] Sehgal, P. B., D. C. Helfgott, U. Santhanam, D. B. Tatter, R. H. Clarick, J. Ghrayeb, and L. T. May. "Regulation of the Acute Phase and Immune Responses in Viral Disease: Enhanced Expression of the 2-Interferon/Hepatocyte-Stimulating Factor/Interleukin-6 Gene in Virus-Infected Human Fibroblasts." *J. Exp. Med.* **167** (1988): 1951–1956.

[175] Schultz, C. L., and R. L. Coffman. "Mechanisms of Murine Isotype Regulations by IL-4." In *IL-4: Structure and Function*, edited by H. Spits, 15–35. Ann Arbor, MI: CRC Press, 1992.

[176] Scuderi, P. "Differential Effects of Copper and Zinc on Human Peripheral Blood Monocyte Cytokine Secretion." *Cell. Immunol.* **126** (1990): 391–405.

[177] Seckinger, P., and J. M. Dayer. "Natural Inhibitors of TNF." In *Tumor Necrosis Factors: Structure, Function, and Mechanism of Action*, edited by B. B. Aggarwal and J. Vilcek, 217–236. New York: Marcel Dekker, 1992.

[178] Seder, R. A., J.-L. Boulay, F. Finkelman, S. Barbier, S. Z. Ben-Sasson, G. Le Gros, and W. E. Paul. "$CD8^+$ T Cells Can Be Primed in vitro to Produce IL-4." *J. Immunol.* **148(6)** (1992): 1652–1656.

[179] Shields, J. G., R. J. Armitage, B. N. Jamieson, P. C. Beverley, and R. E. Callard. "Increased Expression of Surface IgM but Not IgD or IgG on Human B Cells in Response to IL-4." *Immunology* **66** (1989): 224–228.

[180] Smith, K. A. "T-Cell Growth Factor." *Immunol. Rev.* **51** (1980): 337–357.

[181] Snapper, C. M., and W. E. Paul. "Interferon-Gamma and B Cell Stimulatory Factor-1 Reciprocally Regulate Ig Isotype Production." *Science* **236** (1987): 944–947.

[182] Snyder, D. S., and E. R. Unanue. "Corticosteroids Inhibit Murine Macrophage Ia Expression and Interleukin 1 Production." *J. Immunol.* **129** (1982): 1803.

[183] Spits, H., H. Yssel, X. Paliard, R. Kastelein, C. Figdor, and J. E. de Vries. "IL-4 Inhibits IL-2 Mediated Induction of Human Lymphokine Activated Killer Cells, but Not the Generation of Antigen Specific Cytotoxic T Lymphocytes in Mixed Leukocyte Cultures." *J. Immunol.* **141** (1988): 29–35.

[184] Standiford, T. J., R. M. Strieter, S. E. Chensue, J. Westwick, K. Kasahara, and S. L. Kunkel. "IL-4 Inhibits the Expression of IL-8 from Stimulated Human Monocytes." *J. Immunol.* **145** (1990): 1435–1439.

[185] Stepien, H., A. Agro, J. Crossley, I. Padol, C. Richards, and A. Stanisz. "Immunomodulatory Properties of Diazepam-Binding Inhibitor: Effect on Human Interleukin-6 Secretion, Lymphocyte Proliferation and Natural Killer Cell Activity in vitro." *Neuropeptides* **25** (1993): 207–211.

[186] Stewart, J. P., F. G. Behm, J. R. Arand, and C. M. Rooney. "Differential Expression of Viral and Human Interleukin-10 (IL-10) by Primary B-Cell Tumors and B-Cell Lines." *Virology* **200** (1994): 724–732.

[187] Taga, K., B. Cheney, and G. Tosato. "IL-10 Inhibits Apoptotic Cell Death in Human T Cells Starved of IL-2." *Int. Immunol.* **5** (1993): 1599–1608.

[188] Taga, K., J. Chretien, B. Cheney, L. Diaz, M. Brown, and G. Tosato. "Interleukin-10 Inhibits Apoptotic Cell Death in Infectious Mononucleosis T Cells." *J. Clin. Invest.* **94** (1994): 251–260.

[189] Tan, J., B. Deleuran, B. Gesser, H. Maare, M. Deleuran, C. G. Larsen, and K. Thestrup-Pedersen. "Regulation of Human T Lymphocyte Chemotaxis in vitro by T-Cell-Derived Cytokines IL-2, IFN-gamma, IL-4, IL-10, and IL-13." *J. Immunol.* **154(8)** (1995): 3742–3752.

[190] Taylor, M. W., and G. Feng. "Relationship between Interfeon-γ-, Indoleamine 2,3-Dioxygenase, and Tryptophan Catabolism." *FASEB J.* **5** (1991): 2516–2522.

[191] Te Velde, A. A., J. P. G. Klomp, B. A. Yard, J. E. de Vries, and C. Figdor. "Modulation of Phenotypic and Functional Properties of Human Peripheral Blood Monocytes by IL-4." *J. Immunol.* **140** (1988): 1548–1553.

[192] Te Velde, A. A., F. Rousset, C. Peronne, J. E. de Vries, and C. G. Figdor. "IFN-α and IFN-γ Have Different Regulatory Effects on IL-4-Induced Membrane Expression of Fc RIIb and Release of Soluble Fc RIIb by Human Monocytes." *J. Immunol.* **144** (1990): 3052–3059.

[193] Te Velde, A. A., R. J. E. Huijbens, J. E. de Vries, and C. G. Figdor. "IL-4 Decreases Fcγ R Membrane Expression and Fcγ R-Mediated Cytotoxic Activity of Human Monocytes." *J. Immunol.* **144** (1990): 3046–3051.

[194] Te Velde, A. A., M. R. de Waal, R. J. Huijbens, J. E. de Vries, and C. G. Figdor. "IL-10 Stimulates Monocyte Fc Gamma R Surface Expression and Cytotoxic Activity. Distinct Regulation of Antibody-Dependent

Cellular Cytotoxicity by IFN-Gamma, IL-4, and IL-10." *J. Immunol.* **149** (1992): 4048–4052.

[195] Tentori, L., D. L. Longo, J. C. Zuniga Pflucker, C. Wing, and A. M. Kruisbeek. "Essential Role of the Interleukin 2-Interleukin 2 Receptor Pathway in Thymocyte Maturation in vivo." *J. Exp. Med.* **168** (1988): 1741–1747.

[196] Thomas, S. M., L. F. Garrity, C. R. Brandt, C. S. Schobert, G. S. Feng, M. W. Taylor, J. M. Carlin, and G. I. Byrne. "IFN-γ-Mediated Antimicrobial Response." *J. Immunol.* **150** (1993): 5529–5534.

[197] Thompson-Snipes, L., V. Dhar, M. W. Bond, T. R. Mosmann, K. W. Moore, and D. Rennick. "Interleukin-10: A Novel Stimulatory Factor for Mast Cell and Their Progenitors." *J. Exp. Med.* **173** (1991): 507–510.

[198] Trinchieri, G., M. Matsumoto Kobayashi, S. C. Clark, J. Seehra, L. London, and B. Perussia. "Response of Resting Human Peripheral Blood Natural Killer Cells to Interleukin 2." *J. Exp. Med.* **160** (1984): 1147–1169.

[199] Tsudo, M., T. Uchiyama, and H. Uchino. "Expression of Tac Antigen on Activated Normal Human B Cells." *J. Exp. Med.* **160** (1984): 612–617.

[200] Urban, J. L., H. M. Shepard, J. L. Rothstein, B. J. Sugarman, and H. Schreiber. "Tumor Necrosis Factor: A Potent Effect or Effector Molecule for Tumor Cell Killing by Activated Macrophages." *Proc. Natl. Acad. Sci. USA* **83** (1986): 5233–5237.

[201] Valle, A., C. E. Zuber, T. Defrance, O. Djossou, M. De Rie, and J. Banchereau. "Activation of Human B Lymphocytes through CD40 and Interleukin-4." *Eur. J. Immunol.* **19** (1989): 1463–1467.

[202] van der Poll, T., J. Jansen, M. Levi, H. ten Cate, J. W. ten Cate, and S. J. H. Deventer. "Regulation of Interleukin-10 Release by Tumor Necrosis Factor in Humans and Chimpanzees." *J. Exp. Med.* **180** (1994): 1985–1988.

[203] Van Hal, P. Th. W., J. P. M. Hopstaken-Broos, J. J. Wijkhuijs, A. A. Te Velde, C. G. Figdor, and H. C. Hoogsteden. "Regulation of Aminopeptidase-N (CD13) and Fc RIIb (CD23) Expression by IL-4 Depends on the Stage of Maturation of Monocytes/Macrophages." *J. Immunol.* **149** (1992): 1395–1401.

[204] Vercilli, D., H. H. Jabara, B.-W. Lee, N. Woodland, R. S. Geha, and D. Y. M. Leung. "Human Recombinant Interleukin-4 Induces Fc R2/CD23 on Normal Human Monocytes." *J. Exp. Med.* **167** (1988): 1406–1416.

[205] Vieira, P., R. de Waal-Malefyt, M. N. Dang, K. E. Johnson, R. Kastelein, D. F. Fiorentino, J. E. deVries, M. G. Roncarolo, T. R. Mosmann, and K. W. Moore. "Isolation and Expression of Human Cytokine Synthesis Inhibitory Factor (CSIF) cDNA Clones: Homology to Epstein Barr Virus Open Reading Frame BCRFI." *Proc. Natl. Acad. Sci. USA* **88** (1991): 1172–1176.

[206] Waage, A., G. Slupphaug, and R. Shalaby. "Glucocorticoids Inhibit the Production of IL6 from Monocytes, Endothelial Cells and Fibroblasts." *Eur. J. Immunol.* **20** (1990): 2439–2443.

[207] Wanidworanum, C., and W. Strober. "Predominant Role of Tumor Necrosis Factor-Alpha in Human Monocyte IL-10 Synthesis." *J. Immunol.* **151** (1993): 6853–6861.

[208] Weidmann, E., M. Sacchi, S. Plaisance, D. S. Heo, S. Yasumura, W. C. Lin, J. T.Johnson, R. B. Herberman, B. Azzarone, and T. L. Whiteside. "Receptors for Interleukin 2 on Human Squamous Cell Carcinoma Cell Lines and Tumor in situ." *Cancer Resh.* **52** (1992): 5963–5970.

[209] Wheelock, E. F. "Interferon-like Virus-Inhibitor Induced in Human Leukocytes by Photohemagglutinin." *Science* **149** (1965): 310–311.

[210] Wieser, M., R. Boniofer, W. Oster, A. Lindemann, R. Mertelsmann, and F. Herrmann. "Interleukin-4 Induces Secretion of CSF for Granulocytes and CSF for Macrophages by Peripheral Blood Monocytes." *Blood* **73** (1989): 1105–1108.

[211] Willems, F., A. Marchant, J. P. Delville, C. Gerard, A. Delvaux, T. Velu, M. de Boer, and M. Goldman. "Interleukin-10 Inhibits B7 and Intercellular Adhesion Molecule-1 Expression on Human Monocytes." *Eur. J. Immunol.* **24** (1994): 1007–1009.

[212] Wolf, S. F., P. A. Temple, M. Kobayashi, D. Young, M. Dicig, L. Lowe, R. Dzialo, L. Fitz, C. Ferenz, and R. M. Hewick. "Cloning of cDNA for Natural Killer Cell Stimulatory Factor, a Heterodimeric Cytokine with Multiple Biologic Effects on T and Natural Killer Cells." *J. Immunol.* **146** (1991): 3074–3081.

[213] Woloski, B. M. R. N. J., E. M. Smith, W. J. Meyer, III, G. M. Fuller, and J. E. Blalock. "Corticotropin-Releasing Activity of Monokines." *Science* **230** (1985): 1035–1037.

[214] Wong, H. L., D. E. Wilson, J. C. Jenson, P. C. Familletti, D. L. Stremlo, and M. K. Gately. "Characterization of a Factor(s) which Synergizes with Recombinant Interleukin 2 in Promoting Allogeneic Human Cytolytic T-Lymphocyte Responses in vitro." *Cell. Immunol.* **111** (1988): 39–54.

[215] Wu, J., F. Q. Cunha, F. Y. Liew, and W. Y. Weiser. "IL-10 Inhibits the Synthesis of Migration Inhibitory Factor and Migration Inhibitory Factor-Mediated Macrophage Activation." *J. Immunol.* **151** (1993): 4325–4332.

[216] Wu, C.-Y., C. Demeure, M. Kiniwa, M. Gately, and G. Delespesse. "IL-12 Induces the Production of IFN- by Neonatal Human CD4 T Cells." *J. Immunol.* **151** (1993): 1938–1949.

[217] Yamada, Y. K., A. Meager, A. Yamada, and F. A. Ennis. "Human Interferon-Alpha and Gamma Production by Lymphocytes during the Generation of Influenza Virus-Specific Cytotoxic T Lymphocytes." *J. Gen. Virol.* **67** (1986): 2325–2334.

[218] Yokota, T., N. Arai, J. de Vries, H. Spits, J. Banchereau, A. Zlotnik, D. Rennick, M. Howard, Y. Takebe, and S. Miyatake. "Molecular Biology of Interleukin-4 and Interleukin-5 Genes and Biology of Their Products that Stimulate B Cells, T Cells, and Hemopoietic Cells." *J. Immunol. Rev.* **102** (1988): 137–185.

[219] Yssel, H., M. R. De Waal, M. G. Roncarolo, J. S. Abrams, R. Lahesmaa, H. Spits, and J. E. de Vries. "IL-10 is Produced by Subsets of Human CD4+ T Cell Clones and Peripheral Blood T Cells." *J. Immunol.* **149** (1992): 2378–2384.

[220] Zeh, H. J., III, S. Hind, W. J. Storkus, and M. T. Lotze. "Interleukin-12 Promotes the Proliferation and Cytolytic Maturation of Immune Effectors: Implications for the Immunotherapy of Cancer." *J. Immunother.* **14** (1993): 155–161.

[221] Zhang, Y., J. X. Lin, and J. Vilcek. "Synthesis of Interleukin 6 (Interferon-Beta2/B Cell Stimulatory Factor 2) in Human Fibroblasts is Triggered by an Increase in Intercellular Cyclic AMP." *J. Biol. Chem.* **263** (1988): 6177–6182.

[222] Zuniga Pflucker, J. C., K. A. Smith, L. Tentori, D. M. Pardoll, D. L. Longo, and A. M. Kruisbeek. "Are the IL-2 Receptors Expressed in the Murine Fetal Thymus Functional?" *Dev. Immunol.* **1** (1990): 59–66.

Th1/Th2 Effector Choice in the Immune System: A Developmental Program Influenced by Cytokine Signals

Dragana Jankovic
Alan Sher

1 INTRODUCTION

The plasticity of the immune system is perhaps its most important characteristic. In addition to generating antibodies and T cells that recognize specific pathogens, the adaptive immune response can simultaneously choose the appropriate class of effector functions necessary for microbial clearance. For example, the destruction of intracellular bacteria and protozoa requires T-lymphocyte-dependent activation of the host cells in which these intracellular pathogens reside. In contrast, immune elimination of extracellular pathogens usually requires the production of soluble antibodies which trigger their lysis by complement or their phagocytosis by macrophages and/or granulocytes. In the case of viral infections CD8+ cytolytic T lymphocytes are also induced to restrict pathogen spread. Importantly, all of these distinct effector functions depend on the activation of a single type of immune cell, the CD4+ T helper (Th) lymphocyte.

The mechanism by which CD4+ T cells promote different classes of effector responses remained unclear until the mid-1980s when Tim Mosmann and Robert Coffman [38] first demonstrated that mouse CD4+ T-cell clones could be classified into distinct subsets on the basis of their patterns of cytokine production. They termed the two most polarized subsets Th1 and Th2.

TABLE 1 Cytokine secretion profiles of the major Th subsets.

Cytokine	Th1 cells	Th2 cells
IL-2	+ +	−
IFN-γ	+ +	−
TNF-β	+ +	−
IL-4	−	+ +
IL-5	−	+ +
IL-6	−	+ +
IL-9	−	+ +
IL-10[1]	−	+ +
IL-13	−	+ +
GM-CSF	+ +	+
TNF-α	+ +	+
IL-3	+ +	++

[1]Some human Th1 clones can produce IL-10.

Th1 clones were found to produce IL-2, IFN-γ, and TNF-β while Th2 clones instead synthesize IL-4, IL-5, IL-6, IL-9, IL-10, and IL-13 (table 1). CD4+ T cells with similar phenotypes were identified during immune responses in vivo [39, 46], although cells (Th0) with mixed cytokine secretion patterns were also evident [48]. This discovery that Th lymphocytes differ in their production of important cytokine mediators provided an important framework for understanding the diversity of CD4+ T-cell-dependent host effector functions.

While early difficulties in identifying Th1 and Th2 cells among human T lymphocytes generated considerable skepticism about the generality of the Th1/Th2 concept, it is now clear that CD4$^+$ Th subsets similar to those described in mice do exist in humans as well as other species [5, 11, 18]. Moreover, polarized cytokine expression patterns have been documented in other immune cell types. For example, CD8+CD4-$\alpha\beta$+[10] and CD8-CD4-$\gamma\delta$+[14] cells have also been shown to posses Th1-like and Th2-like lymphokine production profiles, suggesting that the overall cytokine response of the host involves the coordination of lymphokine production by distinct lymphocyte lineages.

A crucial aspect of Th1 and Th2 CD4+ T-cell subsets is their distinct immunobiologic functions [1, 39, 46]. The lymphokines secreted by Th1 lymphocytes, by activating macrophages [17] and nonhematopoietic effector cells [53] as well as providing help for cytotoxic CD8+ T-cell responses [41], are important mediators against microbial agents. Conversely, the cytokines secreted by Th2 cells play a role in host resistance against extracellular pathogens by stimulating the growth and differentiation of mast cells and eosinophils as well as the production of antibody isotypes (e.g., IgE) required for their triggering [15]. The biological function of Th1 and Th2 cells is well documented in many infectious diseases where the development of one or the other phenotype may result in resistance or enhance susceptibility to the pathogen involved.

Studies on experimental infection with the protozoan *Leishmania major* have been particularly valuable in establishing this paradigm [35]. Most inbred mouse strains are resistant to the parasite and heal infections by means of a Th1 cytokine-dependent response. In contrast, BALB/c mice develop an uncontrolled and ultimately fatal infection when exposed to the same pathogen. This exacerbation is associated with the development of Th2 cells and is dependent on Th2 lymphokines.

The association of Th1 cells with resistance and Th2 cells with susceptibility is not universal. For example, immune expulsion of intestinal nematodes requires Th2 cytokines and is inhibited by Th1 responses [13]. Thus, the beneficial vs. detrimental function of Th1 and Th2 cells in host resistance is fundamentally dependent on the nature of the infectious agent. Clearly the host requires both responses in order to cope with the different pathogens that it encounters.

In addition to its relevance to infectious disease, the Th1/Th2 paradigm provides a framework for understanding the immunologic basis of other pathologic disorders. Thus, allergic responses to common environmental Ag are clearly dependent on Th2 lymphocytes, whereas Th1 cells have been implicated in the induction of many autoimmune diseases [43]. Th1/Th2 effector choice is therefore a major immunologic determinant of health vs. disease.

Studies utilizing transgenic mice in which all CD4+ T lymphocytes express a single antigen (Ag) receptor of known specificity clearly demonstrate that both Th1 and Th2 cells recognizing the same Ag can be generated from a common precursor population [40, 44]. From this evidence it is clear that Th1 and Th2 cells are not precommitted phenotypes but rather represent endpoint stages of a multistep differentiative process. The key questions are how naïve T cells acquire their distinct cytokine expression profiles and how this decision is directed by the different antigenic stimuli that the immune system encounters.

2 CYTOKINES AS CRITICAL DETERMINANTS OF TH1/TH2 EFFECTOR CHOICE

A number of different factors have been shown to influence the differentiation of CD4+ T lymphocytes into Th1 or Th2 effectors. In particular, the manner in which Ag are presented to T cells can promote the emergence of one or the other subset. Thus, different Ag doses can trigger different Th1/Th2 profiles and the strength of the signal delivered though the MHC-Ag/T-cell receptor interaction can exert a similar influence [9]. In related fashion, "costimulatory" signals provided by the accessory cells which present Ag to T cells have also been shown to affect the outcome of this process [4].

Although the above factors in certain settings can play a dominant role, it is clear from numerous studies that the cytokine environment during initial T-cell activation is the overriding determinant of Th1/Th2 subset selection.

The two pivotal cytokines influencing Th differentiation are IL-12 and IL-4. In vitro, IL-12 drives Ag-stimulated naïve CD4+ T cells into Th1 cells [22], while IL-4 promotes Th2 development [34, 49]. The same requirements for IL-12 or IL-4 have been demonstrated for Th1/Th2 induction in vivo. Thus, knock-out (KO) mice, which lack one or the other cytokine as a result of targeted gene disruption, display severely impaired Th1 or Th2 responses as compared with wild-type littermates [32, 37].

Importantly, to successfully direct Th lymphocyte differentiation, IL-12 or IL-4 must be present at the time of initial T-cell priming by Ag [40, 44]. Because of the latter consideration it was previously assumed that both cytokines must first be induced from non-T cells. In the case of IL-12-driven Th1 responses this requirement is well established with antigen-presenting cells (APC) (e.g., macrophages and dendritic cells) providing the non-T cell source of the cytokine [22, 40]. Indeed, a number of microbial products have been identified that stimulate APC IL-12, thus explaining the ability of the pathogens from which they were derived to promote protective Th1-type immune responses. In the opposite direction, considerable debate persists over the origin of the early IL-4 required for promotion of Th2 responses. A number of different cell types including mast cells, basophils, NK1.1+CD4+ T cells, CD4-CD8- $\gamma\delta$ T cells and eosinophils could potentially serve as an exogenous source of this cytokine [8, 24]. Nevertheless, none of these appear to be obligatory for Th2 differentiation. Instead, since the capacity to produce IL-4 is intrinsic to CD4+ T cells, it has been argued that activated CD4+ lymphocytes may provide their own IL-4 and that this autocrine stimulus may lead to Th2 development. Indeed, there is now abundant evidence from both in vitro and in vivo studies that Th2 differentiation can be supported solely by IL-4 produced by Ag-primed [21, 45] or memory CD4+ lymphocytes [20].

The signaling pathways responsible for the differential effects of IL-12 and IL-4 on T-cell development have now been partially delineated. Upon binding to their receptors IL-12 and IL-4 trigger phosphorylation of specific and distinct *signal transducers and activators of transcription*, Stat-4 and Stat-6, respectively [42]. Genetic disruption of one or the other of these signaling molecules has the same impact on Th1/Th2 development as deletion of the corresponding differentiative cytokines [27, 28, 47, 51, 52].

3 INSTRUCTIONAL VS. STOCHASTIC MODELS OF T-HELPER SUBSET DEVELOPMENT

Based on the concept of positive cytokine influence on development of Th1/Th2 lymphokine secretion profiles two alternative models have emerged involving instructional vs. stochastic differentiation pathways (fig. 1) [36]. In the instructional model (fig. 1(a)), the targets of cytokine action are immature precursors that can differentiate in response to either IL-12 or IL-4 signaling. In contrast, the stochastic model (fig. 1(b)) predicts that the differentiative cytokines act on spontaneously developing T cells expressing random lymphokine patterns selecting those with the appropriate Th1 or Th2 phenotype. Both models predict that mature Th1/Th2 populations should not develop in the absence of IL-12/IL-4 signaling.

Because of the lack of specific cell surface markers for identifying lymphocytes with different cytokine secretion profiles, it is technically difficult to track developing Th lineages. In addition, the frequency of Ag-responding Th cells is very low under physiological conditions. Experiments with transgenic mice in which all CD4+ T lymphocytes express a single Ag receptor of known specificity demonstrated that, in the absence of exogenous IL-12 and IL-4, Ag-primed CD4+ lymphocytes initially produce IL-2 selectively, while synthesis of IFN-γ and IL-4 is detected only in daughter cells [6, 33]. Indeed it would appear that specific derepression of genes encoding these cytokines is required for their expression [42]. Thus, the instructional model postulates that the mechanism of derepression is directly dependent on IL-12 or IL-4[3].

On the other hand, recent in vitro findings suggest that lymphokine expression patterns may evolve independently of direct instruction by differentiative cytokines. Thus, cultures of transgenic T lymphocytes activated under neutral cytokine conditions after expressing IL-2, sequentially develop IFN-γ followed by IL-4-producing cells and this progression is directly related to the number of cell divisions [6, 19]. While addition of cytokines (IL-12 or IL-4) clearly influences the outcome of Th development in the above system, their major effect appears to be the stabilization or positive selection of IFN-γ or IL-4 expressing populations appearing spontaneously in the cultures. Such findings are consistent with a model of Th differentiation in which cytokines play a selective rather than instructional role.

A major hypothesis is that the targets of this selective process are cells expressing random assortments of cytokines generated in a stochastic manner [7, 30, 31]. Indeed, lymphocytes with mixed (Th0) phenotypes can be detected during developing T-cell responses [2, 26, 48]. However, it is not clear whether such phenotypes reflect the cells upon which selection is performed or transitional stages between Th1 and Th2 lymphocytes interconverting under the influence of the cytokines.

4 EMERGENCE OF MATURE TH1/TH2 POPULATIONS IN THE ABSENCE OF CYTOKINE SIGNALING

Further evidence that cast doubts on the classical cytokine instruction model of Th1/Th2 effector choice comes from studies in which the normal IL-12- or IL-4-dependent signaling pathways have been genetically disrupted. Although such animals show impaired subset development, they have been shown by a number of investigators to be capable of generating residual Th1/Th2 lymphocytes, but typically in the presence of a greatly expanded population of cells with the opposing Th phenotype. For example, IL-4 receptor- or Stat-6-KO

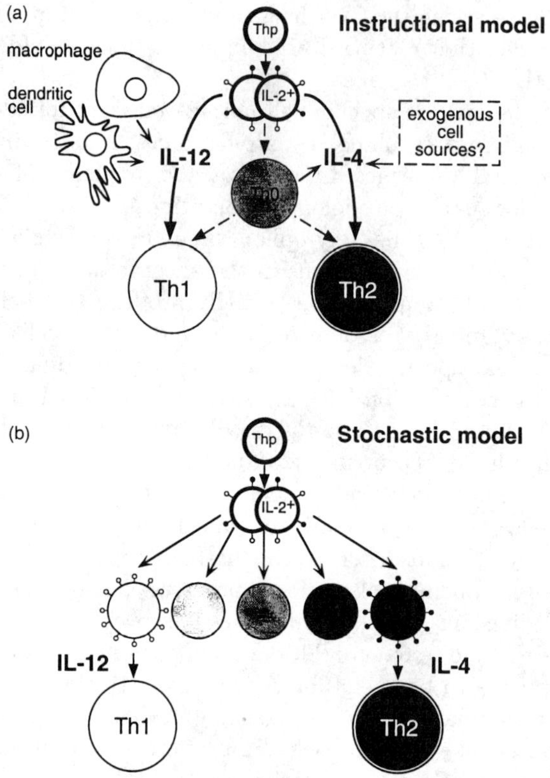

FIGURE 1 (a) The instructional model of Th differentiation. Th precursors (Thp) are stimulated by Ag to produce IL-2 and, in one version of the model, to differentiate into Th0 cells, all of which express a uniform mixed lymphokine secretion pattern. In the presence of differentiative cytokines the activated Thp or Th0 lymphocytes are then instructed to develop into Th1 or Th2 cells. IL-12, which drives Th1 generation, is provided by microbially stimulated dendritic cells and/or macrophages. On the other hand, the IL-4 that drives the Th2 development can originate from CD4+ lymphocytes themselves as well as from a variety of exogenous cell sources. In this model, direct instruction by IL-12/IL-4 is obligatory for the generation of fully differentiated Th1 and Th2 cells. (b) The stochastic model of Th differentiation. Upon Ag encounter, Thp lymphocytes generate a heterogeneous population of cells expressing random combinations of Th1- and Th2-type cytokines. The two extremes are CD4+ lymphocytes synthesizing only Th1- or only Th2-type cytokines and simultaneously expressing the corresponding receptors for IL-12 or IL-4, respectively. In the presence of the appropriate differentiative cytokines, such cells are favored and through positive selection are expanded to generate a homogeneous population of Th1 or Th2 lymphocytes. According to this model, cytokine secretion profiles evolve randomly and IL-12 and IL-4 merely select those with the correct phenotype.

mice exposed to a normally Th2-inducing helminth infectious stimulus exhibit an enhanced Th1 response but nevertheless are able to generate a small, but significant, number of cells with a mature Th2 phenotype [25]. Interestingly and importantly, in these animals, cells with a mixed Th0 phenotype are rare, an observation inconsistent with a stochastic model. Conversely, while Stat-4-KO mice show blunted Th1 and augmented Th2 responses, animals engineered to be deficient in both Stat-4 and Stat-6 display Th1 responses comparable to wild-type animals [29]. Taken together the data from these studies argue that (1) Th subset differentiation is fundamentally an autonomous process that does not require instruction by cytokines, and (2) that instead a principal role of IL-12 and IL-4 may be to negatively select cells expressing the undesired Th2/Th1 cytokine secretion profile while stabilizing those with the desired phenotype.

5 TH SUBSET DIFFERENTIATION AS AN AUTONOMOUS PROCESS MOLDED BY CYTOKINE SIGNALS

The above findings, although inconsistent with an instructional mechanism of Th1/Th2 differentiation, are also incompatible with a purely stochastic model in which the phenotypes emerge as a result of random lymphokine expression and positive selection by IL-12/IL-4. Rather, the combined data appear to support a mechanism involving an autonomous program for generating Th1 and Th2 cells, which are then selected and stabilized by cytokines. In the model that we propose (fig. 2), Ag encounter triggers a predetermined developmental pattern of CD4+ T-lymphocyte differentiation in which a pool of CD4+ cells is generated; these cells are heterogeneous in their lymphokine production but nevertheless homogenous in their expression of receptors for IL-12 and IL-4. Already at this beginning stage, Th cells produce either IFN-γ or IL-4, with lymphocytes synthesizing IFN-γ emerging earlier and in higher number than those expressing IL-4. Importantly, due to the uniform expression of cytokine receptors, all of the activated CD4+ lymphocytes have the potential to develop into mature Th1 or Th2 cells. If exogenous IL-12 or IL-4 is present in high amounts at the time of initial Ag encounter this default pattern is overridden and the emergence of Th1 or Th2 lymphocytes is accelerated.

Depending on the nature of the Ag stimulus (or type of microbial invader) the early activated T cells make an immunologic "decision." In the case of some stimuli (such as intracellular pathogens), IL-12 secretion by APC is triggered, resulting in the stabilization and upregulation of IFN-γ secretion by those T cells already making the cytokine (fig. 2, left). Moreover, the IFN-γ produced results in positive feedback for enhanced IL-12 synthesis. In addition, the few IL-4-producing cells present will be converted by IL-12 to cells making IFN-γ instead. As a consequence of the above processes of stabilization and conversion, the resulting population will be composed

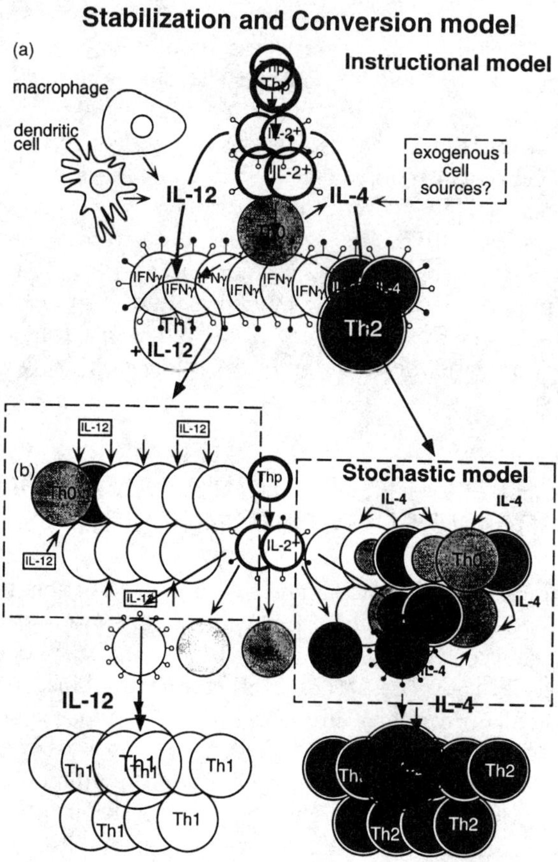

FIGURE 2 The stabilization/conversion model of Th differentiation. After initial encounter with Ag-activated Thp, lymphocytes undergo preprogrammed differentiation into a population consisting predominantly of IFN-γ^+ IL-4^- cells but also containing IFN-γ^- IL-4^+ cells. In the presence of IL-12 induced from APC, signaling through the IL-12 receptor results in the stabilization and upregulation of IFN-γ secretion by IFN-γ^+ IL-4^- cells as well as the induction of the cytokine in IFN-γ^- IL-4^+ producers converting some into Th0 or Th1 cells. Importantly in this model, Th2 cells can be generated in the absence of exogenous signaling by IL-4, utilizing their own autocrine cytokine as a differentiation factor. (Thus, the absence of IL-12 is the key determinant of Th2 differentiation.) When the secretion of IL-4 exceeds its consumption, it acts on neighboring IFN-γ^+ IL-4^- cells by suppressing their IFN-γ production and inducing IL-4 synthesis. As more cells produce IL-4 this process will be accelerated, resulting in a homogenous population of Th2 cells. This model predicts that cells with a mixed Th0 phenotype are unstable and transitory, owing to cytokine cross talk. According to the model IL-12 and IL-4 neither instruct nor select cells with the appropriate phenotype, but instead stabilize Th lymphocytes with the appropriate cytokine secretion pattern while converting cells with the inappropriate phenotype.

predominantly of Th1 cells. These mature Th1 cells maintain their phenotype by expressing an IL-4 receptor that is functionally impaired [23].

In the case of other antigenic stimuli (e.g., helminth infection) which fail to induce significant IL-12 production by APC, IL-4 becomes the dominant differentiative cytokine (fig. 2, right). In this situation the cytokine is produced in an autocrine fashion by the activated T cells themselves and does not depend on exogenous IL-4. The resulting IL-4 receptor/Stat6 triggering leads to induction of more IL-4 as well as the downregulation of IFN-γ production. During the process of conversion of IFN-γ to IL-4 producers, cells with a mixed Th0 phenotype will be generated transiently. (As discussed in the previous section, such cells do not appear in the absence of IL-4 receptor signaling.) Finally, as these cells mature into Th2 lymphocytes they become unresponsive to IL-12 as a result of IL-12 receptor downregulation [50].

Certain features of the above "stabilization/conversion" model are noteworthy in providing an efficient decision-making process for the immune response. First, by rapidly generating lymphocytes expressing either IFN-γ or IL-4, the immune system quickly mobilizes the two basic Th-cell prototypes needed for host defense. Second, the programmed early synthesis of IFN-γ may be functionally important in mediating innate resistance against invading viruses and intracellular pathogens and in limiting their initial spread. Moreover, since the later secretion of IL-4 can result in suppression of both IFN-γ and Th1 cells, the earlier emergence of IFN-γ expressing CD4+ lymphocytes provides a branch point for successfully establishing and stabilizing Th1 responses. A further key element is the role of negative cytokine-mediated cross regulation in the development of homogeneous Th1/Th2 populations. The concept that one Th population can negatively regulate the opposing subset is one of the founding tenets of the Th1/Th2 paradigm [16]. This feature provides a highly economical process for both the generation and preservation of a predominant phenotype and, as emphasized in our model, provides a mechanism for stabilization of Th2 populations when an initial exogenous IL-4 differentiative signal is absent.

6 WHY ONLY TWO MAJOR EFFECTOR CHOICES?

A fundamental aspect of the mechanism of effector choice in the immune system is its binary (Th1/Th2) nature. At face value this can be explained on the basis of a duality (e.g., intracellular vs. helminthic parasites) among the pathogens that the immune response must confront. Nevertheless, it could be argued that the protective functions of the Th2 response (e.g., immunity against helminths) are of limited evolutionary value and that the Th1 response is the primary defense against life-threatening infections. At the same time, uncontrolled Th1 responses are clearly also life-threatening and therefore need to be regulated. Thus, an alternative hypothesis is that the duality of effector choice in the immune system is the result of the need to balance one subset against the other and that, possibly, this balance is asymmetrical with the Th2 arm playing a more regulatory than protective function. Such an asymmetry is also evident in the differential requirement for exogenous IL-12 vs. IL-4 in the development of the two subsets. Indeed, one could argue that in terms of both evolution and ontogeny, the Th2 arm temporally succeeds the Th1 because of the need for this regulation.

7 LESSONS FROM TH1/TH2 DIFFERENTIATION ON THE NATURE OF BIOLOGIC EFFECTOR CHOICE

The decision-making process used by the immune system in confronting different pathogens has broad analogies to other biological systems where external stimuli must be recognized and responded to in a specific manner. The first common element is that of *surveillance*. The immune system handles this problem by generating cells that not only express different receptors for Ag but also exhibit different immunological functions as expressed in their cytokine expression patterns. This feature, as opposed to simple instruction-driven differentiation from a monospecific, inert precursor, provides a mechanism for sampling the nature of the target as well as the capacity to respond rapidly to its presence. The second common element is the process of *signaling*. In the case of the immune system the recognition of a particular class of stimuli is translated into a positive signal which triggers the production of more cells with the same phenotype. (In our model the *absence* of a positive signal also leads to a defined and distinct phenotype.) At the same time, cells which are positively triggered emit negative signals that suppress the emergence of cells of the inappropriate type. Thus, the development of the immune system is regulated by positive and negative signals as well as their absence.

A final common element is *stability*. Once a given type of response is established, it persists long enough to eliminate the stimulus. In the case of the immune system the class of the response is maintained both by genetic stabilization and cytokine cross regulation. One feature of the design of the immune system not shared by all biological effector choice processes is *memory*. Thus, even after a polarized Th1/Th2 response has waned, the more vigorous anamnestic response elicited by the same Ag expresses the original cytokine secretion phenotype [12]. This property of the adaptive immune system results in establishment of more efficient surveillance as the organism continues to be confronted with additional external stimuli.

Overall, the mechanism of effector choice employed by the immune system provides initial flexibility and plasticity, while ensuring an efficient and ultimately stable response. These features have contributed to the immune system's success in securing the survival of higher metazoan species against the different classes of pathogens they encounter while protecting the host against detrimental effects of its own excesses.

ACKNOWLEDGMENTS

We thank M. C. Kullberg and G. Yap for their helpful comments on this chapter.

REFERENCES

[1] Abbas, A. K., K. M. Murphy, and A. Sher. "Functional Diversity of Helper T Lymphocytes." *Nature* **383** (1996): 787–793.

[2] Abehsira-Amar, O., M. Gibert, M. Joliy, J. Theze, and D. Jankovic. "IL-4 Plays a Dominant Role in the Differential Development of Th0 into Th1 and Th2 Cells." *J. Immunol.* **148** (1992): 3820–3829.

[3] Agarwal, S., and A. Rao. "Modulation of Chromatin Structure Regulates Cytokine Gene Expression during T Cell Differentiation." *Immunity* **9** (1998): 765–775.

[4] Anderson, D. E., A. H. Sharpe, and D. A. Hafler. "The B7-CD28/CTLA-4 Costimulatory Pathways in Autoimmune Disease of the Central Nervous System." *Curr. Opin. Immunol.* **11** (1999): 677–683.

[5] Brown, W. C., A. C. Rice-Ficht, and D. M. Estes. "Bovine Type 1 and Type 2 Responses." *Vet. Immunol. Immunopthol.* **63** (1998): 45–55.

[6] Bird, J. J., D. R. Brown, A. C. Mullen, N. H. Moskowitz, M. A. Mahowald, J. R. Sider, T. F. Gajewski, C. R. Wang, and S. L. Reiner. "Helper T Cell Differentiation is Controlled by the Cell Cycle." *Immunity* **9** (1998): 229–237.

[7] Bix, M., and R. M. Locksley. "Independent and Epigenetic Regulation of the Interleukin-4 Alleles in CD4+ T Cells." *Science* **281** (1998): 1352–1354.

[8] Coffman, R. L., and T. von der Weid. "Multiple Pathways for the Initiation of T Helper 2 (Th2) Responses." *J. Exp. Med.* **185** (1997): 373–375.

[9] Constant, S. L., and K. Bottomly. "Induction of Th1 and Th2 CD4+ T Cell Responses: The Alternative Approaches." *Ann. Rev. Immunol.* **15** (1997): 297–322.

[10] Croft, M., L. Carter, S. L. Swain, and R. W. Dutton. "Generation of Polarized Antigen-Specific CD8 Effector Populations: Reciprocal Action of Interleukin (IL)-4 and IL-12 in Promoting Type 2 versus Type 1 Cytokine Profiles." *J. Exp. Med.* **180** (1994): 1715–1728.

[11] Del Prete, G., M. De Carli, C. Mastromauro, D. Macchia, R. Biagiotti, M. Ricci, and S. Romagnani. "Purified Protein Derivative of *Mycobacterium tuberculosis* and Excretory-Secretory Antigen(s) of *Toxocara canis* Expand in vitro Human T Cells with Stable and Opposite (Type 1 T Helper or Type 2 T Helper) Profile of Cytokine Production." *J. Clin. Invest.* **88** (1989): 346–350.

[12] Dutton, R. W., L. M. Bradley, and S. L. Swain. "T Cell Memory." *Ann. Rev. Immunol.* **16** (1998): 201–223.

[13] Else, K. J., F. D. Finkelman, C. R. Maliszewski, and R. K. Grensis. "Cytokine Mediated Regulation of Chronic Helminth Infection." *J. Exp. Med.* **179** (1994): 347–351.

[14] Ferrick, D. A, M. D. Schrenzel, T. Mulvania, B. Hsieh, G. W. Ferlin, and H. Lepper. "Differential Production of Interferon-Gamma and Interleukin-4 in Response to Th1- and Th2-Stimulating Pathogens by Gamma Delta Cells in vivo." *Nature* **373** (1995): 255–257.

[15] Finkelman, F. D., T. Shea-Donohue, J. Goldhill, C. A. Sullivan, S. C. Morris, K. B. Madden, W. C. Gause, and J. F. Urban. "Cytokine Regulation of Host Defense against Parasitic Gastrointestinal Helminths: Lessons from Studies with Rodent Models." *Ann. Rev. Immunol.* **15** (1997): 505–533.

[16] Fiorentino, D. F., M. W. Bond, and T. R. Mosmann. "Two Types of Mouse T Helper Cell. IV. Th2 Clones Secrete a Factor that Inhibits Cytokine Production by Th1 Clones." *J. Exp. Med.* **170** (1989): 2081–2095.

[17] Flesch, I., and S. H. Kaufmann. "Mycobacterial Growth Inhibition by Interferon-Gamma-Activated Bone Marrow Macrophages and Differential Susceptibility among Strains of *Mycobacterium tuberculosis*" *J. Immunol.* **138** (1987): 4408–4413.

[18] Fowell, D., A. J. McKnight, F. Powrie, R. Dyke, and D. Mason. "Subsets of CD4+ T Cells and Their Role in the Induction and Prevention of Autoimmunity." *Immunol. Rev.* **123** (1991): 37–64.

[19] Gett, A. V., and P. H. Hodgkin. "Cell Devision Regulates the T Cell Cytokine Repertoire, Revealing a Mechanism Underlying Immune Class Regulation." *Proc. Natl. Acad. Sci. USA* **95** (1998): 9488–9493.

[20] Gollob, K. J., and R. L. Coffman. "A Minority Subpopulation of CD4+ T Cells Directs the Development of Naïve CD4+ T Cells into IL-4-Secreting Cells." *J. Immunol.* **152** (1994): 5180–5188.

[21] Guery, J. C., F. Galbiati, S. Smiroldo, and L. Adorini. "Selective Development of T Helper (Th)2 Cells Induced by Continuous Administration of Low Dose Soluble Proteins to Normal and β2-Microglobulin-Deficient BALB/c Mice." *J. Exp. Med.* **183** (1996): 485–497.

[22] Hsieh, C. S., S. E. Macatonia, C. S. Tripp, S. F. Wolf, A. O'Garra, and K. M. Murphy. "Development of T_H1 CD4+ T Cells through IL-12 Produced by *Listeria*-Induced Macrophages." *Science* **260** (1993): 547–549.

[23] Huang, H., and W. E. Paul. "Impaired Interleukin 4 Signaling in T Helper Type 1 Cells." *J. Exp. Med.* **187** (1998): 1305–1313.

[24] Jankovic, D., and A. Sher. "Initiation and Regulation of CD4+ T Cell Function in Host-Parasite Models." *Chem. Immunol.* **63** (1996): 51–65.

[25] Jankovic, D., M. C. Kullberg, N. Noben-Trauth, P. Caspar, W. E. Paul, and A. Sher. "Single Cell Analysis Reveals that IL-4R/Stat6 Signaling is not Required for the Development in vivo and in vitro of CD4+ Lymphocytes with a Th2 Cytokine Profile." *J. Immunol.* **164** (2000): 3047–3055.

[26] Kamogawa, Y., L. A. Minasi, S. R. Carding, K. Bottomly, and R. A. Flavell. "The Relationship of IL-4- and IFNγ-Producing T Cells Studied by Lineage Ablation of IL-4-Producing Cells." *Cell* **75** (1993): 985–995.
[27] Kaplan, M. H., Y. L. Sun, T. Hoey, and M. J. Grusby. "Impaired IL-12 Responses and Enhanced Development of Th2 Cells in Stat4-Deficient Mice." *Nature* **382** (1996): 174–177.
[28] Kaplan, M. H., U. Schindler, S. T. Smiley, and M. J. Grusby. "Stat6 is Required for Mediating Responses to IL-4 and for Development of Th2 Cells." *Immunity* **4** (1996): 313–319.
[29] Kaplan, M. H., A. L. Wurster, and M. J. Grusby. "A Signal Transducer and Activator of Transcription (Stat)4-Independent Pathway for the Development of T Helper Type 1 Cells." *J. Exp. Med.* **188** (1998): 1191–1196.
[30] Kelso, A. "Thl and Th2 Subsets: Paradigms Lost?" *Immunol. Today* **16** (1995): 374–379.
[31] Kelso, A., P. Groves, A. B. Troutt, and K. Francis. "Evidence for Stochastic Acquisition of Cytokine Profile by CD4+ T Cells Activated in a T Helper Type 2-Like Response in vivo." *Eur. J. Immunol.* **25** (1995): 1168–1175.
[32] Kopf, M., G. Le Gros, M. Bachmann, M. C. Lamers, H. Bluethmann, and G. Köhler. "Disruption of the Murine IL-4 Gene Blocks Th2 Cytokine Responses." *Nature* **362** (1993): 245–248.
[33] Lederer, J. A., V. L. Perez, L. DesRoches, S. M. Kim, A. K. Abbas, and A. H. Lichtman. "Cytokine Transcriptional Events during Helper T Cell Subset Differentiation." *J. Exp. Med.* **184** (1996): 397–406.
[34] Le Gross, G., S. Ben-Sasson, R. Seder, F. D. Finkelman, and W. E. Paul. "Generation of Interleukin 4 (IL-4)-Producing Cells in vivo and in vitro: IL-2 and IL-4 are Required for in vitro Generation of IL-4 Producing Cells." *J. Exp. Med.* **172** (1990): 921–929.
[35] Locksley, R. M., and S. L. Reiner. "The Regulation of Immunity to *Leishmania major*." *Ann. Rev. Immunol.* **13** (1995): 151–177.
[36] Locksley, R. M., D. J. Fowell, K. Shinkai, A. E. Wakil, D. Lacy, and M. Bix. "Development of CD4+ Effector T Cells and Susceptibility to Infectious Diseases." In *Mechanisms of Lymphocyte Activation and Immune Regulation VII*, edited by S. Gupta, A. Sher, and R. Ahmed, 45–52. New York: Plenum Press, 1998.
[37] Magram, J., S. E. Connaughton, R. R. Warrier, D. M. Carvajal, C. Wu, J. Ferrante, U. Sarmiento, D. A. Faherty, and M. K. Gately. "IL-12-Deficient Mice are Deficient in IFN-γ Production and Type 1 Cytokine Responses." *Immunity* **4** (1996): 471–481.
[38] Mosmann, T. R., H. Cherwinski, M. W. Bond, M. A. Giedlin, and R. L. Coffman. "Two Types of Murine Helper T Cell Clone. I. Definition According to Profiles of Lymphokine Activities and Secreted Proteins." *J. Immunol.* **136** (1986): 2348–2357.

[39] Mosmann, T. R., and R. L. Coffman. "Heterogeneity of Cytokine Secretion Patterns and Functions of Helper T Cell." *Adv. Immunol.* **46** (1989): 111–147.

[40] O'Garra, A. "Cytokines Induce the Development of Functionally Heterogeneous T Helper Cell Subsets." *Immunity* **8** (1998): 275–283.

[41] Ossendorp, F., E. Mengede, M. Camps, R. Filius, and C. J. Melief. "Specific T Helper Cell Requirement for Optimal Induction of Cytotoxic T Lymphocytes against Major Histo-compatibility Complex Class II Negative Tumors." *J. Exp. Med.* **187** (1998): 693–702.

[42] Ricon, M., and R. A. Flavell. "T-Cell Subsets: Transcriptional Control in the Th1/Th2 Decision." *Curr. Biol.* **7** (1997): 729–732.

[43] Romagnani, S. "Lymphokine Production by Human T Cells in Disease States." *Ann. Rev. Immunol.* **12** (1994): 227–257.

[44] Seder, R. A., and W. E. Paul. "Acquisition of Lymphokine-Producing Phenotype by CD4+ T Cells." *Ann. Rev. Immunol.* **12** (1994): 635–973.

[45] Schmitz, J., A. Thiel, R. Kühn, K. Rajewsky, W. Müller, M. Assenmacher, and A. Radbruch. "Induction of Interleukin 4 (IL-4) Expression in T Helper (Th) Cells is not Dependent on IL-4 from non-Th Cells." *J. Exp. Med.* **179** (1994): 1349–1353.

[46] Sher, A., and R. L. Coffman. "Regulation of Immunity to Parasites by T Cells and T Cell-Derived Cytokines." *Ann. Rev. Immunol.* **10** (1992): 385–409.

[47] Shimoda, K., J. van Deursen, M. Y. Sangster, S. R. Sarawar, R. T. Carson, R. A. Tripp, C. Chu, F. W. Quelle, T. Nosaka, D. A. Vignali, P. C. Doherty, G. Grosveld, W. E. Paul, and J. N. Ihle. "Lack of IL-4-Induced Th2 Response and IgE Class Switching in Mice with Disrupted Stat6 Gene." *Nature* **380** (1996): 630–633.

[48] Street, N. E., J. H. Schumacher, T. A. T. Fong, H. Bass, D. F. Fiorentino, J. A. Leverah, and T. R. Mosmann. "Heterogeneity of Mouse Helper T Cells. Evidence from Bulk Cultures and Limiting Dilution Cloning for Precursors of Th1 and Th2 Cells." *J. Immunol.* **144** (1986): 1629–1639.

[49] Swain, S. L., A. D. Weinberg, M. English, and G Huston. "IL-4 Directs the Development of Th2-Like Helper Effectors." *J. Immunol.* **20** (1990): 3796–3806.

[50] Szabo, S. J., A. S. Dighe, U. Gubler, and K. M. Murphy. "Regulation of the Interleukin (IL)-12R Beta 2 Subunit Expression in Developing T Helper 1 (Th1) and Th2 Cells." *J. Exp. Med.* **185** (1997): 817–824.

[51] Thierfelder, W. E., J. M. van Deursen, K. Yamamoto, R. A. Tripp, S. R. Sarawar, R. T. Carson, M. Y. Sangster, D. A. Vignali, P. C. Doherty, G. C. Grosveld, and J. N. Ihle. "Requirement for Stat4 in Interleukin-12-Mediated Responses of Natural Killer and T Cells." *Nature* **382** (1996): 171–174.

[52] Takeda, K., T. Tanaka, W. Shi, M. Matsumoto, M. Minami, S. Kashiwamura, K. Nakanishi, N. Yoshida, T. Kishimoto, and S. Akira. "Essential Role of Stat6 in IL-4 Signalling." *Nature* **380** (1996): 627–630.

[53] Yap, G. S., and A. Sher. "Effector Cells of Both Nonhematopoietic and Hematopoeitic Origin are Required for Interferon (IFN)-γ- and Tumor Necrosis Fator (TNF)-α-dependent Host Resistance to the Intracellular Pathogen, *Toxoplasma gondii.*" *J. Exp. Med.* **189** (1999): 1083–1091.

Oral Tolerance

Howard L. Weiner

1 INTRODUCTION

A majority of the contacts with foreign antigenic materials occur at mucosal surfaces, which is larger than the area of the skin. The mucosal surface is constantly and physiologically exposed to a large variety of antigenic materials. Orally administered antigen encounters the gut-associated lymphoid tissue (GALT[1]) which has the inherent property of not only protecting the host from ingested pathogens but also preventing the host from reacting to ingested proteins. Thus, orally administered antigens induce systemic hyporesponsiveness to the fed proteins and this phenomenon is termed oral tolerance. It was first described in 1911 when Wells fed hen egg proteins to guinea pigs and found them resistant anaphylaxis when challenged [140]. In 1946, Chase fed guinea pigs the contact-sensitizing agent dinitrochlorobenzene (DNCB) and observed that animals had decreased skin reactivity to DNCB [13]. It has also been observed in humans fed and immunized with KLH [42]. There have been many studies trying to elucidate the mechanisms of oral tolerance [85, 139]

[1]Please refer to the abbreviation legend section at end of chapter.

and now it is clear that oral tolerance is mediated by T cells through different mechanisms depending on the dose of antigen fed. Low-dose antigen favors the induction of regulatory T cells which suppress Th1-cell-mediated response and high-dose antigen induces T-cell clonal anergy or deletion. In recent years, oral tolerance has been used successfully to treat autoimmune diseases in animal models and is now being applied to the treatment of human diseases.

2 MECHANISM OF ORAL TOLERANCE

2.1 INDUCTIVE PHASE

The GALT consists of villi that contain epithelial cells, intraepithelial lymphocytes (IELs), lamina propria lymphocytes (LPLs), and Peyer's patches, which are lymphoid nodules interspersed among the villi (fig. 1). Peyer's patches (PP) are one of the primary areas in the GALT where specific immune responses are generated. There have been attempts to use the GALT as a vaccination route, but this is difficult due to the systemic hyporesponsiveness that is generated. Although dietary antigens are degraded by the time they reach the small intestine, studies in humans and rodents have indicated that degradation is partial and that some intact antigen is absorbed into systemic circulation [11, 12, 41]. High-dose oral antigen may result in systemic antigen presentation, which induce hyporesponsiveness either via clonal T-cell anergy or clonal deletion.

It is generally believed that in low-dose antigen-fed animals, oral tolerance is induced in the GALT. Several cells capable of antigen presentation exist in the GALT. These include macrophages, dendritic cells, B cells, and epithelial cells. Macrophage-enriched cells obtained from mice fed ovalbumin (OVA) are able to stimulate antigen-primed lymph node T cells in vitro in an antigen-specific fashion without further exposure to antigen [98]. Dendritic cells have been shown to be the major intestinal antigen-presenting cells (APC) that can acquire and process orally administered antigen [61]. Epithelial cells may preferentially trigger the activation of CD8+ regulatory T cells. In the rat, these epithelial cell-induced CD8+ T cells are antigen specific [9] whereas in the human they were found to be antigen nonspecific [70]. MHC class II positive intestinal epithelial cells from 2,4-dinitrochlorobenzene (DNCB)-fed mice could induce anergy of DNCB-primed T cells [29]. Lamina propria cells (LPC) also may be involved in oral tolerance as antigen-presenting cells. Antigen-pulsed splenic APC stimulated Ag-specific Th0 cytokine production while PP APC in Peyer's patches induced a profile consistent with the provision of T-cell help for IgA production. Presentation of Ag by LPC stimulated a high level of IFN-g and TGF-β, and adoptive transfer of Ag pulsed LPC-induced oral tolerance to that antigen in the recipients [35]. However, the type of APC responsible for the effect of LPC was not clear.

Th2 helper cells are preferentially generated in the GALT [21, 146]. Th2 cell differentiation depends on the cytokine microenvironments or cytokine

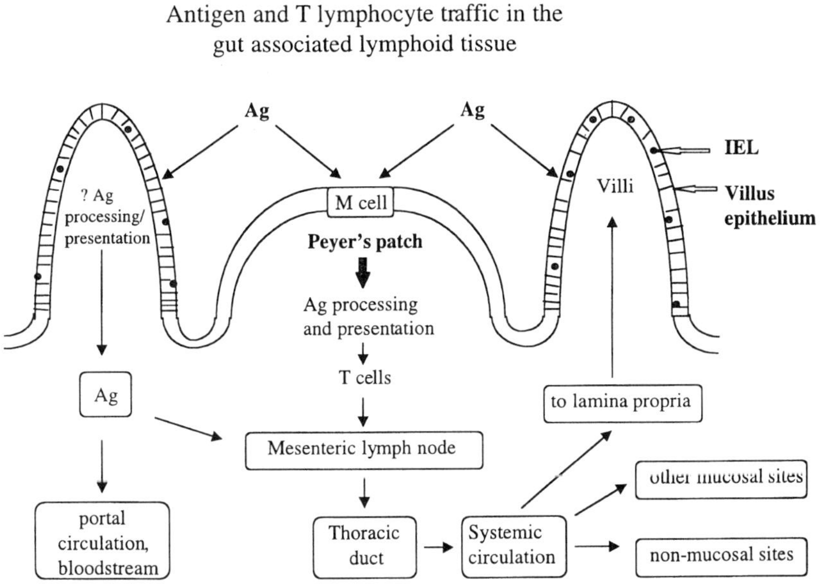

FIGURE 1 Antigen and T-cell traffic in the GALT. Antigen is taken up either via M cells into lymphoid nodules termed Peyer's patches or into the villus epithelium. Particulate antigen is preferentially taken up by M cells and soluble antigen by the villus epithelium. Antigen presentation results in the induction of cells that traffic to the systemic circulation via the mesenteric lymph node and thoracic duct, and then migrate back to the lamina propria and to other mucosal and nonmucosal sites. The villi contain IELs, which are CD8+ T cells unique to the gut. T cells in the lamina propria are in a different state of activation to those in Peyer's patches. Peyer's patches also contain B-cell rich, poorly formed, germinal centers where induction of antibody responses occurs.

milieu in which the Th precursor cells are exposed during their activation [1]. If IL-12 is present during activation, Th1 cells are differentiated while IL-4 induce Th2-cells differentiation. Microenvironment of intestinal mucosa may be crucial for the induction of Th2 or Th3 (TGF-β secreting cells). We have shown that dendritic cells, when exposed to IL-10, can drive Th2 cell differentiation of naive OVA TCR CD4+ transgenic T cells [60]. The influence of cytokine milieu on the antigen presentation by dendritic cells (DC) has also been demonstrated in vivo. Thus, DC exposed to IL-10 in vitro, when injected into footpad of mice, can prime for Th2-cell-type response [22]. It has also been shown in humans that PGE-2-treated DC can produce IL-10 and can prime naive human T cells for Th2-cell differentiation [48]. DC from

Peyer's patches preferentially stimulate Th0 clones to produce huge amounts of IL-4, while DC from spleen induce high IFN-g production [26, 45]. There is also evidence that DC may be involved in oral tolerance induction, in that expansion of DC in vivo with Flt3 ligand enhances oral tolerance [132]. It is possible that dendritic cells, most potent APC in activating resting T cells, under the influence of gut cytokine milieu, present antigen for Th2 or Th3 cell differentiation.

APC provide costimulatory signals required for the activation of T cells. B7.1 and B7.2 are the most important costimulatory molecules. B7.2 has been shown to be critical for Th2-type cell differentiation [27]. To determine the role of costimulatory molecules in the induction of oral tolerance, we have tested the effect of anti-B7.1 or anti-B7.2 mAb on the induction of tolerance by both high- and low-dose antigen feeding [59]. In experimental allergic encephalomyelitis (EAE) model for multiple sclerosis, anti-B7.2 mAb, but not anti-B7.1 mAb, inhibited the induction of oral tolerance induced by low-dose MBP feeding. We also found that CTLA-4 molecules on the APC appear to be critical for the induction of oral tolerance [102]. CD40-CD40 ligand interactions are also important for high-dose oral tolerance [57]. Class II molecules on the APC also appear to be critical for the induction of oral tolerance since oral tolerance can not be induced in class II deficient mice [23].

Recently, it was reported that Cyclooxygenease-2-(Cox-2) dependent arachidonic acid such as PGE2 is produced by lamina propria mononuclear cells and involved in oral tolerance [91]. Although it is antigen nonspecific, it is possible that this mechanism, which is not dependent on the cytokines, is important for generating immunoregulatory T cells.

2.2 EFFECTOR PHASE

It has become clear that there are two primary effector mechanisms of oral tolerance: the induction of regulatory T cells that mediate active suppression and the induction of clonal anergy or deletion. Depending on the dose of the antigen fed, it is determined which form of peripheral tolerance develops following oral administration of antigen. Low doses of antigen favor the generation of active suppression or regulatory-cell-driven tolerance whereas high doses of antigen favor anergy-driven tolerance (fig. 2). However, these two mechanisms are not mutually exclusive; they may occur simultaneously. Moreover, general definitions of "low" and "high" doses need to be established for each antigen.

2.2.1 Active Suppression. Many early studies demonstrated that active suppression is an important mechanism for oral tolerance [85]. After antigens were fed transferable suppression to cell-mediated immune responses was demonstrated using T cells from Peyer's patches, mesenteric lymph node, and spleen as sources of cells for adoptive transfer experiments. Active suppression is believed to be mediated by the induction of regulatory T cells in the GALT, for

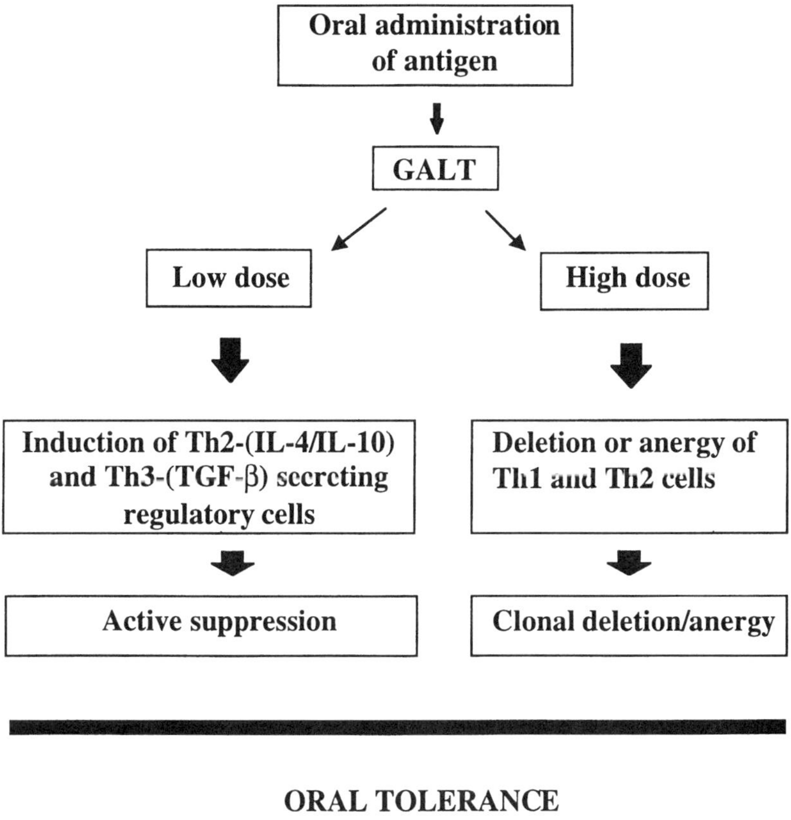

FIGURE 2 The different mechanisms of oral tolerance are determined by the dose of fed antigen. Abbreviations: GALT, gut-associated lymphoid tissue; IL, interleukin; TGF-β, transforming growth factor β; Th, T helper.

example in Peyer's patches [103]; these cells then migrate to the systemic immune system. Several reports indicate that one of the primary mechanisms of active cellular suppression is via the secretion of suppressive cytokines, such as TGF-β, IL-4, IL-10 following antigen specific triggering. TGF-β is produced both by CD4+ and CD8+ GALT-derived T cells and is an important mediator of the active suppression component of oral tolerance [15, 16]. TGF-β secreting CD4+ cells specific for MBP were cloned from the mesenteric lymph nodes of SJL mice [16]. These clones were found to be structurally identical to Th1-disease-inducing clones with respect to TCR usage, major histocompatibility complex (MHC) restriction and epitope recognition, but these clones suppressed rather than induced disease. TGF-β-secreting CD4+ cells were also cloned from MBP-TCR transgenic mice by culturing in the presence of IL-4 but not IL-2 [44]. These clones did not secret IL-2, IFN-g, IL-4, or IL-10. Thus, CD4+ cells that primarily produce TGF-β appear to be a unique T-cell subset that includes mucosal helper-T-cell function and downregulatory properties for Th1 and other immune cells. These cells have been termed Th3 cells. In contrast to Th1 and Th2 cells, Th3 cells provide help for IgA production and primarily secrete TGF-β [16, 83]. Th3-type cells appear distinct from Th2 cells since CD4+ TGF-β-secreting cells that suppress a form of colitis have been generated from IL-4 deficient mice [96]. Studies on rats have also demonstrated an essential role for TGF-β and IL-4 in the prevention of autoimmune thyroiditis by peripheral CD4+ CD45RC− cells and CD4+CD8− thymocytes [107]. Another type of regulatory T cell which is driven by IL-10 and secretes both IL-10 and TGF-β has been proposed, and termed a Tr1 cell [30].

Bystander suppression is a concept that regulatory cells, induced by a fed antigen, can suppress immune response stimulated by an irrelevant antigen, as long as the fed antigen is present in anatomic vicinity. This concept was demonstrated in vitro when it was shown that cells from animals fed MBP suppressed proliferation of OVA-specific cell line across a transwell, but only when triggered by the fed antigen [80]. The soluble factor shown to be responsible for the suppression was TGF-β. Bystander suppression has since been demonstrated in several autoimmune disease models (table 1). Bystander suppression solves a major conceptual problem in the design of antigen- or T-cell-specific therapy for inflammatory autoimmune diseases such as MS, type 1 diabetes, and rheumatoid arthritis (RA), in which the autoantigen is not clear or where there are reactivities to multiple autoantigens in the target tissue. During the course of chronic inflammatory autoimmune processes in animals, there is intra- and inter-antigenic spread of autoreactivity at the target organ [19, 51, 58, 71, 129]. Similarly, in human autoimmune diseases, there are reactivities to multiple autoantigens in the target tissue. For example, in MS, there is immune reactivity to at least three myelin antigens: MBP, PLP, and myelin oligodendrocyte glycoprotein (MOG) [53, 149]. In type 1 diabetes, there are multiple islet-cell antigens that could be the target of autoreactivity, including glutamic acid decarboxylase (GAD), insulin, and heat shock proteins [36]. Because regulatory cells induced by oral antigen secrete antigen-nonspecific cytokines after being triggered by the fed antigen, they suppress inflammation in the microenvironment where the fed antigen is localized. Thus, for a human organ-specific inflammatory disease, it is not necessary to know the specific antigen that is the target of an autoimmune response, but only to administer orally an antigen capable of inducing regulatory cells, which then migrate to the target tissue and suppress inflammation. Bystander suppression has also been shown for IL-10-secreting Tr1 cells in which an OVA-specific Tr1 clone could suppress a murine model of inflammatory bowel disease in vivo when fed [30]. Although bystander suppression was

TABLE 1 Models of Autoimmune and Other Diseases that Demonstrate Bystander Suppression. Abbreviation: BSA, bovine serum albumin; DTH, delayed-type hypersensitivity; EAE, experimental allergic encephalomyelitis; LCMV, lymphocytic choriomeningitis virus; MBP, myelin basic protein; OVA, ovalbumin; PLP, proteolipid; IBD, inflammatory bowel disease.

Autoimmune disease	Immunizing antigen	Oral antigen	Target organ
Arthritis	BSA, mycobacteria	Type II collagen	Joint
EAE	PLP	MBP	Brain
EAE	MBP peptide 71–90	MBP peptide 21–40	Brain
EAE	MBP	OVA	Lymph node, DTH response
Diabetes	LCMV	Insulin	Pancreatic islets
IBD	CD4+CD45RBhi T-cell transfer	OVA	Intestine
Stroke	None	MBP	Brain

initially described in association with regulatory cells induced by oral antigen, the process could in principal be induced by any immune manipulation that induces Th2-Tr1 or Th-3-type regulatory cells. Bystander suppression mediated by TGF-β secretion has also reported in a mouse model of transplantation tolerance [119].

2.2.2 Anergy and Clonal Deletion. Anergy is defined as a state of T-lymphocyte unresponsiveness characterized by the absence of proliferation by IL-2 production, and by diminished expression of IL-2R [106]. Anergy as a mechanism for oral tolerance has been shown indirectly or directly [131, 141]. A single feeding of 20 mg OVA induced a state of anergy in OVA-specific T cells [73]. T-cell clones derived from high-dose MBP-fed rats were characterized. Following several cell divisions in the presence of IL-2, these clones undergo a reversal of unresponsiveness [47]. Anergy as a mechanism has also been shown in a transfer system with OVA-TCR transgenic T cells [131]. All above studies associated with anergy as a tolerance mechanism used relatively high-dose antigen. Studies on the cells rendered anergic have raised the possibility that these cells do not function in a totally passive fashion in the tolerance they evoke. Recent reports have suggested that anergic cells can actively suppress T-cell responses either through modulation of the T-cell activating capacity of the APC (APC/T cell interaction) [116] or by inhibition of T cells recognizing their ligand in close proximity on the same APC ("linked suppression" through T/T cell interactions) [40]. In these cases described above, the so-called anergic cells serve as regulatory cells which mediate tolerance via an active mechanism.

Feeding very high dose antigen induced T-cell clonal deletion using OVA TCR transgenic model [14]. Mowat and coworkers also showed that lymphocytes from these animals die rapidly when cultured in vitro in the absence of antigen [86]. Other investigators did not find deletion in wild-type mice transferred with T cells from OVA TCR transgenic mice when they fed 25 mg OVA [131]. Oral tolerance to high doses of ovalbumin is reported to be normal in fas-deficient lpr mice [81, 84]. It is also reported that IL-12 is required to prevent an induction of Fas-mediated apoptosis after high-dose feeding of OVA to OVA-TCR transgenic mice [69]. Thus, clonal deletion occurs in transgenic mice fed a very high dose of antigen, but its role in high-dose tolerance in normal animals is unclear.

3 MODULATION OF ORAL TOLERANCE

A number of factors have been reported to modulate oral tolerance. As oral tolerance has usually been defined in terms of Th1 responses, anything that suppress Th1 and/or enhances Th2 or Th3 cell development would enhance oral tolerance (table 2). Th3 cells appear to use IL-4 as one of their growth/differentiation factors [44]. Seder also found that IL-4 and TGF-β may serve to promote growth of TGF-β-secreting cells [108]. Thus, intraperitonal (i.p.) IL-4 administration enhances low-dose oral tolerance to myelin basic protein (MBP) in the EAE model and is associated with increased fecal IgA anti-MBP antibodies. Oral IL-10 and IL-4 can also enhance oral tolerance when coadministered with antigen [112]. Cytokines have also been administered by the nasal route [145]. Large doses of IFN-γ given intraperitoneally abrogate oral tolerance [152]. Anti-IL-12 enhances oral tolerance and is associated both with increased TGF-β production and T-cell apoptosis [68], while subcutaneous administration of IL-12 reverses mucosal tolerance [17]. In the uveitis model, intraperitoneal IL-2 potentiates s oral tolerance and is associated with increased production of TGF-β, IL-10, and IL-4 [99]. Oral but not subcutaneous lipopolysaccaride (LPS) enhances oral tolerance to MBP [56] and is associated with increased expression of IL-4 in the brain. Oral IFN-β synergizes with the induction of oral tolerance in SJL/PLJ mice fed low doses of MBP [89].

Cholera toxin (CT) is one of the most potent mucosal adjuvants, and feeding CT abrogates oral tolerance when fed with an unrelated protein antigen [24]. However, when a protein is coupled to recombinant cholera toxin B subunit (CTB) and given orally, there is enhancement of peripheral immune tolerance [115]. Oral administration of corneal epithelial cells coupled to CTB markedly enhanced the corneal allograft survival [65].

Antibody to monocyte chemotactic protein 1 (MCP-1) abrogates oral tolerance [50]. Oral antigen delivery using a multiple emulsion system also enhances oral tolerance [25]. In oral tolerance induction $\gamma\delta$ T cells may have an important role since it is more difficult to induce oral tolerance in animals

TABLE 2 Modulation of oral tolerance. Abbreviations: Ab, antibody; CT, cholera toxin; CTB, choleran toxin B subunit; GVH, graft-versus-host; IFN, interferon; IL, interleukin; LPS, lipopolysaccharide; MCP-1, monocyte chemotactic protein 1.

Augments	Decreases
IL-2	IFN-γ
IL-4	IL-12
IL-10	CT
Anti-IL-12 Ab	Anti-MCP-1
TGF-β	Anti-$\gamma\delta$ Ab
INF-β	GVH
CTB	Anti-B7.2 mAb (low-dose tolerance)
F1t-3 ligand	
LPS	
Multiple emulsions	

depleted of such cells [52, 74] or in delta-chain-deficient animals [113]. The steroid hormone dehydroepiandrosterone (DHEA) breaks intranasally induced tolerance [144] and diesel exhaust particles block induction of oral tolerance in mice [147]. In the arthritis model, administration of TGF-β or dimaprid (a histamine type 2 receptor agonist) i.p.—both of which are believed to promote the development of immunoregulatory cells—enhances the induction of oral tolerance to collagen II even after the onset of arthritis [125].

4 TREATMENT OF AUTOIMMUNE DISEASES IN ANIMALS

Several studies have demonstrated the effectiveness of orally administered myelin antigens in rat and mouse models of autoimmune disease (table 3).

4.1 EXPERIMENTAL ALLERGIC ENCEPHALOMYELITIS

In the Lewis rat, high doses of MBP can suppress EAE via the mechanism of T-cell clonal anergy [46], whereas multiple, lower doses prevent EAE by transferable active cellular suppression [77]. In the nervous system of low-dose-fed animals, inflammatory cytokines such as TNF-α and IFN-γ are downregulated and TGF-β is upregulated [55]. Oral MBP partially suppresses serum antibody responses, especially at higher doses [38]. Administration of myelin to sensitized animals in the chronic guinea pig model or larger doses of MBP in the murine EAE model is protective and does not exacerbate disease [10, 70] and long-term (6 month) administration of myelin in the chronic EAE model was beneficial [101]. EAE can also be suppressed in animals transgenic for an MBP-specific T-cell receptor (TCR) following feeding with MBP [100]. Nasally

TABLE 3 Suppression of autoimmunity by oral tolerance. Abbreviations: AA, adjuvant arthritis; AIA, antigen-induced arthritis; AchR, acetylcholine receptor; CII, type II collagen; CIA, collagen-induced arthritis; EAE, experimental allergic encephalomyelitis; GAD, glutamic acid decarboxylase; gp39, glycoprotein 39; HSP, heat shock protein; IRBP, inter-photoreceptor retinoid-binding protein; MBP, myelin basic protein; MHC, major histocompatibility complex; NOD, nonobese diabetic; PLP, proteolipid protein; S-Ag, S antigen.

(a) Animal models	
Model	**Protein fed**
EAE	MBP, PLP, MOG
Arthritis (CIA, AA, AIA)	CII, HSP, gp39
Uveitis	S-Ag, IRBP
Myasthenia gravis	AchR
Diabetes (NOD mouse)	Insulin, GAD
Transplantation	Alloantigen, MHC peptide
Thyroiditis	Thyroglobulin
Colitis	Haptenized colonic proteins
(b) Human disease trials[84]	
Disease trial	**Protein fed**
Multiple sclerosis	Bovine myelin, glatiramer acetate
Rheumatoid arthritis	Chicken and bovine CII
Uveitis	Bovine S-Ag
Type I diabetes	Human insulin
Systemic sclerosis	Type 1 collagen

administered MBP peptides have been reported to suppress EAE [75]. The latest approach in animal models has been to utilize glatiramer acetate (Cop-1, Copaxone), a drug approved for therapy of multiple sclerosis, which is given to patients by injection. There are reports that Cop-1 suppresses EAE in both mouse and rat [67, 117, 118, 138].

4.2 ARTHRITIS

Oral administration of cartilage antigens, such as type II collagen, suppresses several models of arthritis including collagen-induced arthritis (CIA) [88, 122, 123], adjuvant arthritis [34, 151], pristine arthritis [124], and antigen-induced arthritis [148]. One of the first studies to demonstrate that an orally administered autoantigen can suppress an autoimmune disease was the use of oral type II collagen in CIA [121]. Oral type I collagen has also been shown to suppress adjuvant arthritis via bystander suppression [151]. Oral administration of an immunodominant human collagen peptide modulates CIA in mice [54] and type II collagen peptides given nasally also suppresses CIA in mice [87, 114].

Oral mycobacterial 65-kDA heat shock protein has also been shown to suppress adjuvant arthritis or avridine-induced arthritis [34, 97]. One interesting observation in treating arthritis models with collagen is that the suppression was observed at doses as low as 3 and 30 micrograms, suggesting that the mechanism involved is the generation of suppressive regulatory T cell rather than clonal anergy.

4.3 DIABETES

Oral insulin has been shown to delay and, in some instances, prevent diabetes in the nonobese diabetic (NOD) mouse model. Such suppression is transferable [150], primarily with $CD4^+$ cells [7]. Immunohistochemistry of pancreatic islets of Langerhans isolated from insulin-fed animals demonstrates decreased insulitis associated with decreased IFN-γ, as well as increased expression of TNF, IL-4, IL-10, TGF-β, and prostaglandin (PGE2) [33]. Recently, it was also reported that nasal administration of the insulin B chain or GAD and aerosol insulin suppresses diabetes in the NOD mouse [20, 37, 128]. Under special experimental conditions, large doses of OVA given to OVA double transgenic mice resulted in diabetes mediated by OVA-specific CTL [8]. These animals expressed OVA on the islets under the rat insulin promoter and were made chimeric to enrich for OVA-specific transgenic TCR CTL. Oral insulin suppressed diabetes in a viral induced model of diabetes in which LCMV was expressed under the insulin promoter and animals were infected with LCMV to induce diabetes [133]. Protection was associated with protective cytokine shifts (IL-4/IL-10, TGF-β) in the islets. Oral administration of the B-chain of insulin, a 30-amino-acid peptide, slowed the development of diabetes and prevented diabetes in some animals [95]. This effect was associated with a decrease in IFN-γ and an increase in IL-4, TGF-β, and IL-10 expression. Oral administration of recombinant GAD from plants suppresses the development of diabetes in the NOD mouse [66] as does oral administration of a plant-based CTB-insulin fusion protein [2].

4.4 UVEITIS

Oral administration of S antigen (S-Ag), a retinal autoantigen that induces experimental autoimmune uveitis (EAU), or of S-Ag peptides, prevents or markedly diminishes the clinical appearance of S-Ag-induced disease as measured by ocular inflammation [92, 111, 134]. S-Ag-induced EAU can also be suppressed by feeding an HLA peptide [142]. Feeding interphotoreceptor binding protein (IRBP) suppresses IRBP-induced disease and is potentiated by IL-2 [143]. Oral feeding of retinal antigen can not only prevent acute disease but can also effectively suppress a second attack in chronic-relapsing EAU, showing that oral tolerance may have practical clinical implications in uveitis, which is predominantly a chronic-relapsing condition in humans [126, 127].

4.5 OTHER MODELS

Although myasthenia gravis is an antibody-mediated disease, oral and nasal administration of the Torpedo acetylcholine receptor (AchR) to Lewis rats prevented or delayed the onset of myasthenia gravis [63, 136, 137]. Purified AchR was found more effective than an unpurified mixture [94]. Experimental autoimmune myasthenia gravis (EAMG) can also be suppressed by nasally administered AchR [63, 64], AchR peptides [49], and human AchR fragments [4]. It has recently been reported that nasal tolerance for EAMG by AchR is shown to be mediated by TGF-β-secreting CD4 cells using CD8 knockout mice [109].

It has been shown that oral feeding of haptenized colonic protein (HCP) effectively prevents 2,4,6-trinitrobenzene sulfonic acid (TNBS) induced granulomatous colitis via the generation of TGF-β-secreting T cells [90]. Oral administration of allogeneic cells prevents sensitization by skin grafts and changes accelerated rejection of vascularized cardiac allografts to an acute form that is typical of unsensitized recipients [105]. Orally administered allopeptides in the Lewis rat reduces delayed-type hypersensitivity (DTH) responses to the peptide [104]. Oral, but not intravenous, alloantigen was accompanied by elevation of intragraft levels of IL-4 [32]. Oral alloantigen enhanced corneal allograft survival even in preimmune hosts [65]. Oral thyroglobulin has been shown to suppress autoimmune thyroiditis [31] and feeding peptides of the Der p I allergen suppressed responses to the whole allergen [39]. Experimental granulomatous colitis in mice is abrogated by TGF-β-mediated oral tolerance after administration of haptenized colonic proteins [90]. Oral tolerization to adenoviral antigens permitted long-term gene expression using recombinant adenoviral vectors [43].

5 TREATMENT OF AUTOIMMUNE DISEASES IN HUMANS

Investigators have shown that exposure of a contact-sensitizing agent via the mucosa prior to subsequent skin challenge led to unresponsiveness in a portion of patients studied [62]. KLH administered orally to human subjects has been reported to decrease subsequent cell-mediated immune responses although antibody responses were not affected [42]. Nasal KLH has also been reported to induce tolerance in humans [135].

On the basis of the long history of oral tolerance and the safety of the approach, human trials have been initiated in MS, RA, uveitis, and diabetes (table 3(b)). These initial trials suggest that there has been no systemic toxicity or exacerbation of disease, although reproducible clinical efficacy has yet to be demonstrated. Results in humans, however, have paralleled several aspects of what has been observed in animals.

In MS patients, MBP- and PLP-specific TGF-β-secreting Th3-type cells have been observed in the peripheral blood of patients treated with an oral bovine myelin preparation and not in patients who were untreated [28]. There was no increase in MBP- or PLP-specific IFN-γ-secreting cells in treated patients. These results demonstrate that it is possible to immunize via the gut for autoantigen-specific TGF-β-secreting cells in a human autoimmune disease by oral administration of the autoantigen. However, a recently completed 515-patient, placebo-controlled, double-blind Phase III trial of single-dose bovine myelin in relapsing-remitting MS did not show differences in the number of relapses between placebo and treated groups. A large placebo effect was observed (AutoImmune, Inc., Lexington, MA, USA). The dose of myelin was 300 mg given in capsule form and contained 8 mg MBP and 15 mg PLP. Preliminary analysis of magnetic resonance imaging data showed significant changes favoring oral myelin in certain patient subgroups. A new trial of oral tolerance in MS is being undertaken with glatiramer acetate (Cop-1), an MBP analogue, which is currently given by injection to MS patients but which has been shown to be effective orally in animals and to induce regulatory cells that mediate bystander suppression [118, 138].

In RA, a 280-patient double-blind phase II dosing trial of chicken type II collagen in doses ranging from 20 mg to 2500 mg for six months demonstrated statistically significant positive effects in the group treated with the lowest dose [6]. Oral administration of larger doses of bovine type II collagen (1–10 mg) did not show a significant difference between tested and placebo groups, although a higher prevalence of responders was reported for the groups treated with type II collagen [110]. These results are consistent with animal studies of orally administered type II collagen in which protection against adjuvant- and antigen-induced arthritis and bystander suppression was observed only at the lower doses [148, 151]. An open-label pilot study of oral collagen in juvenile RA gave positive results with no toxicity [5]. This lack of systemic toxicity is an important feature for the clinical use of oral tolerance, especially in children for whom the long-term effects of immunosuppressive drugs is unknown. Recently completed several Phase II trials of oral collagen, which involved 805 patients treated with CII and 296 treated with placebo, showed that 60 μg was the most significant dose compared to other doses [6, 130]. Using linear logistic regression, a statistically significant effect was found in patients treated on oral CII versus those on the placebo. However, integrated efficacy analysis of response predictors including HLA, CII antibodies, tender and swollen joint count showed no predictors. Based on these data, a phase III trial has been carried out administering 60 μg vs. placebo. No differences were observed between groups; however, a 51% response rate occurred in the placebo group.

In uveitis, a pilot trial of S-Ag and an S-Ag mixture has been completed at the National Eye Institute (Bethesda, MD, USA) and showed positive trends with oral bovine S-Ag but not the retinal mixture [93]. Feeding of peptide derived from the patient's own HLA antigen appeared to have an effect on uveitis in that patients could discontinue their steroids because of reduced intraocular inflammation mediated by oral tolerance [127].

Trials have been initiated in new-onset diabetes in which recombinant human insulin is administered orally, and trials are underway in subjects at risk for diabetes as part of the diabetes prevention trial (DPT-1). Preliminary analysis of a randomized double-blind placebo-controlled study of oral insulin in newly diagnosed type 1 diabetes demonstrated preserved beta-cell function as measured by endogenous C-peptide insulin responses in adult new-onset diabetics fed 10 mg of recombinant human insulin as compared to those fed placebo [18].

Oral desensitization to nickel allergy in humans induces a decrease in nickel-specific T cells and affects cutaneous eczema [3]. Positive effects were reported in an open-label pilot study of oral type 1 collagen in patients with systemic sclerosis [72]. A pilot immunological study of oral MHC peptides has been initiated in transplantation patients. Based on results to date in humans, it appears that the clinical application of oral antigen for the treatment of human conditions will depend on the specific disease and the nature and dosages of proteins administered. The use of synergists or mucosal adjuvants may be required to enhance biologic effects. Also, recombinant human proteins may be more efficacious than animals proteins [79].

6 FUTURE DIRECTIONS

Although it is clear that oral antigen can suppress autoimmunity and inflammatory diseases in animals, much remains to be learned. Under certain experimental conditions, oral antigen has worsened autoimmune diseases in animals [8, 76, 78, 120]. Cell surface molecules and cytokines associated with inductive events in the gut that generate and modulate oral tolerance are not completely understood. Important areas of investigation include cytokine and chemokine milieu, antigen presentation and costimulation requirements, routes of antigen processing, form of the antigen, role of the liver, the effect or oral antigens on antibody and IgE responses and on CTLs, and the role of $\gamma\delta$ T cells or CD4+CD25+ cells. As the molecular events associated with the generation and modulation of oral tolerance are better understood, the ability to apply mucosal tolerance successfully for the treatment of human autoimmune and other diseases will be further enhanced.

ABBREVIATION LEGEND

AChR	Acetylcholine receptor
APC	Antigen-presenting cells
CIA	Collagen-induced arthritis
CTB	Cholera toxin B subunit
DC	Dendritic cell
DHEA	Dehydroepiandrosterone
DNCB	Dinitrochlorobenzene
DPT-1	Diabetes prevention trial
EAMG	Experimental autoimmune myasthenia gravis
EAU	Experimental autoimmune uveitis
GAD	Glutamic acid decarboxylase
GALT	Gut-associated lymphoid tissue
HCP	Haptenized colonic protein
IELs	Intraepithelial lymphocytes
IFN-g	Interferon gamma
IRBP	Interphotoreceptor binding protein
KLH	Keyhole limpet hemocyanin
LCMV	Lymphocytic choreomeningitis virus
LPC	Lamina propria cells
LPL	Lamina propria lymphocytes
MBP	Myelin basic protein
MCP-1	Monocyte chemotactic protein 1
MHC	Major histocompatibility complex
MOG	Myelin oligodendrocyte glycoprotein
NOD	Nonobese diabetic
OVA	Ova albumin
PGE2	Prostaglandin
PP	Peyer's patches
RA	Rheumatoid arthritis
TCR	T-cell receptor
TGF-β	Transforming growth factor beta

REFERENCES

[1] Abbas, A. K., K. M. Murphy, and A. Sher. "Functional Diversity of Helper T lymphocytes." *Nature* **383** (1996): 787–793.

[2] Arakawa, T., J. Yu, D. K. Chong, J. Hough, P. C. Engen, and W. H. Langridge. "A Plant-Based Cholera Toxin B Subunit-Insulin Fusion Protein Protects against the Development of Autoimmune Diabetes." *Nat. Biotechnol.* **16** (1998): 934–938.

[3] Bagot, M., D. Charue, M. L. Flechet, N. Terki, A. Toma, and J. Revuz. "Oral Desensitization in Nickel Allergy Induces a Decrease in Nickel-Specific T Cells." *Eur. J. Dermatol.* **5** (1995): 614–617.

[4] Barchan, D., M. C. Souroujon, S. H. Im, C. Antozzi, and S. Fuchs. "Antigen-Specific Modulation of Experimental Myasthenia Gravis: Nasal Tolerization with Recombinant Fragments of the Human Acetylcholine Receptor Alpha-Subunit." *Proc. Natl. Acad. Sci. USA* **96** (1999): 8086–8091.

[5] Barnett, M. L., D. Combitchi, and D. E. Trentham. "A Pilot Trial of Oral Type II Collagen in the Treatment of Juvenile Rheumatoid Arthritis." *Arthritis Rheum.* **39** (1996): 623–628.

[6] Barnett, M. L., J. M. Kremer, E. W. St. Clair, D. O. Clegg, D. Furst, M. Weisman, M. J. F. Fletcher, P. T. Lavin, E. Finger, A. Morales, C. H. Le, and D. E. Trentham. "Treatment of Rheumatoid Arthritis with Oral Type II Collagen: Results of a Multicenter, Double-Blind, Placebo-Controlled Trial." *Arthritis Rheum.* **41** (1998): 290–297.

[7] Bergerot, J., N. Fabien, V. Maguer, and C. Thivolet. "Oral Administration of Human Insulin to NOD Mice Generates CD4+ T Cells that Suppress Adoptive Transfer of Diabetes." *J. Autoimmun.* **7** (1994): 655–663.

[8] Blanas, E., F. R. Carbone, J. Allison, J. F. A. P. Miller, and W. R. Heath. "Induction of Autoimmune Diabetes by Oral Administration of Autoantigen." *Science* **274** (1996): 1707–1709.

[9] Bland, P. W., and L. G. Warren. "Antigen Presentation by Epithelial Cells of the Rat Small Intestine. II. Selective Induction of Suppressor T Cells." *Immunology* **58** (1986): 9–14.

[10] Brod, S. A., A. al-Sabbagh, R. A. Sobel, D. A. Hafler, and H. L. Weiner. "Suppression of Experimental Autoimmune Encephalomyelitis by Oral Administration of Myelin Antigens. IV. Suppression of Chronic Relapsing Disease in the Lewis Rat and Strain 13 Guinea Pig." *Ann. Neurol.* **29** (1991): 615–622.

[11] Bruce, M. G., and A. Ferguson. "The Influence of Intestinal Processing on the Immunogenicity and Molecular Size of Absorbed, Circulating Ovalbumin in Mice." *Immunology* **59** (1986): 295–300.

[12] Bruce, M. G., and A. Ferguson. "Oral Tolerance Induced by Gut-Processed Antigen." *Adv. Exp. Med. Biol.* **216A** (1987): 721–731.

[13] Chase, M. "Inhibition of Experimental Drug Allergy by Prior Feeding of the Sensitizing Agent." *Proc. Soc. Exp. Biol. Med.* **61** (1946): 257–259.
[14] Chen, Y., J. Inobe, R. Marks, P. Gonnella, V. K. Kuchroo, and H. L. Weiner. "Peripheral Deletion of Antigen-Reactive T Cells in Oral Tolerance." *Nature* **376** (1995): 177–180.
[15] Chen, Y., J. Inobe, and H. L. Weiner. "Induction of Oral Tolerance to Myelin Basic Protein in CD8-Depleted Mice: Both CD4+ and CD8+ Cells Mediate Active Suppression." *J. Immunol.* **155** (1995): 910–916.
[16] Chen, Y., V. K. Kuchroo, J.-I. Inobe, D. A. Hafler, and H. L. Weiner. "Regulatory T-Cell Clones Induced by Oral Tolerance: Suppression of Autoimmune Encephalomyelitis." *Science* **265** (1994): 1237–1240.
[17] Claessen, A. M., B. M. von Blomberg, J. De Groot, D. A. Wolvers, G. Kraal, and R. J. Scheper. "Reversal of Mucosal Tolerance by Subcutaneous Administration of Interleukin-12 at the Site of Attempted Sensitization." *Immunology* **88** (1996): 363–367.
[18] Coutant, R., A. Zeidler, R. Rappaport, D. Schatz, S. Schwartz, P. Raskin, D. Rogers, B. Bode, S. Crockett, J. Marks, L. Deeb, S. Chalew, and N. MacLaren. "Oral Insulin Therapy in Newly Diagnosed Immune-Mediated (Type I) Diabetes. Preliminary Analysis of a Randomized Double-Blind Placebo-Controlled Study." *Diabetes* **47** (Suppl 1) (1998): A97.
[19] Cross, A. H., V. K. Tuohy, and C. S. Raine. "Development of Reactivity to New Myelin Antigens during Chronic Relapsing Autoimmune Demyelination." *Cell. Immunol.* **146** (1993): 261–270.
[20] Daniel, D., and D. R. Wegmann. "Protection of Nonobese Diabetic Mice from Diabetics by Intranasal or Subcutaneous Administration of Insulin Peptide B-(9-23)." *Proc. Natl. Acad. Sci. USA* **93** (1996): 956–960.
[21] Daynes, R., B. Araneo, T. Dowell, K. Huang, and D. Dudley. "Regulation of Murine Lymphokine Production in vivo. III. The Lymphoid Tissue Microenvironment Exerts Regulatory Influences over T-Helper Cell Function." *J. Exp. Med.* **171** (1990): 979–996.
[22] DeSmedt, T., M. Van Mechelen, G. De Becker, J. Urbain, O. Leo, and M. Moser. "Effect of Interleukin-10 on Dendritic Cell Maturation and Function." *Eur. J. Immunol.* **27** (1997): 1229–1235.
[23] Desvignes, C., H. Bour, J. F. Nicolas, and D. Kaiserlian. "Lack of Oral Tolerance but Oral Priming for Contact Sensitivity to Dinitrofluorobenzene in Major Histocompatibility Complex Class II-Deficient Mice and in CD4+ T-Cell-Depleted Mice." *Eur. J. Immunol.* **26** (1996): 1756–1761.
[24] Elson, C. O., and W. Ealding. "Cholera Toxin Feeding Did Not Induce Oral Tolerance in Mice and Abrogated Oral Tolerance to an Unrelated Protein Antigen." *J. Immunol.* **133** (1984): 2892–2897.
[25] Elson, C. O., M. Tomasi, M. T. Dertzbaugh, G. Thaggard, R. Hunter, and C. Weaver. "Oral Antigen Delivery by Way of a Multiple Emulsion

System Enhances Oral Tolerance." *Ann. NY Acad. Sci.* **778** (1996): 156–162.

[26] Everson, M. P., D. G. Lemak, J. R. McGhee, and K. W. Beagley. "FACS-Sorted Spleen and Peyer's Patch Dendritic Cells Induce Different Responses in Th0 Clones." *Adv. Exp. Med. Biol.* **417** (1997): 357–362.

[27] Freeman, G. J., V. A. Boussiotis, A., Anumanthan, G. M. Bernstein, K.-Y. Ke, P. D. Rennert, G. S. Gray, J. G. Gribben, and L. M. Nadler. "B7-1 and B7-2 Do Not Deliver Identical Costimulatory Signals, since B7-2 but Not B7-1 Preferentially Costimulates the Initial Production of IL-4." *Immunity* **2** (1995): 523–532.

[28] Fukaura, H., S. C. Kent, M. J. Pietrusewicz, S. J. Khoury, H. L.Weiner, and D. A. Hafler. "Induction of Circulating Myelin Basic Protein and Proteolipid Protein-Specific Transforming Growth Factor-Beta1-Secreting Th3 T Cells by Oral Administration of Myelin in Multiple Sclerosis Patients." *J. Clin. Invest.* **98** (1996): 70–77.

[29] Galliaerde, V., C. Desvignes, E. Peyron, and D. Kaiserlian. "Oral Tolerance to Haptens: Intestinal Epithelial Cells from 2,4-Dinitrochlorobenzene-Fed Mice Inhibit Hapten-Specific T-Cell Activation in vitro." *Eur. J. Immunol.* **25** (1995): 1385–1390.

[30] Groux, H., A. O'Garra, M. Bigler, M. Rouleau, S. Antonenko, J. E. de Vries, and M. G. Roncarolo. "A CD4+ T-Cell Subset Inhibits Antigen-Specific T-Cell Responses and Prevents Colitis." *Nature* **389** (1997): 737–742.

[31] Guimaraes, V. C., J. Quintans, M.-E. Fisfalen, F. H. Straus, K. Wilhelm, G. A. Medeiros-Neto, and L. J. DeGroot. "Suppression of Experimental Autoimmune Thyroiditis by Oral Administration of Thyroglobulin." *Endocrinology* **136** (1995): 3353–3359.

[32] Hancock, W., M. Sayegh, C. Kwok, H. Weiner, and C. Carpenter. "Oral but Not Intravenous, Alloantigen Prevents Accelerated Allograft Rejection by Selective Intragraft Th2 Cell Activation." *Transplantion* **55** (1993): 1112–1118.

[33] Hancock, W. W., M. Polanski, Z. J. Zhang, N. Blogg, and H. L. Weiner. "Suppression of Insulitis in NOD Mice by Oral Insulin Administration is Associated with Selective Expression of IL-4, IL-10, TGF-β, and Prostaglandin-E." *Am. J. Pathol.* **147** (1995): 1193–1199.

[34] Haque, M. A., S. Yoshino, S. Inada, H. Nomaguchi, O. Tokunaga, and O. Kohashi. "Suppression of Adjuvant Arthritis in Rats by Induction of Oral Tolerance to Mycobacterial 65-kDa Heat Shock Protein." *Eur. J. Immunol.* **26** (1996): 2650–2656.

[35] Harper, H. M., L. Cochrane, and N. A. Williams. "The Role of Small Intestinal Antigen-Presenting Cells in the Induction of T-Cell Reactivity to Soluble Protein Antigens: Association between Aberrant Presentation in the Lamina Propria and Oral Tolerance." *Immunology* **89** (1996): 449–456.

[36] Harrison, L. C. "Islet-Cell Antigens in Insulin-Dependent Diabetes: Pandora's Box Revisited." *Immunol. Today* **13** (1992): 348–352.

[37] Harrison, L. C., M. Dempsey-Collier, D. R. Kramer, and K. Takahashi. "Aerosol Insulin Induces Regulatory CD8 $\gamma\delta$ T Cells that Prevent Murine Insulin-Dependent Diabetes." *J. Exp. Med.* **184** (1996): 2167–2174.

[38] Higgins, P., and H. L. Weiner. "Suppression of Experimental Autoimmune Encephalomyelitis by Oral Administration of Myelin Basic Protein and Its Fragments." *J. Immunol.* **140** (1988): 440–445.

[39] Hoyne, G. F., M. G. Callow, M. C. Kuo, and W. R. Thomas. "Inhibition of T-Cell Responses by Feeding Peptides Containing Major and Cryptic Epitopes: Studies with the Der p I Allergen." *Immunology* **83** (1994): 190–195.

[40] Hoyne, G. F., and J. R. Lamb. "Regulation of T-Cell Function in Mucosal Tolerance." *Immunol. Cell Biol.* **75** (1997): 197–201.

[41] Husby, S., J. C. Jensenius, and S.-E. Svehag. "Passage of Undergraded Dietary Antigen into the Blood of Healthy Adults. Further Characterization of the Kinetics of Uptake and the Size Distribution of the Antigen." *Scand. J. Immunol.* **24** (1986): 447–452.

[42] Husby, S., J. Mestecky, Z. Moldoveanu, S. Holland, and C. O. Elson. "Oral Tolerance in Humans. T Cell but Not B-Cell Tolerance after Antigen Feeding." *J. Immunol.* **152** (1994): 4663–4670.

[43] Ilan, Y., R. Prakash, A. Davidson, V. Jona, G. Droguett, M. S. Horwitz, N. R. Chowdhury, and J. R. Chowdhury. "Oral Tolerization to Adenoviral Antigens Permits Long-Term Gene Expression using Recombinant Adenoviral Vectors." *J. Clin. Invest.* **99** (1997): 1098–1106.

[44] Inobe, J., A. J. Slavin, Y. Komagata, Y. Chen, L. Liu, and H. L. Weiner. "IL-4 is a Differentiation Factor for Transforming Growth Factor-Beta Secreting Th3 Cells and Oral Administration of IL-4 Enhances Oral Tolerance in Experimental Allergic Encephalomyelitis." *Eur. J. Immunol.* **28** (1998): 2780–2790.

[45] Iwasaki, A., and B. L. Kelsall. "Freshly Isolated Peyer's Patch, but Not Spleen, Dendritic Cells Produce Interleukin 10 and Induce the Differentiation of T Helper Type 2 Cells." *J. Exp. Med.* **190** (1999): 229–239.

[46] Javed, N. H., I. E. Gienapp, K. L. Cox, and C. C. Whitacre. "Exquisite Peptide Specificity of Oral Tolerance in Experimental Autoimmune Encephalomyelitis." *J. Immunol.* **155** (1995): 1599–1605.

[47] Jewell, S., J. Dierksheide, A. Curry, A. Shrestha, and J. Waldman. "Suppression of Experimental Autoimmune Encephalomyelitis (EAE) by Portal Vein (PV) Injection of Myelin Basic Protein (MBP)." *FASEB J.* **12** (1998): A600.

[48] Kalinski, P., C. M. Hilkens, A. Snijders, F. G. Snijdewint, and M. L. Kapsenberg. "IL-12-Deficient Dendritic Cells, Generated in the Presence of Prostaglandin E2, Promote Type 2 Cytokine Production in Maturing Human Naive T Helper Cells." *J. Immunol.* **159** (1997): 28–35.

[49] Karachunski, P. I., N. S. Ostlie, D. K.Okita, and B. M. Conti-Fine. "Prevention of Experimental Myasthenia Gravis by Nasal Administration of Synthetic Acetylcholine Receptor T Epitope Sequences." *J. Clin. Invest.* **100** (1997): 3027–3035.

[50] Karpus, W. J., K. J. Kennedy, S. L. Kunkel, and N. W. Lukacs. "Monocyte Chemotactic Protein 1 Regulates Oral Tolerance Induction by Inhibition of T Helper Cell-Related Cytokines." *J. Exp. Med.* **187** (1998): 733–741.

[51] Kaufman, D. I., M. Clare-Salzler, J. Tian, T. Forsthuber, G. S. P. Ting, P. Robinson, M. A. Atkinson, E. E. Sercaz, A. J. Tobin, and P. V. Lehmann. "Spontaneous Loss of T-Cell Tolerance to Glutamic Acid Decarboxylase in Murine Insulin-Dependent Diabetes." *Nature* **366** (1993): 69–72.

[52] Ke, Y., K. Pearce, J. P. Lake, H. K. Ziegler, and J. A. Kapp. "Gamma Delta T Lymphocytes Regulate the Induction and Maintenance of Oral Tolerance." *J. Immunol.* **158** (1997): 3610–3618.

[53] Kerlero de Rosbo, N., R. Milo, M. B. Lees, D. Burger, C. C. A. Bernard, and A. Ben-Nun. "Reactivity to Myelin Antigens in Multiple Sclerosis: Peripheral Blood Lymphocytes Respond Predominantly to Myelin Oligodendrocyte Glycoprotein." *J. Clin. Invest.* **92** (1993): 2602–2608.

[54] Khare, S. D., C. J. Krco, M. M. Griffiths, H. S. Luthra, and C. S. David. "Oral Administration of an Immunodominant Human Collagen Peptide Modulates Collagen-Induced Arthritis." *J. Immunol.* **155** (1995): 3653–3659.

[55] Khoury, S. J., W. W. Hancock, and H. L. Weiner. "Oral Tolerance to Myelin Basic Protein and Natural Recovery from Experimental Autoimmune Encephalomyelitis as Associated with Downregulation of Inflammatory Cytokines and Differential Upregulation of Transforming Growth Factor β, Interleukin 4, and Prostaglandin Expression in the Brain." *J. Exp. Med.* **176** (1992): 1355–1364.

[56] Khoury, S. J., O. Lider, A. al-Sabbagh, and H. L. Weiner. "Suppression of Experimental Autoimmune Encephalomyelitis by Oral Administration of Myelin Basic Protein. III. Synergistic Effect of Lipopolysaccharide." *Cell. Immunol.* **131** (1990): 302–310.

[57] Kweon, M. N., K. Fujihashi, Y. Wakatsuki, T. Koga, M. Yamamoto, J. R. McGhee, and H. Kiyono. "Mucosally Induced Systemic T-Cell Unresponsiveness to Ovalbumin Requires CD40 Ligand-CD40 Interactions." *J. Immunol.* **162** (1999): 1904–1909.

[58] Lehmann, P., T. Forsthuber, A. Miller, and E. Sercarz. "Spreading of T-Cell Autoimmunity to Cryptic Determinants of an Autoantigen." *Nature* **358** (1992): 155.

[59] Liu, L., V. K. Kuchroo, and H. L. Weiner. "B7.2 but Not B7.1 Costimulation is Required for the Induction of Low-Dose Oral Tolerance." *FASEB J.* **I** (1998): A597.

[60] Liu, L., B. E. Rich, J.-I. Inobe, W. Chen, and H. L. Weiner. "Induction of T Helper 2 Cell Differentiation in the Primary Immune Response: Dendritic Cells Isolated from Adherent Cell Culture Treated with Interleukin-10 Prime Naive CD4+ T Cells to Secrete Interleukin-4." *Int. Immunol.* **10** (1998): 1017–1026.

[61] Liu, L. M., and G. G. MacPherson. "Antigen Acquisition by Dendritic Cells: Intestinal Dendritic Cells Acquire Antigen Administered Orally and Can Prime Naive T Cells in vivo." *J. Exp. Med.* **177** (1993): 1299–1307.

[62] Lowney, E. D. "Immunologic Unresponsiveness to a Contact Sensitizer in Man." *J. Invest. Dermatol.* **51** (1968): 411–417.

[63] Ma, C.-G., G.-X. Zhang, B.-G. Xiao, J. Link, T. Olsson, and H. Link. "Suppression of Experimental Autoimmune Myasthenia Gravis by Nasal Administration of Acetylcholine Receptor." *J. Neuroimmunol.* **58** (1995): 51–60.

[64] Ma, C. G., G.-X. Zhang, B. G. Xiao, and H. Link. "Cellular mRNA Expression of Interferon-Gamma (IFN-γ), IL-4 and Transforming Growth Factor-Beta (TGF-β) in Rats Nasally Tolerized Against Experimental Autoimmune Myasthenia Gravis (EAMG)." *Clin. Exp. Immunol.* **104** (1996): 509–516.

[65] Ma, D., J. Mellon, and J. Y. Niederkorn. "Oral Administration as a Strategy for Enhancing Corneal Allograft Survival." *Brit. J. Ophthalmol.* **81** (1997): 778–784.

[66] Ma, S. W., D. L. Zhao, Z. Q. Yin, R. Mukherjee, B. Singh, H. Y. Qin, C. R. Stiller, and A. M. Jevnikar. "Transgenic Plants Expressing Autoantigens Fed to Mice to Induce Oral Immune Tolerance." *Nature Med.* **3** (1997): 793–796.

[67] Maron, R., A. Slavin, and H. L. Weiner. "Oral Tolerance to Glatiramer Acetate (Copl, Copaxone) in MBPT Cell Receptor Transgenic Mice." *J. Neuroimmunol.* **90** (1998): 82.

[68] Marth, T., W. Strober, and B. L. Kelsall. "High-Dose Oral Tolerance in Ovalbumin TCR-Transgenic Mice: Systemic Neutralization of IL-12 Augments TGF-β Secretion and T-Cell Apoptosis." *J. Immunol.* **157** (1996): 2348–2357.

[69] Marth, T., M. Zeitz, B. R. Ludviksson, W. Strober, and B. L. Kelsall. "Extinction of IL-12 Signaling Promotes Fas-Mediated Apoptosis of Antigen-Specific T Cells." *J. Immunol.* **162** (1999): 7233–7240.

[70] Mayer, L., and R. Shlien. "Evidence for Function of Ia Molecules on Gut Epithelial Cells in Man." *J. Exp. Med.* **166** (1987): 1471–1483.

[71] McCarron, R., R. Fallis, and D. McFarlin. "Alterations in T-Cell Antigen Specificity and Class II Restriction during the Course of Chronic Relapsing Experimental Allergic Encephlomyelitis." *J. Neuroimmunol.* **29** (1990): 73–79.

[72] McKown, K. M., L. D. Carbone, J. Bustillo, J. M. Sever, A. H. Kang, and A. E. Postlethwaite. "Open Trial of Oral Type I Collagen in Patients with Systemic Sclerosis." *Arthritis Rheum.* **40** (1997): S100.

[73] Melamed, D., and A. Friedman. "Direct Evidence for Anergy in T Lymphocytes Tolerized by Oral Administration of Ovalbumin." *Eur. J. Immunol.* **23** (1993): 935–942.

[74] Mengel, J., F. Cardillo, L. S. Aroeira, O. Williams, M. Russo, and N. M. Vaz. "Anti-$\gamma\delta$ T-Cell Antibody Blocks the Induction and Maintenance of Oral Tolerance to Ovalbumin in Mice." *Immunol. Lett.* **48** (1995): 97–102.

[75] Metzler, B., and D. C. Wraith. "Inhibition of Experimental Autoimmune Encephalomyelitis by Inhalation but Not Oral Administration of the Encephalitogenic Peptide: Influence of MHC Binding Affinity." *Int. Immunol.* **5** (1993): 1159–1165.

[76] Meyer, A. L., J. M. Benson, I. E. Gienapp, K. L. Cox, and C. C. Whitacre. "Suppression of Murine Chronic Relapsing Experimental Autoimmune Encephalomyelitis by the Oral Administration of Myelin Basic Protein." *J. Immunol.* **157** (1996): 4230–4238.

[77] Miller, A., A. al-Sabbagh, L. Santos, M. P. Das, and H. L. Weiner. "Epitopes of Myelin Basic Protein that Trigger TGF-β Release Following Oral Tolerization are Distinct from Encephalitogenic Epitopes and Mediate Epitope Driven Bystander Suppression." *J. Immunol.* **151** (1993): 7307–7315.

[78] Miller, A., O. Lider, O. Abramsky, and H. L. Weiner. "Orally Administered Myelin Basic Protein in Neonates Primes for Immune Responses and Enhances Experimental Autoimmune Encephalomyelitis in Adult Animals." *Eur. J. Immunol.* **24** (1994): 1026–1032.

[79] Miller, A., O. Lider, A. al-Sabbagh, and H. L. Weiner. "Suppression of Experimental Autoimmune Encephalomyelitis by Oral Administration of Myelin Basic Protein. V. Hierarchy of Suppression by Myelin Basic Protein from Different Species." *J. Neuroimmunol.* **39** (1992): 243–250.

[80] Miller, A., O. Lider, and H. L. Weiner. "Antigen-Driven Bystander Suppression Following Oral Administration of Antigens." *J. Exp. Med.* **174** (1991): 791–798.

[81] Miller, M. L., J. S. Cowdery, C. A. Laskin, M. Curtin, Jr., and A. D. Steinberg. "Heterogeneity of Oral Tolerance Defects in Autoimmune Mice." *Clin. Immunol. Immunopathol.* **31** (1984): 231–240.

[82] Moingeon, P., H. C. Chang, B. P. Wallner, C. Stebbins, A. Z. Frey, and E. L. Reinherz. "CD2-Mediated Adhesion Facilitates T Lymphocyte Antigen Recognition Function." *Nature* **339** (1989): 312–314.

[83] Mosmann, T. R., and S. Sad. "The Expanding Universe of T-Cell Subsets: Th1, Th2, and More." *Immunol. Today* **17** (1996): 138–146.

[84] Mowat, A. "Putative Role of p55 TNF Receptor, but Not Fas in Oral Tolerance." *FASEB J.* **12** (1998): A598.

[85] Mowat, A. M. "The Regulation of Immune Responses to Dietary Protein Antigens." *Immunol. Today* **8** (1987): 93–98.

[86] Mowat, A. M., M. Steel, E. A. Worthy, P. J. Kewin, and P. Garside. "Inactivation of Th1 and Th2 Cells by Feeding Ovalbumin." *Ann. NY Acad. Sci.* **778** (1996): 122–132.

[87] Myers, L. K., J. M. Seyer, J. M. Stuart, and A. H. Kang. "Suppression of Murine Collagen-Induced Arthritis by Nasal Administration of Collagen." *Immunology* **90** (1997): 161–164.

[88] Nagler-Anderson, C., L. A. Bober, M. E. Robinson, G. W. Siskind, and F. J. Thorbeke. "Suppression of Type II Collagen-Induced Arthritis by Intragastric Administration of Soluble Type II Collagen." *Proc. Natl. Acad. Sci. USA* **83** (1986): 7443–7446.

[89] Nelson, P. A., Y. Akselband, S. M. Dearborn, A. al-Sabbagh, Z. J. Tian, P. A. Gonnella, S. S. Zamvil, Y. Chen, and H. L. Weiner. "Effect of Oral Beta Interferon on Subsequent Immune Responsiveness." *Ann. NY Acad. Sci.* **778** (1996): 145–155.

[90] Neurath, M. F., I. Fuss, B. L. Kelsall, D. H. Presky, W. Waegell, and W. Strober. "Experimental Granulomatous Colitis in Mice is Abrogated by Induction of TGF-β-Mediated Oral Tolerance." *J. Exp. Med.* **183** (1996): 2605–2616.

[91] Newberry, R. D., W. F. Stenton, and R. G. Lorenz. "Cyclooxygenase-2-Dependent Arachidonic Acid Metabolites are Essential Modulators of the Intestinal Immune Response to Dietary Antigen." *Nature Med.* **5** (1999): 900–906.

[92] Nussenblatt, R. B., R. R. Caspi, R. Mahdi, C. C. Chan, F. Roberge, O. Lider, and H. L. Weiner. "Inhibition of S-Antigen-Induced Experimental Autoimmune Uveoretinitis by Oral Induction of Tolerance with S-Antigen." *J. Immunol.* **144** (1990): 1689–1695.

[93] Nussenblatt, R. B., I. Gery, H. L. Weiner, F. Ferris, J. Shiloach, N. Ramaley, C. Perry, R. Caspi, D. A. Hafler, S. Foster, and S. M. Whitcup. "Treatment of Uveitis by Oral Administration of Retinal Antigens: Results of a Phase I/II Randomized Masked Trial." *Am. J. Ophthalmol.* **123** (1997): 583–592.

[94] Okumura, S., K. McIntosh, and D. B. Drachman. "Oral Administration of Acetylcholine Receptor: Effects on Experimental Myasthenia Gravis." *Ann. Neurol.* **36** (1994): 704–713.

[95] Polanski, M., N. S. Blogg, J. Zhang, and H. L. Weiner. "Oral Administration of the Immunodominant B-Chain of Insulin Suppresses Diabetes in NOD Mice and is Associated with a Switch from Th1 to Th2 Cytokines." *J. Autoimmun.* **10** (1997): 339–346.

[96] Powrie, F., J. Carlino, M. W. Leach, S. Mauze, and R. L. Coffman. "A Critical Role for Transforming Growth Factor-Beta but Not Interleukin 4 in the Suppression of T Helper Type 1-Mediated Colitis by CD45RB (low) CD4+ T Cells." *J. Exp. Med.* **183** (1996): 2669–2674.

[97] Prakken, B. J., R. van der Zee, S. M. Anderton, P. J. S. van Kooten, W. Kuis, and W. van Eden. "Peptide-Induced Nasal Tolerance for a Mycobacterial Heat Shock Protein 60 T-Cell Epitope in Rats Suppresses Both Adjuvant Arthritis and Nonmicrobially Induced Experimental Arthritis." *Proc. Natl. Acad. Sci. USA* **94** (1997): 3284–3289.

[98] Richman, L. K., A. S. Graeff, and W. Strober. "Antigen Presentation by Macrophage-Enriched Cells from the Mouse Peyer's Patch." *Cell. Immunol.* **62** (1981): 110–118.

[99] Rizzo, L. V., N. E. Miller-Rivero, C.-C. Chan, B. Wiggert, R. B. Nussenblatt, and R. R. Caspi. "Interleukin-2 Treatment Potentiates Induction of Oral Tolerance in a Murine Model of Autoimmunity." *J. Clin. Invest.* **94** (1994): 1668–1672.

[100] al-Sabbagh, A. M., G. Garcia, A. J. Slavin, H. L. Weiner, and P. A. Nelson. "Combination Therapy with Oral Myelin Basic Protein and Oral Methotrexate Enhances Suppression of Experimental Autoimmune Encephalomyelitis." *Neurology* **48** (1997): A421.

[101] al-Sabbagh, A. M., E. P. Goad, H. L. Weiner, and P. A. Nelson. "Decreased CNS Inflammation and Absence of Clinical Exacerbation of Disease after Six Months' Oral Administration of Bovine Myelin in Diseased SJL/J Mice with Chronic Relapsing Experimental Autoimmune Encephalomyelitis." *J. Neurosci. Resh.* **45** (1996): 424–429.

[102] Samoilova, E. B., J. L. Horton, H. Zhang, S. J. Khoury, H. L. Weiner, and Y. Chen. "CTLA4 is Required for the Induction of High-Dose Oral Tolerance." *Int. Immunol.* **10** (1998): 491–498.

[103] Santos, L. M. B., A. al-Sabbagh, A. Londono, and H. L. Weiner. "Oral Tolerance to Myelin Basic Protein Induces Regulatory TGF-β-Secreting T Cells in Peyer's Patches of SJL Mice." *Cell. Immunol.* **157** (1994): 439–447.

[104] Sayegh, M. H., S. J. Khoury, W. H. Hancock, H. L. Weiner, and C. B. Carpenter. "Induction of Immunity and Oral Tolerance with Polymorphic Class II Major Histocompatability Complex Allopeptides in the Rat." *Proc. Natl. Acad. Sci. USA* **89** (1992): 7762–7766.

[105] Sayegh, M. H., Z. J. Zhang, W. W. Hancock, C. A. Kwok, C. B. Carpenter, and H. L. Weiner. "Down-Regulation of the Immune Response to Histocompatibility Antigen and Prevention of Sensitization by Skin Allografts by Orally Administered Alloantigen." *Transplantion* **53** (1992): 163–166.

[106] Schwartz, R. H. "A Cell Culture Model for T Lymphocyte Clonal Anergy." *Science* **248** (1990): 1349–1356.

[107] Seddon, B., and D. Mason. "Regulatory T Cells in the Control of Autoimmunity: The Essential Role of Transforming Growth Factor Beta and Interleukin 4 in the Prevention of Autoimmune Thyroiditis in Rats by Peripheral CD4(+)CD45RC- Cells and CD4(+)CD8(−) Thymocytes." *J. Exp. Med.* **189** (1999): 279–288.

[108] Seder, R. A., T. Marth, M. C. Sieve, W. Strober, J. J. Letterio, A. B. Roberts, and B. Kelsall. "Factors Involved in the Differentiation of TGF-β-Producing Cells from Naive CD4+ T Cells: IL-4 and IFN-β Have Opposing Effects, while TGF-β Positively Regulates Its Own Production." *J. Immunol.* **160** (1998): 5719–5728.

[109] Shi, F. D., H. L. Li, H. B. Wang, X. F. Bai, P. H. van der Meide, H. Link, and H. G. Ljunggren. "Mechanisms of Nasal Tolerance Induction in Experimental Autoimmune Myasthenia Gravis: Identification of Regulatory Cells." *J. Immunol.* **162** (1999): 5757–5763.

[110] Sieper, J., S. Kary, H. Sörensen, R. Alten, U. Eggens, W. Hüge, F. Hiepe, A. Kühne, J. Listing, N. Ulbrich, J. Braun, A. Zink, and N. A. Mitchison. "Oral Type II Collagen Treatment in Early Rheumatoid Arthritis." *Arthritis Rheum.* **39** (1996): 41–51.

[111] Singh, V. K., H. K. Kalra, K. Yamaki, and T. Shinohara. "Suppression of Experimental Autoimmune Uveitis in Rats by the Oral Administration of the Uveitopathogenic S-Antigen Fragment and a Cross-Reactive Homologous Peptide." *Cell. Immunol.* **139** (1992): 81–90.

[112] Slavin, A. J., R. Maron, G. Garcia, P. Gonnella, and H. L. Weiner. "Oral Administration of IL-4 and IL-10 Enhance the Induction of Low-Dose Oral Tolerance." *FASEB J.* **II** (1998): A599.

[113] Spahn, T. W., and H. L. Weiner. "$\gamma\delta$ T Cells are Necessary for Low-Dose but Not High-Dose Oral Tolerance." *FASEB J.* **II** (1998): A597, 3464.

[114] Staines, N. A., N. Harper, F. J. Ward, V. Malmström, R. Holmdahl, and S. Bansal. "Mucosal Tolerance and Suppression of Collagen-Induced Arthritis (CIA) Induced by Nasal Inhalation of Synthetic Peptide 184–198 of Bovine Type II Collagen (CII) Expressing a Dominant T-Cell Epitope." *Clin. Exp. Immunol.* **103** (1996): 368–375.

[115] Sun, J.-B., C. Holmgren, and C. Czerkinsky. "Cholera Toxin B Subunit: An Efficient Transmucosal Carrier-Delivery System for Induction of Peripheral Immunological Tolerance." *Proc. Natl. Acad. Sci. USA* **91** (1994): 10795–10799.

[116] Taams, L. S., A. J. M. L. van Rensen, M. C. M. Poelen, C. A. C. M. van Els, A. C. Besseling, J. P. A. Wagenaar, W. van Eden, and M. H. M. Wauben. "Anergic T-Cells Actively Suppress T-Cell Responses via the Antigen-Presenting Cell." *Eur. J. Immunol.* **28** (1998): 2902–2912.

[117] Teitelbaum, D., R. Arnon, and M. Sela. "Immunomodulation of Experimental Allergic Encephalomyelitis by Oral Administration of Copolymer 1 (Copaxone)." *J. Neuroimmunol.* **90** (1998): 85.

[118] Teitelbaum, D., R. Arnon, and M. Sela. "Immunomodulation of Experimental Autoimmune Encephalomyelitis by Oral Administration of Copolymer 1." *Proc. Natl. Acad. Sci. USA* **96** (1999): 3842–3847.

[119] Teng, Y., R. Gorczynski, and N. Hozumi. "The Function of TGF-Beta-Mediated Innocent Bystander Suppression Associated with Physiological Self-Tolerance in vivo." *Cell. Immunol.* **190** (1998): 51–60.

[120] Terato, K., J. Y. Xiu, H. Miyahara, M. A. Cremer, and M. M. Griffiths. "Induction by Chronic Autoimmune Arthritis in DBA/1 Mice by Oral Administration of Type II Collagen and *Escherichia coli* Lipopolysaccharide." *Brit. J. Rheumatol.* **35** (1996): 828–838.

[121] Thompson, H. S., and N. A. Staines. "Suppression of Collagen-Induced Arthritis with Pregastrically or Intravenously Administered Type II Collagen." *Agents Actions* **19** (1986): 318–319.

[122] Thompson, H. S. G., N. Harper, D. J. Bevan, and N. A. Staines. "Suppression of Collagen-Induced Arthritis by Oral Administration of Type II Collagen: Changes in Immune and Arthritic Responses Mediated by Active Peripheral Suppression." *Autoimmunity* **16** (1993): 189–199.

[123] Thompson, H. S. G., and N. A. Staines. "Gastric Administration of Type II Collagen Delays the Onset and Severity of Collagen-Induced Arthritis in Rats." *Clin. Exp. Immunol.* **64** (1986): 581–586.

[124] Thompson, S. J., H. S. G. Thompson, N. Harper, M. J. Day, A. J. Coad, C. J., Elson, and N. A. Staines. "Prevention of Pristane-Induced Arthritis by the Oral Administration of Type II Collagen." *Immunology* **79** (1993): 152–157.

[125] Thorbecke, G. J., R. Schwarcz, J. Leu, C. Huang, and W. J. Simmons. "Modulation by Cytokines of Induction of Oral Tolerance to Type II Collagen." *Arthritis Rheum.* **42** (1999): 110–118.

[126] Thurau, S. R., C. C. Chan, R. B. Nussenblatt, and R. R. Caspi. "Oral Tolerance in a Murine Model of Relapsing Experimental Autoimmune Uveoretinitis (EAU): Induction of Protective Tolerance in Primed Animals." *Clin. Exp. Immunol.* **109** (1997): 370–376.

[127] Thurau, S. R., M. Diedrichs-Mohring, H. Fricke, S. Arbogast, and G. Wildner. "Molecular Mimicry as a Therapeutic Approach for an Autoimmune Disease: Oral Treatment of Uveitis Patients with an MHC-Peptide Crossreactive with Autoantigen—First Results." *Immunol. Lett.* **57** (1997): 193–201.

[128] Tian, J., M. A. Atkinson, M. Clare-Salzler, A. Herschenfeld, T. Forsthuber, P. V. Lehmann, and D. L. Kaufman. "Nasal Administration of Glutamate Decarboxylase (GAD65) Peptides Induces Th2 Responses and Prevents Murine Insulin-Dependent Diabetes." *J. Exp. Med.* **183** (1996): 1561–1567.

[129] Tisch, R., X.-D.Yang, S. M. Singer, R. S. Liblau, L. Fugger, and H. O. McDevitt. "Immune Response to Glutamic Acid Decarboxylase Correlates with Insulitis in Nonobese Diabetic Mice." *Nature* **366** (1993): 72–75.

[130] Trentham, D., R. Dynesius-Trentham, E. Orav, D. Combitchi, C. Lorenzo, K. Sewell, D. Hafler, and H. Weiner. "Effects of Oral Administration of Type II Collagen on Rheumatoid Arthritis." *Science* **261** (1993): 1727–1730.

[131] Van Houten, N., and S. F. Blake. "Direct Measurement of Anergy of Antigen-Specific T Cells Following Oral Tolerance Induction." *J. Immunol.* **157** (1996): 1337–1341.

[132] Viney, J. L., A. M. Mowat, J. M., O'Malley, E. Williamson, and N. A. Fanger. "Expanding Dendritic Cells in vivo Enhances the Induction of Oral Tolerance." *J. Immunol.* **160** (1998): 5815–5825.

[133] Von Herrath, M. G., T. Dyrberg, and M. B. A. Oldstone. "Oral Insulin Treatment Suppresses Virus-Induced Antigen-Specific Destruction of Beta Cells and Prevents Autoimmune Diabetes in Transgenic Mice." *J. Clin. Invest.* **98** (1996): 1324–1331.

[134] Vrabec, T. R., D. S. Gregerson, H. S. Dua, and L. A. Donoso. "Inhibition of Experimental Autoimmune Uveoretinitis by Oral Administration of S-Antigen and Synthetic Peptides." *Autoimmunity* **12** (1992): 175–184.

[135] Waldo, F. B., A. W. L. Van Den Wall Bake, J. Mestecky, and S. Husby. "Suppression of the Immune Response by Nasal Immunization." *Clin. Immunol. Immunopathol.* **72** (1994): 30–34.

[136] Wang, H.-M., and K. A. Smith. "The Interleukin-2 Receptor: Functional Consequences of Its Bimolecular Structure." *J. Exp. Med.* **166** (1987): 1055.

[137] Wang, Z. Y., J. Qiao, and H. Link. "Suppression of Experimental Autoimmune Myasthenia Gravis by Oral Administration of Acetylcholine Receptor." *J. Neuroimmunol.* **44** (1993): 209–214.

[138] Weiner, H. L. "Oral Tolerance with Copolymer 1 for the Treatment of Multiple Sclerosis." *Proc. Natl. Acad. Sci. USA* **96** (1999): 3333–3335.

[139] Weiner, H. L. "Oral Tolerance: Immune Mechanisms and Treatment of Autoimmune Diseases." *Immunol. Today* **18** (1997): 335–343.

[140] Wells, H. G. "Studies on the Chemistry of Anaphylaxis (III). Experiments with Isolated Proteins, Especially Those of the Hen's Egg." *J. Infect. Dis.* **8** (1911): 147–171.

[141] Whitacre, C. C., I. E. Gienapp, C. G. Orosz, and D. Bitar. "Oral Tolerance in Experimental Autoimmune Encephalomyelitis. III. Evidence for Clonal Anergy." *J. Immunol.* **147** (1991): 2155–2163.

[142] Wildner, G., and S. R. Thurau. "Cross-Reactivity between an HLA-B27-Derived Peptide and a Retinal Autoantigen Peptide: A Clue to Major Histocompatibility Complex Association with Autoimmune Disease." *Eur. J. Immunol.* **24** (1994): 2579–2585.

[143] Wildner, G., and S. R. Thurau. "Orally Induced Bystander Suppression in Experimental Autoimmune Uveoretinitis Occurs Only in the Periphery and Not in the Eye." *Eur. J. Immunol.* **25** (1995): 1292–1297.

[144] Wolvers, D. A., J. M. Bakker, W. M. Bagchus, and G. Kraal. "The Steroid Hormone Dehydroepiandrosterone (DHEA) Breaks Intranasally Induced Tolerance, when Administered at Time of Systemic Immunization." *J. Immunol.* **89** (1998): 19–25.

[145] Xiao, B. G., X. F. Bai, G. X. Zhang, and H. Link. "Suppression of Acute and Protracted-Relapsing Experimental Allergic Encephalomyelitis by

Nasal Administration of Low-Dose IL-10 in Rats." *J. Neuroimmunol.* **84** (1998): 230–237.

[146] Xu-Amano, J., W. K. Aicher, T. Taguchi, H. Kiyono, and J. R. McGhee. "Selective Induction of Th_2 Cells in Murine Peyer's Patches by Oral Immunization." *Int. Immunol.* **4** (1992): 433–445.

[147] Yoshino, S., M. Ohsawa, and M. Sagai. "Diesel Exhaust Particles Block Induction of Oral Tolerance in Mice." *J. Pharmacol. Exp. Theor.* **287** (1998): 679–683.

[148] Yoshino, S., E. Quattrocchi, and H. L. Weiner. "Oral Administration of Type II Collagen Suppresses Antigen-Induced Arthritis in Lewis Rats." *Arthritis Rheum.* **38** (1995): 1092–1096.

[149] Zhang, J., S. Markovic, J. Raus, B. Lacet, H. L. Weiner, and D. A. Hafler. "Increased Frequency of IL-2 Responsive T-Cells Specific for Myelin Basic Protein and Proteolipid Protein in Peripheral Blood and Cerebrospinal Fluid of Patients with Multiple Sclerosis." *J. Exp. Med.* **179** (1993): 973–984.

[150] Zhang, J. Z., L. Davidson, G. Eisenbarth, and H. L. Weiner. "Suppression of Diabetes in NOD Mice by Oral Administration of Porcine Insulin." *Proc. Natl. Acad. Sci. USA* **88** (1991): 10252–10256.

[151] Zhang, J. Z., C. S. Lee, O. Y., Lider, and H. L. Weiner. "Suppression of Adjuvant Arthritis in Lewis Rats by Oral Administration of Type II Collagen." *J. Immunol.* **145** (1990): 2489–2493.

[152] Zhang, Z., and J. G. Michael. "Orally Inducible Immune Unresponsiveness is Abrogated by IFN-γ Treatment." *J. Immunol.* **144** (1990): 4163–4165.

Part III: Design Principles for the Immune System

An Introduction to Immuno-ecology and Immuno-informatics

Charles G. Orosz

1 INTRODUCTION

Most immunologists accept the premise that the immune system is a complex network of interactive elements, yet few have given serious consideration to the biologic implications of this fact. This is unfortunate, since few elemental features impact the immune system as profoundly as does its networked character. By design, the immune network is an extremely large collection of decentralized, diverse components with extensive, nonlinear connections. Yet it is highly dynamic, and inherently capable of maintaining coherence under stress, of adapting to change, and of learning. How does it do each of these things? What keeps this complex network from plunging into chaos when perturbed? How does the immune system deal with its own complexity? These seminal questions have yet to be addressed by immunologists. Until they are, immunology will remain a discipline that is adept at accumulating biologic facts, but incapable of deciphering the in situ mechanisms of biologic function. Any real progress toward this goal will require a deep appreciation for the networked character of the immune system, an area that has received little study to date. This area, referred to herein as *immuno-ecology*, is the study of the immunologic principles that permit effective immunologic function within the context of the immensely complex immunologic network.

The purpose of this communication is to discuss the key principles of *immuno-ecology* in detail. In so doing, this discussion will provide the conceptual framework for the related topic *immuno-informatics*, which is the study of the immune system as a cognitive, decision-making device. *immuno-informatics* addresses the mechanisms by which the immune system converts stimuli into information, how it processes and communicates that information, and how the information is used to promote an effective *immuno-ecology*. In general, *immuno-informatics* is a complex, sophisticated topic that is beyond the scope of this communication. However, it will be referred to periodically to demonstrate the interdependence of *immuno-ecology* and *immuno-informatics*.

To appreciate the concept of *immuno-ecology*, certain basic features of the immune network need to be discussed. Humans inherently think in a linear, stepwise (logical) fashion, and unconsciously impose linear, stepwise patterns on the things that they study. Unfortunately, networks are multifactorial, nonlinear arrays of interactive elements, making them confusing (illogical) to humans. Humans also tend to be anthropomorphic, and to impart human qualities and values to the things that they study. These human tendencies often obscure the basic nature of the immune network.

It is important to remember that *even human leukocytes are inhuman*. Leukocytes are brainless automatons that accomplish things by mass action, not by individual effort. As individuals, they are insignificant and expendable, careless about the needs or rights of others, and incapable of thought, reason, or creativity. Each leukocyte independently responds in a predictable, genetically preprogrammed fashion to integrated signals from its immediate environment. There are no leader leukocytes. The immediate micro environment defines and coordinates all leukocyte behavior. In this regard, leukocytes are more like ants than humans, and *immune responses are basically leukocyte swarm functions*. It is these swarm functions, and not the individual leukocytes, that are unpredictable, because the swarm functions occur under changing local conditions (some of which may be generated by the swarm itself).

Leukocyte swarms, like colonies of ants, can reliably accomplish surprisingly complicated tasks, even under a variety of adverse conditions. This resiliency stems from the fact that swarms are composed of very large numbers of diverse individuals charged with redundant tasks. As a consequence, the swarms are effective, but highly inefficient, and individual swarm members are unimportant and expendable. Inefficiency is abhorrent to modern humans, but is quite common in biology, where it works to insure successful task performance, even when many individuals die or malfunction. Obviously, such an organizational scheme imparts high biologic survival value. Given these considerations, the challenge for immuno-ecologists is to determine (a) the range of tasks that can be performed by leukocyte swarms, (b) the environmental signals that determine whether any of these tasks is performed, and (c) how leukocyte swarms influence the local environment.

Another challenge for immuno-ecologists is to learn how the immune system controls the evolution of its swarm functions, and thus avoids disorder, confusion, and chaos. Again, it is important to remember that immunologic swarm functions are leaderless, so humanized paradigms based on an immunologic orchestra or an immune militia, which are so pervasive in immunology, are not useful here. Instead, a premise that is new to immunology will be explored. This premise holds that the *control options available to immune networks are provided by the innate principles of network design and function.* The purpose of this communication is to discuss this premise in detail.

During this discussion, examples of immunologic processes that reflect network design principles will be provided. The reader will note that most of these examples are drawn from the field of transplant immunology, which merely reflects the specific research interests of the author. Nevertheless, transplantation provides an excellent platform for these discussions, since the three primary forms of graft rejection (hyperacute, acute, and chronic rejection) reflect pathologic developments at different points along the damage/repair axis. This axis, representing the biologic continuum of inflammation, immunity, and tissue repair, constitutes an integrated set of biologic networks charged with damage control and protection from invasion by opportunistic pathogens. An overview of transplant immunology is beyond the scope of this discussion, but readers may be interested in recent reviews on this topic [34, 45].

2 PRINCIPLES OF NETWORK DESIGN

There are at least four basic principles of network design that provide a wide range of control options for the immune system. These are (a) phylogenic layering, (b) parallel processing, (c) dynamic engagement, and (d) variable connectivity. Because each of these is extremely important, but rarely (if ever) considered by immunologists, each will be discussed in considerable detail below.

2.1 PHYLOGENIC LAYERING

Stated simply, *phylogenic layering* refers to the time-honored biologic principle that *new processes are built on top of older, less-effective processes.* While simple in concept, the impact of this principle is quite profound. Since the less-effective, older processes are often retained, *phylogenic layering provides the immune system with a series of back-up responses.* Complete response failures are actually quite rare. Further, *phylogenic layering is the principle behind scaffolding, a critical design feature of immune responses.* In general, early steps in the immune response provide the specific conditions, or scaffold, required for later steps in the response. *Scaffolding serves to limit and organize immune responses*, and is a network design feature that contributes significantly to the preservation of order as immune responses unfold.

The biologic value of the back-up function provided by *phylogenic layering* is intuitively obvious. Tiers of evolutionary developments have provided a panorama of response options, all of which are generally effective, though not all equally efficient. The presence of multiple response options implies that there is no single, prototypic mechanism of immune response to any given set of foreign antigens. For example, there are multiple mechanisms of acute allograft rejection (contrary to the current premise of experimental transplantation). Although most textbooks credit the acute rejection of allografts to CD8-positive cytolytic T cells, experimental data indicates that isolated CD4-positive T cells [24], and populations of noncytolytic T cells [44], can also reject allografts. The actual mechanism, or combination of mechanisms, employed to reject an allograft by any given individual depends on the unique panorama of immunologic and physiologic conditions associated with both the allograft and its recipient (as will be discussed below). For example, our studies with inbred strains of mice have demonstrated that cardiac allografts transplanted from strain 1(Balb/c) to strain 2 (C57Bl/6) are rejected exclusively by CD4-positive T cells, whereas allografts transplanted in the reciprocal direction require both CD4-positive and CD8-positive T cells for rejection. Thus, the same degree of antigenic disparity can evoke a different panorama of immune responses in different individuals. By extension, the mechanisms of acute allograft rejection in outbred populations must be highly individualized and quite unpredictable.

Given these unpredictable, individualized acute rejection processes, it is improbable that a single, standardized immunosuppressive strategy can operate effectively in all patients (contrary to the current premise of clinical transplantation). This conjecture is supported by several experimental and clinical observations. Graft rejection is generally thought to require specific, proinflammatory cytokines (the hormones of the immune system), such as interleukin 2 (IL2), gamma interferon (IFNg), and tumor necrosis factor (TNF). Indeed, a major principle of clinical immunosuppression is the pharmacologic elimination of cytokine production or function (for review, see MacGregor and Bradley [29]). Nevertheless, experimental allograft studies with genetically engineered mice have demonstrated that cardiac allograft rejection develops efficiently despite the deletion of genes encoding the supposedly critical proinflammatory cytokines, IL2 [28] or IFNg [39].

Similarly, there are function-impairing polymorphisms in all human cytokine genes tested to date [37], including genes encoding the proinflammatory cytokines IL2, TNF, and IFNg, and genes encoding the anti-inflammatory cytokines interleukin 10 (IL10) and transforming growth factor b (TGFb). This results in different cytokine production capacities among individual humans. Because of the Mendelian distribution of these polymorphic cytokine genes, there is a broad continuum of cytokine production patterns among humans, resulting in some individuals with a high proinflammatory predisposition (high TNF, IFNg/low IL10, TGFb production), some with a high anti-inflammatory predisposition (high IL10, TGFb/low TNF, IFNg production), and a majority of individuals who fall somewhere in between. This clearly influences the design options for an immune response in any given individual, and therefore the efficacy of various immunosuppressants in populations of such individuals. Cytokine gene polymorphism helps to explain why many patients respond poorly to immunosuppressive strategies that appear to work well in others [20].

If there is no universal therapeutic strategy that works in all patients, then physicians have only two choices: continue to use the existing therapeutics and tolerate the large number of patients for whom it is relatively ineffective, or develop individually tailored therapeutics. This alternative is currently considered to be impractical, because it requires an *understanding* of immune functions in a given individual, and an *understanding* of immunosuppressive drug effects in that same individual. Both of these requirements are well beyond the reach of current medicine, which has an understanding of neither. *It is this very need for a deep understanding of immune function in individuals, rather than the superficial appreciation afforded by anecdotal cataloging of immune phenomena in humans, in experimental animals, and in vitro, that has precipitated this discussion of immune regulation in the context of network theory.*

The initial steps in the development of such an understanding will involve the identification of critical immunologic and physiologic factors operating in both the graft donor and recipient that influence the ultimate immune relationship that develops between the graft and the recipient. In other words, for each transplant event, it will be necessary to appreciate (a) the individual immune capacities of transplant recipient (for example, cytokine genotype), (b) the immunologic features of the allograft that contribute to its immunogenicity (type and degree of alloantigenicity, i.e., disparity among MHC-encoded proteins), (c) physiologic features of the allograft that impact local immune reactivity ("privileged" sites, etc.), and (d) the type and degree of proinflammatory physiologic insult endured by the graft during the peritransplant period (donor brain death, ischemia time, reperfusion injury). Also required is a much better appreciation for the full range of immune relationships that can develop between the graft and its recipient (i.e., acute rejection, chronic rejection, tolerance), and for the impact of the critical physiologic and immune factors on the development of this immune relationship. Finally, it will be necessary to develop an arsenal of conditioning or corrective strategies with which to influence the developing immune relationship in individual allograft recipients.

Ultimately, physicians will monitor early immune developments in transplant patients, using immunologic and physiologic mile markers, and guide them pharmacologically toward allograft tolerance, i.e., drug-free allograft acceptance without attendant pathology. This approach represents a profound change in the philosophy of transplant patient care, which currently relies on cursory pretransplant immunologic evaluation of the recipient, semistandardized posttransplant immunotherapy, and reflexive, semistandardized corrective responses to unfavorable posttransplant developments.

A second important consequence of *phylogenic layering* is the fact that it *imposes scaffolding requirements on the development of immune responses.* In general, later developments in the evolution of an immune response depend on the earlier placement of critical chaperone elements. This is a direct application of the broad principle that the environment profoundly influences the behavior of the leukocyte swarm. The scaffolding principle needs to be appreciated at two levels. First, it is the principle which links the grand processes of tissue physiology, inflammation, immunity and tissue repair into a seamless and flexible homeostatic mechanism. More specifically, it suggests that *inflammation provides some of the scaffold elements that are essential for the development of acute allograft rejection, and thus provides important control options during the development of the host-graft relationship.* The scaffold principle is generally overlooked by transplant immunologists, who favor, instead, a very narrow focus on the direct control of alloreactive T-cell activation. So much so that the control of acute rejection through manipulation of broader proinflammatory processes remains almost completely unexplored.

The exception to this is studies on the contributions of extracellular matrix molecules to the acute rejection process. Among immunologists, the extracellular matrix is often considered to be an inert "back fill" for tissue parenchymal cells. However, this view is extremely naive. Those who study tissue repair mechanisms have long known that cutaneous tissue repair depends on the early development of a "provisional matrix" whose unique elements control the initiation and development of tissue remodeling and repair processes [8]. One such specialized matrix element is fibronectin, a circulating molecule that polymerizes within the extracellular matrix at sites of traumatic or immune-induced vascular leakage [26]. The fibronectin molecule bears ligands for cell surface integrins, like VLA-4 and VLA-5, expressed by infiltrating leukocytes [15, 30], allowing fibronectin-guided, integrin-mediated leukocyte movement through the tissues. Furthermore, alternate splice variants of fibronectin can be produced by these infiltrating leukocytes [9], providing discrete new trails for subsequently infiltrating cell types. The importance of this system is highlighted by the experimental observation that treatment of rodent cardiac allograft recipients with a peptide called CS-1 blocks the development of acute allograft rejection responses [10]. CS-1 is the peptide segment of a fibronectin splice variant that serves as the specific ligand for VLA-4 [15], an integrin displayed by activated T cells. Thus, competitive elimination of an environmental signal (delivered by fibronectin) that is generated during early inflammation (as the provisional matrix) can disrupt the function of the local proinflammatory scaffold, and subvert the subsequent development of alloimmunity.

This example of provisional matrix function belies the extremely important, but grossly underappreciated role of the matrix as the physical and biochemical scaffold upon which inflammatory processes, such as acute rejection, are built. Not only does matrix composition change radically under various conditions [25], but many immunologically active molecules, such as TGFb and numerous chemokines, attach themselves to matrix molecules [43]. It is intriguing to consider the possibility that the extracellular matrix actually serves as a biochemical bulletin board upon which is posted not only important information about local tissue function, but also information about the progression of local inflammatory and immune responses. Thus, the matrix provides constantly updated local news to the parenchymal cells and to infiltrating immune cells. This represents a clear interface between *immuno-ecology* and *immuno-informatics*. It also underscores the importance of *immuno-informatics* and *immuno-ecology* as fields of study. Immunologists should learn to read the matrix, as do the local parenchymal cells and visiting leukocytes.

At a second level, the scaffolding principle implies the *ordered, conditional use of resources during the development of immune responses*. This extremely important feature of *phylogenic layering* provides the network principle that keeps the enormously complex immune system from plunging into chaos during any given immune response. Like chess pieces, all of the available immunologic components are not employed at once by the immune system. Rather, specific patterns of component use unfold as earlier responses create the specific conditions that permit (or preclude) the development of later response patterns. These later responses, in turn, create a new set of immune conditions, and perpetuate the on-site evolutionary process of immunity. Thus, *even when available, a particular immune response element is used only when and if the immune conditions develop which allow its use*. This *conditional use principle* is a critical design feature of the immune response. It contributes significantly to a broader design principle, the micro-environmental control of response development.

On a broad scale, the *conditional use principle* underlies the orderly progression from tissue damage, through inflamation, immunity, and finally to tissue repair. It allows the safe accumulation of biologically active response elements that are not needed until later phases of the response, and thus eliminates the difficult logistical problem of providing specific response elements to a particular site exactly when they are needed. Further, the *conditional use principle* permits a form of biologic conservation, the use of the same element at different times for different purposes. This would explain the perplexing ability of cytokines to have different effects under different conditions [33].

In networks, context is everything. For any given response element, its order of use, or its combinatorial pattern of use, may be much more important than its simple availability. Based on experience with in vitro studies, this concept is generally appreciated by immunologists. However, the overwhelming number of possibilities provided by combinatorial cytokine function is simply too complicated to consider. This has stalled the incorporation of this principle into mechanistic models of immune function.

Because of the *scaffolding* and *conditional use principles*, immune resources, such as leukocyte subpopulations and cytokines, can be stockpiled at an inflammatory site for later use. This stockpiling reduces the danger of inadequate resource provision as the immune response evolves. It must be remembered that the immune system cannot know in advance the nature of the agents responsible for inflammation, and must be prepared to respond to virtually any pathogen. Resource stockpiling allows the immune system to initiate any of a wide range of response options by simply choosing from the stockpile those elements that it needs. Given this immune strategy, the common experimental practice of cataloging immune response elements, especially cytokines, that accumulate at inflammatory sites will probably not help to understand the immune response that will subsequently develop, and merely adds to the overall confusion in the field. Indeed, studies on intragraft cytokine production have taught us only that virtually all cytokines accumulate to some degree in allografts within the first week of the response [32], but they have taught us nothing about the importance of these cytokines to the developing immune response. The scaffolding principle of the networked immunity theory suggests that it would be much more informative to identify the *cytokine receptors* that are expressed by infiltrating leukocytes at graft sites. Alternatively, the expression of immune functions that develop as downstream consequences of selective cytokine utilization, such as chemokine receptor expression [41], would be informative. This information would identify the specific cytokines that are being selected for use from the resource stockpile at any given time, and thus define the nature of the developing immune responses. This conceptualization is new to transplant immunology, and needs to be tested. While chemokines are now under study [13, 19], the patterns of cytokine receptors that are expressed in allografts have yet to be investigated, or even discussed. Indeed, this vacuum of information extends beyond alloimmunity, and includes autoimmunity and immunity to infectious agents, as well.

In summary, the first principle of network design, *phylogenic layering*, provides the immune system with an immunologic insurance policy in the form of back-up response options. It also contributes an important organizational element to the immune system—scaffolding—which ensures an orderly progression of events during the evolution of immune responses.

2.2 PARALLEL PROCESSING

In the context of immunity, *parallel processing* refers to two strategic maneuvers that operate on different scales, (a) *the simultaneous effort by multiple, similar individuals toward the same task*, and (b) *the simultaneous use of multiple, different mechanisms for the same task.* The former accounts for the perplexing redundancy in the immune system, which permits overall success despite individual failures. This concept has already been discussed.

The second tactical maneuver permitted by *parallel processing*, simultaneous initiation of multiple responses, allows the rapid development of a generalized, broad-spectrum response. This is advantageous because the immune system cannot know in advance the number or identity of the pathogens that it will encounter at a site of inflammation. An alternative strategy, to send only pathogen-specific response elements to the site, requires systems of pathogen identification and information transfer that are beyond the capabilities of the immune system. A second alternative, to scout the site and then call in pathogen-specific response elements, wastes valuable time, risks improper or incomplete pathogen identification, and limits options to unexpected countermeasures mounted by the pathogens. Instead, *the immune system builds a tailored immune response de novo to whatever antigens are encountered at the site of inflammation.* The process begins as the simultaneous initiation of an array of basic responses, which is edited with time under the influence of antigen. It is interesting to note that the local stockpiling of immune resources at the inflammatory site, permitted by the scaffolding principle, clearly facilitates this response strategy.

The concurrent employment of multiple responses in a networked system is inherently unstable, and cannot persist for long. This is due to the often contradictory agendas of the different immune processes. The potential for the development of competing processes is an inherent problem of all networks. Thus, immune responses must mature toward a selective use of only the most effective, compatible processes. In networks, this does not necessarily result in the development of the best possible response. Rather, it results in the *best response possible under existing conditions.*

For most immune responses, including allograft rejection, this maturation process occurs under *four response constraints*: (a) *the nature and amount of the foreign antigens*, (b) *the profile of immune resources available to the individual*, (c) *tissue-specific restrictions imposed on immune response options*, and (d) *prior immune experience with the specific pathogens.* As the immune response matures, those mechanisms most responsive to antigens under these constraints are favored. The other mechanisms fail to develop, either due to lack of support, or to negative feedback signals produced by the more responsive mechanisms. This refinement of defense strategy may appear to humans as an effort by the immune system to optimize its efficiency. However, inefficiency is not a serious concern for the immune system. *A truly serious concern is the compromise of an immune response by the concurrent utilization of contradictory immune processes.*

It is important to appreciate that most local negative feedback mechanisms are primarily housekeeping operations. They are engendered by immune decisions about *how to best eliminate antigens.* For example, negative feedback can cause T cells to become functionally paralyzed (anergic) or to die (apoptosis). T cells are plentiful, available, and expendable, so T-cell crippling or destruction is an acceptable and effective strategy for the local regulation of immune responses. This minor housekeeping function should not be confused with the far more important cognitive processes by which the immune system decides *whether to accept or eliminate an antigen.* This life-and-death decision-making process, which can lead to either immunologic tolerance or immunologic aggression, represents another important interface between *immuno-ecology* and *immuno-informatics.*

Given their central importance to the evolution of immune responses, it is worthwhile to consider the *four response constraints* in some detail. The *first constraint*, the *nature and amount of antigen*, is a source of extreme design variability in immune responses. For example, there are three ways for the immune system to recognize alloantigens of the graft: (a) a relatively small number of recipient CD4+ T cells can recognize peptides derived from polymorphic graft proteins displayed by antigen-presenting cells (APC) of the recipient via their "self" MHC class II molecules, (b) a much larger number of recipient CD4+ T cells can recognize peptides derived from either polymorphic or monomorphic proteins that are displayed by graft-derived APC (allografts often carry passenger leukocytes into the recipient) via their foreign MHC class II molecules, and (c) a large number of recipient CD8+ T cells can recognize peptides from either polymorphic or monomorphic proteins displayed by graft-derived APC via their foreign MHC class I molecules. Each of these pathways generates an unique set of immune responses which are more or less effective at graft rejection. Further, any combination of responses can be activated, depending on whether allografts display foreign MHC class I molecules, MHC class II molecules, neither, or both. In general, allograft rejection responses can be weak or strong, and mediated by any combination of numerous rejection mechanisms, depending of the nature and amount of graft alloantigens. This is the reason that some transplant centers attempt to match the MHC molecules of graft donors and recipients, thus controlling to some degree the nature and amount of alloantigens delivered to the recipient. Of course, grafts are a somewhat unusual immune experience. Normally, antigen type and amount are totally uncontrollable for immune responses to infectious agents and for autoimmunity.

To date, there has been little appreciation for the broad *variation in the immune capacities of individuals (constraint #2)*. However, it is now clear that a wide range of cytokine production capacities exists among humans [37], resulting in the random distribution of different sets of immune network options among individuals. One can envision that function-affecting polymorphisms exist among many or most of the genes encoding immune response elements (including chemokines, tissue growth factors, and their receptors), and that each individual has a somewhat unique combination of these polymorphisms. This forces the construction of individualized immune responses which have accommodated the genetically encoded strengths and weaknesses of their immune systems. While therapeutic strategies routinely ignore this consideration, physicians are well aware of individualized physiologies, and have learned to accommodate them empirically. Investigators tend to overlook this consideration. Indeed, most preclinical animal studies, which utilize genetically homogenous inbred strains of mice or rats, to have effectively neutralized the constraint imposed by immunologic individuality.

Tissue-specific constraints on immune options (constraint #3) also tend to be overlooked. Nevertheless, it is generally accepted that the location of antigen deposition is a major factor in the design of immune responses. For example, "privileged sites" like the eye and testes [3, 48], where immune responses are forbidden, are well known. Similarly, it is generally accepted that antigens placed subcutaneously tend to drive elimination responses (the premise for vaccination), whereas the same antigens placed intravenously [23] or in the gut [6] tend to drive acceptance responses. However, some manifestations of this principle are more subtle. In this regard, it is interesting to note that the process of acute allograft rejection displays different patterns of pathology in heart (myocardial infarction), kidney (tubular malfunction), liver (biliary remodeling), intestine (epithelial loss), lung (bronchiolar remodeling) and skin (vascular failure) allografts. Further, the incidences of rejection, the rates of rejection, and the relative susceptibilities to immunosuppression all differ among these tissues, suggesting that the immunologic mechanisms of acute rejection may also differ to some degree among these tissues. The basis for these tissue-specific differences has yet to be integrated with current immunologic conceptualization, and the important issue of tissue-specific constraints on local immune response options remains largely unexplored.

It should be noted that these tissue-specific responses are generally *un*related to the presence of tissue-specific antigens. Such antigens clearly exist [21], but do not operate at this level. Rather, tissue-specific immune responses reflect the *special agreements negotiated between the physiologic network and the immune network of a given tissue during the local development of an immune response.* These agreements are made because of competing network agendas. The physiologic network attempts to maintain homeostasis, despite the severe perturbation caused by the visiting immune network. In turn, the immune network attempts to do its job within the somewhat unique and changing biochemical environment of the specific tissue. Each network must adjust to accommodate the other, resulting in both altered tissue biochemistry and restricted immune response options. Given these considerations, one might predict that immunologic principles generated, for example, from studies with rodent cardiac allografts may not apply to human renal allograft recipients.

Finally, the *fourth constraint, prior immune experience* (immunologic memory), provides a rapid response component that gives the immune system a significant advantage over common environmental pathogens. The development of and utilization of immunologic memory represents another important interface between *immuno-ecology* and *immuno-informatics.* Immunologic memory is well appreciated by immunologists and physicians, who routinely exploit immunologic memory by vaccinating humans to protect them from specific pathogens. In contrast, prior experience with alloantigens is generally considered to be detrimental to allograft survival. So much so that each patient is clinically tested prior to transplantation for preexisting donor-reactive alloantibodies, a consequence of prior humoral allosensitization, and a primary contraindication to transplantation. Interestingly, no patient is ever tested for donor-reactive T-cell allosensitization, due primarily to a lack of accepted testing methodologies. Instead, cellular allosensitization is generally inferred from humoral allosensitization.

One of the primary tasks of the immune system is to determine whether to accept or reject any given foreign element. We have learned to influence this immunologic decision by process of vaccination, prior provision of antigen under defined conditions that promote antigen elimination. It should also be possible to influence this decision toward antigen acceptance by an alternative form of vaccination. There is little call for this approach, with the exception of transplantation. There, it has been established clinically that the process of donor-specific transfusion, the IV delivery of donor alloantigens to graft recipients a few weeks prior to transplantation, can reduce posttransplant immunologic complications [17]. Thus, immunologic memory can deviate immune reactivity either toward acceptance or rejection of foreign elements, highlighting the role that prior antigenic experience plays in the design of immune responses.

The utilization by the immune system of the *parallel-processing* principle again suggest that cataloging the accumulated immune response elements during the early phases of an immune response is relatively uninformative. While it is a popular way to generate large amounts of publishable data, it does little to promote an understanding of immune response mechanisms. Unfortunately, this experimental strategy has been showcased by the recent development of gene chip technology, and other "high throughput" experimental methodologies associated with "discovery-based research" (the euphemism for a style of science based on serendipitous observation rather than hypothesis-and-test). These methodologies have a role in modern science, since they can exponentially increase the amount of *information* that can be obtained from a given experimental system. However, this information should not be confused with *understanding*. Identification of all the nodes and connectors in a network does not provide much information regarding the pattern of network function in response to a particular perturbation. This is the classic scientific dilemma posed by Einstein regarding the watchmaker's son, who dismantled his father's watch in a vain effort to understand time. In this respect, there is little difference between physics and biology. Thus, the theory of networked immunity suggests that discovery-based research is unlikely to contribute significantly to an understanding of immune function, and may only add further confusion to the field. Despite the powerfully enhanced information gathering capacity of these methodologies, the roulettelike nature of the *four response constraints* make it virtually impossible to predict the nature of the immune response pattern that will emerge from the network as the consequence of a specific antigenic perturbation.

This important problem is generally underappreciated by investigators. In vitro studies, a time-honored approach to experimentation, are designed to minimize experimental variables to an extreme. This is done in an effort to generate reproducible results, the cornerstone of scientific fact. (The unacknowledged principle is that immune responses develop unpredictably under normal conditions, an indirect admission of its complex character.) Under such minimal conditions, the impact of the *four response constraints* is essentially eliminated, and investigators tend to lose sight of their importance. Unfortunately, any minimization of immune complexity changes the conditions of the immune response, making the results as artificial as the experimental conditions. Obviously, all such results should be considered as conditional, but too often they are considered universal.

The *four response constraints* are often minimized during in vivo studies, as well, especially those involving rodent experimental models. Such studies commonly involve inbred rodent strains, which tends to freeze the variables of the *four response constraints* into a reproducible pattern. Commonly, the same studies conducted in different rodent strains will yield different results. For example, infection of mice from inbred strain 1 (C3H) with Leishmania leads to pathogen clearance and immunity, whereas infection of mice from strain 2 (Balb/c) leads to fatal chronic disease [4, 40] This is a vague reflection of the problems that would be encountered if immunologists performed their studies with outbred mice, where the *four response constraints* would operate unfettered. Such studies are virtually never performed (for this very reason).

In contrast, the consequences of the *four response constraints* are commonly encountered clinically during the treatment of outbred humans. This has lead to a disparity between experimental and clinical observations, and to the charge by clinical investigators that information obtained in rodent preclinical studies may not be applicable to humans. Indeed, they are absolutely correct, not because the basic immune mechanisms of rodents and humans are fundamentally different (in fact, the *scaffolding principle* implies that they are fundamentally similar), but because the *conditions* of immune responses in humans are far more variable than those of rodent preclinical studies. Immunologic conditions that have been inadvertently selected in murine studies may rarely develop in humans (or even in outbred mice), resulting in significant "bench to bedside" translational problems. For example, anti-CD4 mAb is a potent and reliable immunosuppressant that permits long-term, drug-free cardiac allograft acceptance in mice [31, 36], but clinical trials with anti-CD4 mAb in renal transplant patients revealed little or no efficacy [11]. Many other examples of such discrepancies exist, and have called into question the value of preclinical studies in rodent models. The theory of networked immunity suggests that studies with rodent models are perfectly valid and applicable to humans, *provided that the conditions of the immune responses are matched with respect to the four constraints of response development.*

In summary, the immune system has employed the strategy of *parallel processing* as a broad-based, rapid response mechanism to improve its chances for successful competition with invading pathogens. The advantage of this strategy is that it provides speed, flexibility, and multiple response options to the immune system. The price of this strategy is the wasteful, inefficient use of immune resources, which means little to the immune system, but bothers humans. The advantages of parallel immune processing confer high survival value to the individual. However, the fact of *parallel processing* has confused and frustrated investigators, who generally expect uniformity, predictability, and a purity of purpose from the immune system. Both the theory of networked immunity and general observations suggest that these are unreasonable expectations.

2.3 DYNAMIC ENGAGEMENT

The phenomenon of *dynamic engagement* is a primary leukocyte swarm function, and refers to the fact that *immune responses are mediated by very large numbers of leukocytes, each of which operates only briefly, and is then replaced by other similar leukocytes.* In effect, this provides nonsynchronous, consecutive waves of self-limiting immune processes. The process of tissue inflammation is characteristically associated with the local accumulation of leukocytes within the damaged tissues. Most immunologists consider this to be a static, rather than dynamic, process. However, leukocytes are highly mobile within tissues, and it is an anatomic fact that the leukocyte exit portals, the lymphatics, are usually only a few cell diameters away from their entry portals, the postcapillary venules. Thus, it is quite possible that leukocytes accumulate at inflammatory sites only because their entry rate exceeds their exit rate.

While the complex leukocyte-endothelial interactions that promote leukocyte entry into tissues are well defined [42], the mechanisms of leukocyte emigration from tissues are virtually unknown. Further, the transit times of individual infiltrating T cells or macrophages through inflammatory sites have not yet been determined. Thus, there is little direct evidence either for or against the principle of *dynamic leukocyte engagement* during inflammation. This should not denigrate the principle. It merely reflects the fact that to date, no immunologic conceptualization has suggested its existence or importance. However, *dynamic engagement* is a clear inference of the theory of networked immunity, and it now deserves critical evaluation.

The principle of *dynamic engagement* suggests that leukocytes arrive continuously at inflammatory sites, function briefly, and then depart. In this situation, departure may include both functional crippling (anergy) and cell death (apoptosis), as well as actual emigration from the site. Again, this seems inherently wasteful to humans but, in fact, the supply of leukocytes is essentially inexhaustible. In return for this seemingly inefficient use of immune resources, the immune system acquires two important capabilities. *Dynamic engagement allows the continuous monitoring of the site for persistence of antigen, and it allows the local immune response to be self-limiting.*

In this context, it is useful to think of macrophages as environmental sampling devices, and of T cells as detectors of local "foreignness." Macrophages continuously obtain, process, and display elements of the local environment. When a T cell enters an inflammatory site, it seeks out the nearest macrophages and queries it via TcR/MHC interactions to determine if any of the displayed elements are antigenic (foreign). In the absence of macrophages, the T cells are functionally blind to antigens. As macrophages sample the environment, they also determine whether "danger" exists in the form of local tissue damage, which they can sense through a series of receptors for tissue damage products [7]. Thus, by interacting with T cells they can detect coincident "foreignness" and "danger." (Such flows of processed information represent another interface between *immuno-ecology* and *immuno-informatics.*) The macrophages then instruct the T cell to behave appropriately, i.e., by developing antigen elimination programs. This instruction is transmitted via well-recognized costimulation molecules, like CD40 and B7 [5, 18], which in conjunction with TcR engagement, induce the T cell to produce proinflammatory cytokines, like IFNg. These cytokines, in turn, provide alarm signals to all the other local cell types. Should the T cell remain in the area producing cytokines, the value of this alarm is diminished. Similar to a household smoke alarm, its value is enhanced if the alarm is interrupted. Then if it resumes, it represents new evidence that the antigen still persists. In an effort to utilize this principle, the immune system simply clears or deactivates the first T cells and replaces them with others.

For this system to be effective, several criteria must be met. First, the alarm devices (T cells) and the sampling devices (macrophages) must be numerous and widely distributed. In fact, macrophages are randomly dispersed throughout rejecting allografts, and constitute about half of the infiltrating leukocytes. Most of the remaining leukocytes are T cells. Second, both the alarm devices (T cells) and their signals (cytokines) must be transient, to avoid prolonged, inappropriate signaling. It has long been known that most cytokines are highly labile in vivo, a fact that complicates clinical cytokine-based therapies. Many cytokines have simultaneously produced antagonists, such as soluble cytokine receptors [16] and receptor antagonists [46]. The need to extinguish cytokine signals explains the perplexing coproduction of cytokines and their antagonists during inflammation, a fact that has no satisfactory explanation in current paradigms of immune function.

The T cells must also be transient. Indeed, T cells not only have the option to leave, but they have two failsafe devices in case they cannot: they can either shut down functionally (anergy) or they can die (apoptosis). These devices protect the process of *dynamic engagement* in cases of severe inflammation. It should be noted that in this context, T-cell anergy and apoptosis are housekeeping devices. Since both mechanisms are inherently anti-inflammatory, and since there is no other explanation for them in current paradigms of immunity, immunologists have implicated them as mechanisms of immune tolerance. Network considerations offer an alternative purpose, and suggest that anergy and apoptosis are mechanisms of signal dampening, not mechanisms of immunologic tolerance, at least under these conditions.

Third, the environmental sampling devices (macrophages) must be transient, in order to avoid prolonged, inappropriate antigen presentation. In fact, macrophages, like T cells, have the option to leave, but it is not known if they do so. It is also not known if they can enter an anergic state in which, for example, they limit their expression of the costimulatory molecules necessary for effective antigen presentation to T cells. Clearly, macrophages can undergo apoptosis, and indeed, may be instructed to do so by cytolytic CD4+ T cells [47].

In summary, it appears that the immune system displays many features that support the hypothesis of *dynamic engagement* as an important design principle of immune responses. *Dynamic engagement* would provide important control options and significant order to immune responses. Again, the immune system may sacrifice efficient utilization of resources to gain the benefits of *dynamic engagement*: the ability to instantaneously and continuously monitor the inflammatory site for the persistence of foreign antigens, and the ability to generate self-limiting responses without the need for elaborate control mechanisms. Indeed, it may be forced into this strategy to accommodate its nature as a networked swarm function.

2.4 VARIABLE CONNECTIVITY

By means of the principle of *variable connectivity, the immune system can take advantage of the most basic characteristic of networks, nonlinear connectivity, to control its functions.* At the heart of this principle is the issue of homeostasis, the maintenance of stability in the face of change. Homeostasis is the cornerstone of physiology, which uses many network principles for the design and function of tissues. The basic design of networks, i.e., large groups of multiply interconnected elements, automatically imparts a strong tendency toward homeostasis. By means of their interconnections, each network element is more or less related to all the others, so dangerous perturbations are those that can alter network function, not just the function of individual network elements. Thus, networks can easily withstand "minor" perturbations, i.e., those perturbations that fail to interfere with network function, even though they may drastically influence one or more of the network elements. This inherent resistance to change is reinforced by another network feature, the ability to adjust the degree of interconnectivity among elements. Since each element defines its homeostatic identity by referencing its neighbors, increased referencing during a perturbation adds rigidity to the network. This combination of inherent resistance to change and adjustable interconnectivity provides a built-in shock absorber to networked systems, i.e., an inherent tendency toward homeostasis.

Occasionally, overwhelming perturbations occur, and the network is forced to employ its self-correcting mechanisms. Inflammation and immunity are, in essence, corrective tools of physiology, designed to promote the reestablishment of homeostasis after major or minor tissue perturbations. However, immunity can be dangerous tool, since it is inherently capable of causing severe tissue damage, and further compromising physiologic homeostasis. Further, the immune system is a network unto itself, so immunity, which always takes place within the context of a perturbed physiology, actually represents a clash between the interests of two temporarily overlapping networks. How networks interact is a sophisticated problem that lies beyond the scope of this discussion. However, it is important to note that the immune system never operates in a vacuum, and that the networked processes of immunity and physiology, when merged during inflammation, always have some influence on each other.

Another interesting feature of the immune system is that it is capable of building a functional network de novo at almost any place in the body, and of using this network to accomplish a task, such as pathogen eradication. What is often overlooked is the fact that *once its task is accomplished, the local immune network is rapidly demolished.* This is somewhat unusual, since most biologic networks (except some that are operative during ontogeny and maturation) are established and then maintained. In contrast, the temporary immune networks that are constructed at sites of inflammation do not strive for homeostasis, they strive for dissolution. Indeed, the development of immune network homeostasis at an inflammatory site is pathologic, and results in chronic inflammatory diseases. It is not yet clear if immune deconstruction is an active or a passive process. The principle of *dynamic engagement* would support a passive process whereby the local immune network spontaneously deconstructs when the antigenic stimulus is eliminated. However, observations regarding the biologic effects of anti-inflammatory immune mediators, such as IL10 [12, 22] and TGFb [22, 27], suggest that active mechanisms of deconstruction can also be deployed.

Unfortunately, few immunologists consider the immune response as an exercise in temporary network assembly. As such, it is a fairly sophisticated exercise. For example, as the immune system constructs a local network, it must not only assemble the necessary immune response components at the site, but it must be cautious about their interconnections. Too few interconnections approach a linear relationship among network elements, which imparts susceptibility (ease of disruption) and minimizes network options. Too many interconnections freeze the network into unresponsiveness. Although these two conditions may have biologic utility in certain situations, resilience and flexibility are the goals of most biologic networks, and this requires them to develop an intermediate level of interconnectivity. Biologic networks often turn this constraint into an advantage, because it offers an opportunity for network control. In essence, the *network can adjust its function by varying the degree of connectivity among its elements.* By employing this functional rheostat, the immune system can adjust, to some degree, the evolution (and devolution) of its own response options.

If cells of the hematopoietic lineage represent the immune network elements that can be assembled at an inflammatory site, then their products must represent the network connectors. In this context, immunologic connectors are a large and varied family of molecules, including secreted products (cytokines, chemokines, antibodies, etc.) and cell surface molecules (adhesion molecules, costimulation molecules, receptors for secreted products and matrix molecules, etc.). The degree of connectivity among network elements is regulated by their controlled, conditional expression of these connectors, a process known as "cell activation" to immunologists who study the behavior of T cells and macrophages. The degree of connectivity is also regulated by the production of a series of disconnectors. Immunologists have long been perplexed by the apparent simultaneous production of both immune mediators and their inhibitors during inflammatory responses. Immune elements like soluble HLA molecules [38], soluble cytokine receptors [16], and soluble receptor antagonists [46] represent anomalies that do not fit well into current paradigms of immunity. However, they are easily accommodated by the theory of networked immunity. Thus, for most biologic networks, it appears that homeostasis can be accomplished by rheostatic adjustments in network function, and that these adjustments are accomplished by controlled fluctuations in network interconnectivity. These fluctuations represent a constant rebalancing of connectors and disconnectors. In this regard, homeostasis is a highly dynamic process.

The immune network principles of *variable connectivity* and *dynamic engagement* may overlap at the level of immune connectors and disconnectors. As sensor cells move through the inflammatory site, they respond to local stimuli by producing signals. If these signals were long lasting, they would undermine both the value of their own information, and the strategy of *dynamic engagement.* Thus, the immune system employs a balanced production of immune connectors and disconnectors. This limits the impact of the connectors to short periods and distances, thus enhancing the information value of signals produced by sequentially arriving sensor cells. Unfortunately, the mechanisms that are responsible for the dynamic rebalancing of connectors and disconnectors have yet to be studied, and this represents an important, new area for immunologic investigation, as suggested by the theory of networked immunity.

The connector/disconnector processes would look quite different at different scales. From the individual sensor's (T cell) point of view, its own stimulus-induced (TcR-mediated) activation and signal (cytokine) production are rapidly counterbalanced by its loss of function (departure, anergy, apoptosis) and the extinction of its signals (via disconnectors and proteases). In general, the T-cell achieves its 15 minutes of fame. In contrast, from the infiltrate point of view, the connector/disconnector process would appear somewhat similar to the effect produced by fireflies in the night, since only a small subset of T cells distributed randomly throughout an inflammatory site actually experiences TcR ligation. Most have inappropriate TcR, see no antigen, and go away unfulfilled. Finally, from the viewpoint of an external observer indiscriminately monitoring the entire inflammatory site, the connector/disconnector process appears to be a confusing, continuous production of multiple mediators with conflicting functions. The latter represents the current state of immunologic understanding.

While the temporary immune network that develops at a site of inflammation is not interested in homeostasis, it can still utilize the principle of *variable connectivity* to its advantage. In this case, *variable connectivity* allows the de novo construction of immune responses that are tailored to fit the *four immune response constraints*, i.e., antigen characteristics, local conditions, available immune resources, and prior antigenic experience. The accommodation of the *four response constraints* during the construction of immune responses implies that each immune response is an unique entity. This is far from current immunologic conceptualization, which holds that prototypic immune responses exist and are deployed when necessary. The fact that each immune response is a novel exercise in network construction provides a significant advantage to the immune system (although it is a clear disadvantage to human investigators), since it allows the immune system to accommodate "surprises" offered by environmental pathogens that are constantly evolving strategies to subvert the immune system. However, this advantage is not without price, since the immune system must now accommodate several problems. First, without blueprints for construction, how does the immune system build coherent responses? Second, how does it know whether the response pattern being built is effective or ineffective? Third, how does it keep from developing a set of responses with contradictory agendas? Suffice it to say that the immune system is capable of learning and adaptation, and therefore has cognitive abilities. These cognitive abilities are poorly understood, but badly in need of study, since they hold the real keys to the control of immune system function. Such studies should be designed to determine how the immune system generates, posts, processes, and stores information about itself and its environment, i.e., *immuno-informatics.* While the principles of immuno-ecology serve mainly to provide an infrastructure to the immune system, the motor that drives and shapes immune responses is *immuno-informatics.*

These considerations regarding *variable connectivity* have significant impact on clinical and experimental transplantation. In theory, there are critical network elements whose loss would change the character of the network response. There would also be many elements whose loss would be inconsequential. In allogeneic responses, critical elements include T cells and macrophages. Other elements like B cells and NK cells are apparently unnecessary, although they may not be inconsequential. Still other elements, like eosinophils, are indeed inconsequential. Targeting the critical elements should have interesting, but not always desirable outcomes. The immunotherapeutic anti-CD3 monoclonal antibody, OKT3, is a potent immunosuppressant, but has been associated with the cytokine storm syndrome, an explosive, life-threatening release of proinflammatory cytokines into the systemic circulation [1]. Even the "wonder drug," cyclosporine, which blocks cytokine (connector) production after T-cell activation [35], displays various tissue toxicities [14]. In this context, it is interesting to regard immunosuppressive drugs as artificial network disconnectors. However, these pharmacologic disconnectors create artificial connector/disconnector imbalances that may be responsible for subsequent pathologic developments. Perhaps it would be wiser to identify and utilize the natural connectors available to the system, i.e., soluble cytokine receptors, etc. Unfortunately, such "disconnector therapy" would have to be given continuously and indefinitely (like current pharmacologic therapy), since the immune stimulus (graft alloantigens) remains in place and continuously drives immune network construction forward.

As an alternative approach, one could attempt to influence the immune network during the process by which it decides whether to reject or accept the graft antigens, rather than after the immune system has committed to eradicating the alloantigens. This, of course, refers to the induction of allogeneic tolerance. Until recently, few immunologists acknowledged that the adult immune system had an option to accept foreign antigens, let alone the ability to decide whether or not to do so. However, it is now apparent that the immune system can and does make such decisions regularly, and that it apparently uses the *four immune response constraints* discussed previously as the basis for these decisions. Further discussion of this decision-making process moves into the topic of immune cognition, and thus into *immuno-informatics.*

Presumably, the final arbitrator of immune cell function is the genome, which can coordinately express a variety related gene products after integrating information obtained from inside or outside of the cell. Thus, the most effective way to shepherd the immune response in a desired direction may be to influence the delivery of this information. Since the genetic elements responsible for the integration of external signals that lend to changes in cell behavior are the various promoter-binding nuclear proteins, agonists or antagonists of selected nuclear proteins, such as NF-kB or NF-AT, may prove useful for immunomodulation [2]. At the very least, such agents would alter the display of immune connectors and disconnectors. At best, they may push the immune decision-making process in a desired direction. This strategy needs to be evaluated in various models of immunity.

In summary, *variable connectivity* provides the immune system with a rheostatlike mechanism of immune response control. This endows immune responses with flexibility and resiliency. These are important advantages, since the immune system must not only be able to build an immune network de novo at virtually any site in the body, but it must build one that is effective against virtually any pathogen. *Variable connectivity*, along with *phylogenic layering*, *parallel processing*, and *dynamic engagement*, allow this process to proceed effectively much more often than not.

3 CONCLUDING REMARKS

Immunologists are perplexed and confused by the complexity of the immune system. Traditionally, they have dealt with this problem by ignoring it. They employ a time-honored analytic approach that depends on the rigorous minimization of variables to study immune responses, believing that the immune system is but the sum of its parts, and that understanding the parts will lead to an understanding of the whole. What it has led to is more confusion. In fact, the immune system is much more than the sum of its parts. It is a complex, adaptive network, capable of cognition, conditional responses, and adaptation to change. The complexity associated with its networked character is an indispensable feature on which the immune system depends for its critical advantage over environmental pathogens. The network can not only accommodate error and missing elements, but it provides multiple response options which can be exercised in a flexible, conditional fashion. For this, it accepts some degree of inefficiency.

How the immune system keeps from being as confused as immunologists by its own complexity is a problem worth studying. This area, termed *immuno-ecology*, is a new experimental frontier for immunologists, and an essential first step toward a better understanding of immune systems and their functions. Such studies will undoubtedly require new and unusual patterns of thought on the part of investigators. They will have to learn systems thinking rather than analytic thinking. This diametrically opposed approach will inevitably lead to problems among immunologists, but these are unavoidable.

Early efforts at systems thinking have already suggested that immune network functions are far from chaotic, and instead represent an ordered evolution of events and a conditional selection of response options under a kaleidoscopic variety of constraints imposed by genetics, physiology, and the nature of the invading pathogen. Complexity merely acts to supply more options to the system. Understanding that such options exist leads directly to the critical question of how individual options are selected and evaluated for efficacy by the immune system. Studies in this area, termed *immuno-informatics*, remain rudimentary, at best.

What is already clear is that studies in *immuno-ecology* and *immuno-informatics* will require new tools and new experimental approaches. The preferred approach of most immunologists, in vitro investigations, will be useless for this purpose, since in vitro studies rely on the absolute minimization of experimental variables. While this may insure reproducibility of results, it does so by eliminating response options, and it is not possible to study the selection of response options under conditions where there are none.

The alternative, in vivo investigations, is equally problematic. While in vivo studies preserve the rich complexity of the immune network, they rarely provide more than a demonstration of the endpoint of the complex immune processes that evolve under the conditions of the experiment. Understanding how this endpoint was reached requires some understanding of how complex adaptive networks, including the immune system, function. It may be time for immunologists to develop such an understanding, perhaps by first using computer simulations to study the behavior of complex networks in action, and then by probing animal models of immunity for similar behavior. At the moment, the method of approach is not as important as the fact that such approaches are under development. What should no longer be acceptable is the general dismissal of complexity as anything but an integral and formative feature of the immune system.

REFERENCES

[1] Abramowicz, D., L. Schandene, M. Goldman, A. Crusiaux, P. Vereerstraeten, L. D. Pauw, J. Wybran, P. Kinnaert, E. Dupont, and C. Toussaint. "Release of Tumor Necrosis Factor, Interleukin-2, and Gamma-Interferon in Serum after Injection of OKT3 Monoclonal Antibody in Kidney Transplant Recipients." *Transplantation* **47** (1989): 606.

[2] Baeuerle, P. A., and V. R. Baichwal. "NF-kB as a Frequent Target for Immunosuppressive and Anti-inflammatory Molecules." *Adv. Immunol.* **65** (1997): 111–137.

[3] Bellgrau, D., D. Gold, H. Selawry, J. Moore, A. Franzusoff, and R. C. Duke. "A Role for CD95 Ligand in Preventing Graft Rejection." *Nature* **377** (1995): 630–632.

[4] Belosevic, M., D. S. Finbloom, P. H. Van Der Meide, M. V. Slayter, and C. A. Nacy. "Administration of Monocloal Anti-IFNg Antibodies in vivo Abrogates Natural Resisitance of C3H/HeN Mice to Infection with Leishmania Major." *J. Immunol.* **143** (1989): 266–274.

[5] Bluestone, J. A. "Costimulation and Its Role in Organ Transplantation." *Clin. Transplants* **10** (1996): 104–109.

[6] Chen, Y., V. K. Kuchroo, J.-I. Inobe, D. A. Hafler, and H. L. Weiner. "Regulatory T-Cell Clones Induced by Oral Tolerance: Suppression of Autoimmune Encephalomyelitis." *Science* **265** (1994): 1237–1240.

[7] Chen, W., U. Syldath, K. Bellmann, V. Burkart, and H. Kolb. "Human 60-kDa Heat-Shock Protein: A Danger Signal to the Innate Immune System." *J. Immunol.* **162** (1999): 3212–3219.

[8] Clark, R. A. "Fibronectin Matrix Deposition and Fibronectin Receptor Expression in Healing and Normal Skin." *J. Invest. Dermatol.* **94** (1990): 128S–134S.

[9] Coito, A. J., L. F. Brown, J. H. Peters, J. W. Kupiec-Weglinski, and L. Van De Water. "Expression of Fibronectin Splicing Variants in Organ Transplantation. A Differential Pattern between Rat Cardiac Allografts and Isografts." *Am. J. Pathol.* **150** (1997): 1757–1772.

[10] Coito, A. J., S. Korom, E. Graser, H.-D. Volk, L. Van De Water, and J. W. Kupiec-Weglinski. "Blockade of Very Late Antigen-4 Integrin Binding to Fibronectin in Allograft Recipients. I. Treatment with Connecting Segment-1 Peptides Prevents Acute Rejection by Suppressing Intragraft Mononuclear Cell Accumulation, Endothelial Activation, and Cytokine Expression." *Transplantation* **65** (1998): 699–706.

[11] Cooperative Clinical Trials in Transplantation Research Group. "Murine OKT4A Immunosuppression in Cadaver Donor Renal Allograft Recip-

ients: A Cooperative Clinical Trials in Transplantation Pilot Study." *Transplantation* **63** (1997): 1243–1251.

[12] Ding, L., P. S. Linsley, L.-Y. Huang, R. N. Germain, and E. M. Shevach. "IL-10 Inhibits Macrophage Costimulatory Activity by Selectively Inhibiting the Up-regulation of B7 Expression." *J. Immunol.* **151** (1993): 1224–1234.

[13] Fairchild, R. L., A. M. VanBuskirk, T. Kondo, M. E. Wakely, and C. G. Orosz. "Expression of Chemokine Genes during Rejection and Long-Term Acceptance of Cardiac Allografts." *Transplantation* **663** (1997): 1807–1812.

[14] Ferguson, R. M. "A Multicenter Experience with Sequential ALG/Cyclosporine Therapy in Renal Transplantation." *Clin. Transplants* **2** (1988): 285–294.

[15] Ferguson, T. A., H. Mizutani, and T. S. Kupper. "Two Integrin-Binding Peptides Abrogate T Cell-Mediated Immune Resposes in vivo." *Immunology* **88** (1991): 8072–8076.

[16] Fernandez-Botran, R., P. M. Chilton, and Y. Ma. "Soluble Cytokine Receptors: Their Roles in Immunoregulation, Disease, and Therapy." In *Advances in Immunology*, edited by F. J. Dixon, vol. 63, 269–336. San Diego, CA: Academic Press, 1996.

[17] Francis, D. M. A., P. J. Masendycz, L. J. Dumble, and G. J. A. Clunie. "Enhancement of Renal Allografts by Simultaneous Cyclosporine and Donor-Specific Blood Transfusion." *Transplantation Proc.* **19** (1987): 1464–1466.

[18] Grewal, I. S., and R. A. Flavell. "A Central Role of CD40 Ligand in the Regulation of CD4+ T-Cell Responses." *Immunol. Today* **17** (1996): 410–414.

[19] Hancock, W. W. "Chemokines and the Pathogenesis of T Cell-Dependent Immune Responses." *Am. J. Pathol.* **148** (1996): 681–684.

[20] Hutchinson, I. V., V. Pravica, and P. J. Sinnott. "Genetic Regulation of Cytokine Synthesis: Consequences for Acute and Chronic Organ Allograft Rejection." *Graft* **1** (1998): 186–192.

[21] Joyce, S., J. M. Mathew, M. W. Flye, and T. Mohanakumar. "A Polymorphic Human Kidney-Specific Non-MHC Alloantigen. Its Possible Role in Tissue-Specific Allograft Immunity." *Transplantation* **53** (1992): 1119–1127.

[22] Jungi, T. W., M. Brcic, S. Eperon, and S. Albrecht. "Transforming Growth Factor-Beta and Interleukin-10, but Not Interleukin-4, Down-Regulate Procoagulant Activity and Tissue Factor Expression in Human Monocyte-Derived Macrophages." *Thrombosis Resh.* **76** (1994): 463–474.

[23] Kloke, O., and E. Kolsch. "Modulation of the Antigraft Response by Preimmunization." *Transplantation* **38** (1984): 526–531.

[24] Krieger, N. R., D. Yin, and C. G. Fathman. "CD4+ but Not CD8+ Cells are Essential for Allorejection." *J. Exp. Med.* **184** (1996): 2013–2018.

[25] Kupiec-Weglinski, J. W., A. J. Coito, J. Binder, and M. de Sousa. "The Expression of Extracellular Matrix Proteins Represents an Integral Part of the Host Immune Response in Organ Transplantation." *Surgical Forum* (1995): 415–418.

[26] Kupiec-Weglinski, J. W., A. J. Coito, A. Gorski, and M. de Sousa. "Lymphocyte Migration and Tissue Positioning in Allograft Recipients: The Role Played by Extracellular Matrix Proteins." *Transplantation Rev.* **9** (1995): 29–40.

[27] Lee, Y. J., Y. Han, H. T. Lu, V. Nguyen, H. Qin, P. H. Howe, B. A. Hocevar, J. M. Boss, R. M. Ransohoff, and E. N. Benveniste. "TGF-beta Suppresses IFN-gamma Induction of Class II MHC Gene Expression by Inhibiting Class II Transactivator Messenger RNA Expression." *J. Immunol.* **158** (1997): 2065–2075.

[28] Li, X. C., P. Roy-Chaudhury, W. W. Hancock, R. Manfro, M. S. Zand, Y. Li, X. X. Zheng, P. W. Nickerson, J. Steiger, T. R. Malek, and T. B. Strom. "IL-2 and IL-4 Double Knockout Mice Reject Islet Allografts: A Role of Novel T Cell Growth Factors in Allograft Rejection." *J. Immunol.* **161** (1998): 890–896.

[29] MacGregor, M. S., and J. A. Bradley. "Overview of Immunosuppressive Therapy in Organ Transplantation." *Brit. J. Hosp. Med.* **54** (1995): 276–284.

[30] Mackay, C. R., and B. Imhof. "A. Cell Adhesion in the Immune System." *Immunol. Today* **14** (1993): 99–102.

[31] Madsen, J. C., W. N. Peugh, K. J. Wood, and P. J. Morris. "The Effect of Anti-L3T4 Monoclonal Antibody Treatment on First-Set Rejection of Murine Cardiac Allografts." *Transplantation* **44** (1987): 849–852.

[32] Morgan, C. J., R. P. Pelletier, J. T. Hernandez, E. Huang, R. Ohye, R. M. Ferguson, and C. G. Orosz. "Cytokine mRNA Expression during Development of Acute Rejection in Murine Cardiac Allografts." *Transplantation Proc.* **25** (1993): 114–116.

[33] Nathan, C., and M. Sporn. "Cytokines in Context." *J. Cell Biol.* **113** (1991): 981–986.

[34] Orosz, C. G., S. D. Bergese, E. Wakely, D. Xia, G. M. Gordillo, and A. M. VanBuskirk. "Acute versus Chronic Graft Rejection: Related Manifestations of Allosensitization in Graft Recipients." *Transplantation Rev.* **11** (1997): 38–50.

[35] Orosz, C. G., D. C. Roopenian, and F. H. Bach. "Analysis of Cloned T-Cell Function. II. Differential Blockade of Various Cloned T-Cell Functions by Cyclosporine." *Transplantation* **36** (1983): 706–711.

[36] Orosz, C. G., E. Wakely, S. D. Bergese, A. M. VanBuskirk, R. M. Ferguson, D. Mullet, G. Apseloff, and N. Gerber. "Prevention of Murine Cardiac Allograft Rejection with Gallium Nitrate: Comparison with Anti-CD4 mAb." *Transplantation* **61** (1996): 783–791.

[37] Perrey, C., V. Pravica, P. J. Sinnott, and I. V. Hutchinson. "Genotyping for Polymorphisms in Interferon-g, Interleukin-10, Transforming Growth

Factor-b1 and Tumour Necrosis Factor-a Genes: A Technical Report." *Transpl. Immunol.* **6** (1998): 193–197.

[38] Puppo, F., M. Scudeletti, F. Indiveri, and S. Ferrone. "Serum HLA Class I Antigen: Markers and Modulators of an Immune Response?" *Immunol. Today* **16** (1995): 124–127.

[39] Raisanen-Sokolowski, A., T. Glysing-Jensen, J. Koglin, and M. E. Russell. "Reduced Transplant Arteriosclerosis in Murine Cardiac Allografts Placed in Interferon-g Knockout Recipients." *Am. J. Pathol.* **152** (1998): 359–365.

[40] Sadick, M. D., F. P. Heinzel, B. J. Holaday, R. T. Pu, R. S. Dawkins, and R. M. Locksley. "Cure of Murine Leishmaniasis with Anti-Interleukin 4 Monoclonal Antibody." *J. Exp. Med.* **171** (1990): 115–127.

[41] Sallusto, F., A. Lanzavecchia, and C. R. Mackay. "Chemokines and Chemokine Receptors in T-Cell Priming and Th1/Th2-Mediated Responses." *Immunol. Today* **19** (1998): 568–574.

[42] Shimizu, A., W. Walter, Y. Tanaka, and S. Shaw. "Lymphocyte Interactions with Endothelial Cells." *Immunol. Today* **13** (1992): 106–112.

[43] Taipale, J., and J. Keski-oja. "Growth Factors in the Extracellular Matrix." *FASEB J.* **11** (1997): 51–59.

[44] VanBuskirk, A. M., M. E. Wakely, and C. G. Orosz. "Acute Rejection of Cardiac Allografts by Noncytolytic CD4+ T Cell Populations." *Transplantation* **62** (1996): 300–302.

[45] VanBuskirk, A., D. Pidwell, P. W. Adams, and C. G. Orosz. "Transplantation Immunology." *JAMA* **278** (1997): 1993–1999.

[46] Wahl, S. M., G. L. Costa, M. Corcoran, L. M. Wahl, and A. E. Berger. "Tranforming Growth Factor-b Mediates IL-1-Dependent Induction of IL-1 Receptor Antagonist." *J. Immunol.* **150** (1993): 3553–3560.

[47] Wang, R., A. M. Rogers, T. L. Ratliff, and J. H. Russell. "CD95-Dependent Bystander Lysis caused by CD4+ T Helper 1 Effectors." *J. Immunol.*. **157** (1996): 2961–2968.

[48] Wilbanks, G. A., M. Mammolenti, and J. W. Streilein. "Studies on the Induction of Anterior Chamber-Associated Immune Deviation (ACAID). III. Induction of ACAID Depends upon Intraocular Transforming Growth Factor-Beta." *Eur. J. Immunol.* **22** (1998): 165–173.

The Creation of Immune Specificity

Irun R. Cohen

1 REPRESENTATIONS

Physiological systems such as the immune system and the nervous system are adaptive systems; they *sense* and *respond* to the changing environment. Sensing and responding are most successful when they fit the requirements for life in the environment; *fit* responses mean *fitness*. Hence, it can be said that adaptive systems provide a creature with a functional representation of its reality. The response, in a sense, can be viewed as an internal image, a type of mirror image, of the stimulus that triggered the response [7]. Each of the systems composing the body specializes in mapping a particular aspect of the individual's niche in the world. The nervous system, for example, senses and responds to a wide array of information or energy originating from within and without the body as light, sound, pressure, gravity, chemical signals, and so forth. The immune system is tuned to a more limited world of sensation: molecular shape. Molecular shape is sensed by receptors that bind ligands through complementary, noncovalent interactions.

Immune receptors are of two types: somatically generated *antigen receptors* and *innate receptors*. The innate receptors are encoded in the germline of the species and bind both internal body signals (such as cytokines, chemokines, and others) and extrinsic signals (derived from bacteria, viruses, and other parasites). The output of the immune system in response to its receptor input can be detected as changes in the states and activities of immune cells.

2 INFLAMMATION

The immune response is a complex reaction marked by the production of antibodies and other effector molecules such as cytokines, chemokines, and adhesion molecules, and by the activation of various types of cells: T cells, B cells, macrophages, neutrophils, dendritic cells, endothelial cells, mast cells, and others. These immune cells and molecules mediate a great many effects; they cause cells to grow, to die, to migrate, to adhere, and they activate all kinds of genes. Immune agents control pathogens and kill abnormal cells, but they also trigger the healing and repair of body tissues. The details are beyond our present scope, and information about immune agents and their effects is accessible in text books. For the needs of the present discussion, we may designate the multitude of immune processes by the term *inflammation.* Inflammation relates to the impact of the immune system on the body. There are many different types and degrees of inflammation, and that's what the immune system is about. Let us return to immune representations.

The immune receptors that sense molecular shape and the ensuing inflammation together constitute the *representations* of the individual's immune environment. Immune wresentations that fit the individual's reality generate types of inflammation that facilitate well-being, procreation, and survival, all the rest permitting. Immune representations that do not fit the situation may be felt as undesirable or ineffective inflammation: allergies (damaging inflammatory responses to otherwise harmless antigens), toxic shock syndromes (life-threatening, systemic inflammatory responses triggered by foreign invader molecules), autoimmune diseases (persistent inflammation triggered by otherwise normal antigens of the body), or immune responses to pathogenic invaders that are ineffective in controlling the invaders.

3 IMMUNE PHYSIOLOGY

It has been proposed that the immune system functions to achieve certain ends: to discriminate between the self and the nonself [11], to protect the body from infectious diseases [10], or to ward off "danger" [15]. However, some may think that the assignment of *ends*, what is known as *teleology*, is not really compatible with the empirical method of science [7]. Science can only hope to understand *how* things work; *why* things work is a matter of opinion, not of experiment. Empirical thinkers might feel more at ease characterizing the output of the immune system in physiological terms. The output of the immune system, in physiological terms, is not *protection* or *discrimination*; it is *inflammation*. The function of the immune system is to produce inflammation commensurate with the body's need to keep itself fit for living.

4 SPECIFICITY GIVEN

The fidelity of a representation often can be measured by its power of discrimination; to how fine a degree can the immune system discriminate between different molecular shapes and between different concrete needs for different types of inflammation. Functional representation, in a word, emerges from *specificity*. Wherein lies immune specificity?

The clonal selection paradigm, the classic paradigm of immunology, teaches that specificity is an intrinsic property of immune receptors. Antigen specificity is insured automatically through the act of clonal selection; an antigen will activate only those clones of lymphocytes that *recognize* it. Recognition means specific binding. Specific recognition is the initiating act that generates the response [2].

5 REDUCTION TO CHEMISTRY

Specificity, according to classical clonal selection, not only initiates the immune response; specificity also regulates the immune response. It goes like this: The immune response is turned on by the binding between an antigen and a lymphocyte clone bearing an antigen receptor specific for the antigen. The response, likewise, is turned off by either the lack of the antigen or the lack of the specific antigen receptor [2]. The immune response, much like a chemical reaction, is regulated, all other factors permitting, by the concentrations of the specific reactants—the antigens and the receptors. The entry of an antigen activates a response, and the destruction of the antigen by the inflammatory products of the response feeds back to shut off the response: no antigen, no response. Classical clonal selection enjoys a compelling logic and a commendable reduction to underlying chemistry. Specificity runs the show. If you see an antigen, kill it.

6 AUTOIMMUNITY

If the immune system is designed to kill whatever it can see, then the immune system must never be allowed to see the self. According to clonal selection, self-recognition, autoimmunity, is "forbidden" [3]. The mechanism envisioned to prevent autoimmunity is the deletion of every lymphocyte clone that happens to bear an antigen receptor that can bind a self-antigen.

Strangely, the deletion of self-binding clones is still taught today to be the major factor that prevents autoimmune disease, despite the fact that healthy individuals are demonstrably populated with many clones of autoimmune lymphocytes [6, 12]. Logic and tradition constitute rules; a puzzling observation, such as the prevalence of healthy autoimmunity, can always be dismissed as an "exception" to the standing rule. An observation may evolve into an accepted fact only if the observation makes sense. Hence, the transformation of an unexpected observation into the status of an accepted fact is slow; logic and tradition retard the birth of unforeseen facts. Autoimmunity, in the logic of classical clonal selection, never made any sense.

7 IMMUNE MAINTENANCE

Healthy autoimmunity, however, makes sense when we view the output of the immune system as *inflammation* (physiology) rather than as *rejection of dangerous* pathogens (teleology). Changing a word makes a difference. No immunologist would deny that inflammation is needed to repair and maintain the body, and not only to defend the body against pathogens. Immune cells and their molecules of inflammation such as cytokines are essential to the processes of wound healing, angiogenesis, regeneration, tissue modeling, and waste disposal [7]. Maintaining the body means recognizing the state of the tissues and supplying whatever type and measure of inflammation is required. The immune system, by producing and regulating inflammation, is a key agent in the physiology of body maintenance. Self-maintenance, in a word, is compatible with measured autoimmunity; self-maintenance requires autoimmunity, at least of some types. My colleague Michal Schwartz, myself, and other coinvestigators have found that populations of potentially dangerous autoimmune T cells can help heal the traumatized central nervous system [17, 16].

The essence of immune regulation is not the specific deletion of autoimmune lymphocytes, but the necessity to provide the type and measure of inflammation that fits the circumstances, whether of body defense or of body maintenance. The requirements for immune specificity do not end at discriminating danger, or of self from nonself; the immune system must specifically diagnose varied situations and dynamically adjust specific inflammatory response phenotypes appropriately [9, 18]. Our present problem is to understand how this can be done with the requisite specificity, despite receptor degeneracy, molecular pleiotropism, and redundancy.

8 DEGENERACY

All biologic receptors are intrinsically degenerate; any receptor can bind more than one ligand [13]. The reasons for receptor degeneracy are inherent in the noncovalent chemistry of ligand binding; a full discussion of receptor degeneracy can be seen elsewhere [7]. The bottom line is that there is no way that a receptor can be specific for only one antigen or signal molecule. From whence is the specificity of the immune response, if not from the immune receptor?

9 PLEIOTROPISM

The specificity problem is compounded by the fact that the effector molecules produced by the immune system are pleiotropic; any cytokine, chemokine, or other cell-interaction molecule produces more than one effect. Interferon (IFN)γ, for example, has been shown to activate more than 200 different genes [4]. The effects of a single molecule may even be logically contradictory; tumor necrosis factor (TNF)α, for example, kills some cells while stimulating the growth of others. The specificity of the immune-response phenotype cannot be reduced to a simple one-to-one relationship between a molecule and an effect.

10 REDUNDANCY

Another challenge to specificity is redundancy; different immune recognition and effector molecules may produce very similar effects [7]. The components of the immune system are redundant. One-to-one specificity is quite rare.

11 SPECIFICITY TAKEN

Specificity implies a one-to-one relationship between a receptor and a ligand and between a cause and an effect. Despite the teachings of clonal selection, immune specificity is not given automatically by one-to-one interactions between receptor and ligand molecules. Immune specificity is not reducible to chemical specificity. Immune specificity has to be created by immune physiology. Specificity is not given; it has to be taken. How, indeed, can the immune system create immune specificity out of degeneracy, pleiotropism, and redundancy?

12 DECONSTRUCTION AND CO-RESPONDENCE

An unasked question has no answer; an asked question has, at the very least, an idea. A clone's antigen receptor, and any innate receptor too, is not chemically specific for a single ligand; but a degree of specificity, perhaps a high degree of specificity, can be seen to emerge from multiclonal *cooperativity* and from *deconstruction.* An example of both cooperativity and deconstruction is visible in what I have called *co-respondence* [7, 8].

Clonal selection could not have anticipated the fact that the immune system usually perceives an antigen redundantly using two chemically independent sets of antigen receptors, and that each set of receptors sees a different aspect of the antigen. B cells, and their antibodies, see the *conformation*, usually the *native* conformation of a segment of the antigen, while T cells see a denatured, processed fragment of the peptide sequence of the antigen integrated onto the surface conformation of an MHC molecule. Moreover, each set of receptor interactions is facilitated (or inhibited) by antigen-presenting cells (APC) responsive to the *context* of ancillary signals in which the antigen makes its appearance. In other words, the immune system does not perceive the antigen as it exists unto itself; the immune system perceives aspects of the antigen and its context *deconstructed* according to immune rules [8].

The response to the deconstructed antigen is not the sum of the responses of each individual agent (B cell, T cell, and APC); the immune response is the cooperative outcome of the mutual interactions of the different agents and their diverse perceptions (B cell, T cell, and APC). Each agent persists in seeing its own unique deconstructed world; conformation, processed sequence-in-MHC, context. Yet, each agent modifies its own response behavior under the influence of what each of its fellow agents sees and reports (by way of cytokines, chemokines, antibodies, and other cell-interaction molecules). Thus, a composite picture of the antigen is created by synergy: the mutual cooperativity of semiautonomous agents, each perceiving a different molecular world. This cooperative reconstruction of the immune object is generated by the process I have called co-respondence [7]. A *higher order* specificity, more precise than the degenerate world view of the system's individual receptors, is thus created by the cooperative interactions of the agents comprising the immune system. A single antigen receptor on a single clone satisfied the naive chemical paradigm of clonal selection. The complexity of the real world is represented by the unanticipated complexity of the immune apparatus. Of course, we now need to understand immune reconstruction at the molecular level. We need computer models, perhaps even a novel computer language, to perceive the functional images created by co-respondence.

13 PATTERNS

Patterns of elements are another way of achieving cooperative specificity. The specificity of a pattern can extend beyond the limits of the specificity (or degeneracy) of the discrete elements comprising the pattern. Consider, for example, that our prodigious ability to discriminate between various colors and shades of colors is the work of only three distinct types of color receptors (cones). The types of cones in our retinas are each most sensitive to red, green, or blue; that's all. How do we manage to discriminate so well between so many colors if we have only three types of cones? Fortunately, the three color receptors are degenerate; the perceptive fields of the receptors overlap so that each receptor can respond to some extent to different photons. Thus a single wavelength of light may activate two or three types of cones, each to a different degree. Different wavelengths of light can activate different patterns of responses from the various types of cones. Our ability to discriminate between so many shades of color emerges from the *patterns* of the signals generated by the various cones responding to the incoming light [7]. Color discrimination, therefore, is created by the cooperativity between patterns of *degenerate* receptors. Immune patterns are not composed of photons; they are composed of populations of responding clones, polyclonal antibodies, mixtures of cytokines, and more. Nevertheless, populations of degenerate immune cells and molecules, like retinal cones, produce distinct patterns of activity. The patterns are specific, even if the individual cells and molecules are degenerate. The principle of specificity through cooperative interaction is a common strategy in biology. Note that degeneracy, according to this point of view, is not an impediment to specificity; degeneracy is a requirement for specificity.

14 IMMUNE DECISIONS

In the light of the foregoing, how does the immune system make its decisions? One can think about decision making in at least two different ways: transformational and reactive [14]. A transformational system transforms an input of information (or energy) into a different form of information (or energy). For example, the information borne by an antigen (including the antigen's context) is transformed into a specific immune response that fits the situation. A transformational representation of the antigen proceeds serially step-by-step according to a chain of discrete signals, one signal at a time. A transformational system makes discrete decisions at each point in its path.

A reactive system, in contrast to a transformational system, creates a dynamic representation of the environment by responding in parallel to many signals all at once. The reactive system does not deliberate step-by-step; it reacts simultaneously to the barrage of signals striking the system. A reactive system navigates through a sea of often conflicting signals. The central nervous system exemplifies a reactive system; it gets the individual through life by just reacting. A reactive system can be seen to have made decisions only through hindsight. You can look back and trace a path of (seeming) decisions that got you where you are now. In the process of arriving, however, it is hard to pinpoint discrete, isolated decisions.

Does the immune system proceed by making isolated decisions as a transformational system, or does the immune system just respond continuously as a reactive system? At present, I believe that the immune system is a true reactive system. It is continuously busy maintaining the body and defending against potential pathogens, all at once. It does not have the leisure to wait idly for this or that discrete signal. The immune system builds, without respite, a dynamic representation of its field of operations. This field includes both the body of the individual and invaders of that body. Immune decisions are made through an ongoing dialogue of interacting agents [1]. This dialogue reconstructs the immune system's representation of its world.

15 COGNITION

I think the analogy between the immune system and the central nervous system is apt; both use a self-organizing, representational strategy for defining and maintaining the self [7]. The representations in both systems include images of the individual body. I have termed the immune representation of the self the immunological homunculus [5]. The immunological homunculus represents natural autoimmunity directed to a particular set of self-antigens. Immune maintenance requires the immune representation of the self [7].

ACKNOWLEDGMENTS

I am the incumbent of the Mauerberger Chair in Immunology, Director of the Robert Koch-Minerva Center for Research in Autoimmune Disease, and Director of the Center for the Study of Emerging Diseases.

REFERENCES

[1] Atlan, H., and I. R. Cohen. "Immune Information, Self-Organization, and Meaning." *Int. Immunol.* **10** (1998): 711–717.
[2] Burnet, F. M. *The Clonal Selection Theory of Acquired Immunity.* Cambridge: Cambridge University Press, 1959.
[3] Burnet, F. M. *Self and Not-Self.* Cambridge: Cambridge University Press, 1969.
[4] Boehm, U., T. Klamp, M. Groot, and J. C. Howard. "Cellular Responses to Interferon-γ." *Ann. Rev. Immunol.* **15** (1997): 749-795.
[5] Cohen, I. R. "The Cognitive Paradigm and the Immunological Homunculus." *Immunol. Today* **13** (1992): 490–494.
[6] Cohen, I. R. "The Cognitive Principle Challenges Clonal Selection." *Immunol. Today* **13** (1992): 441–444.
[7] Cohen, I. R. *Tending Adam's Garden: Evolving the Cognitive Immune Self.* San Diego, CA: Academic Press, 2000.
[8] Cohen, I. R. *Discrimination and Dialogue in the Immune System, Seminars in Immunology,* in press.
[9] Grossman, Z. "Contextual Discrimination of Antigens by the Immune System: Towards a Unifying Hypothesis." *Theoretical and Experimental*

Insights into Immunology, edited by A. Perelson and G. Weisbuch, 71–89. Berlin: Springer-Verlag, 1992.

[10] Hoffmann, J. A., F. C. Kafatos, C. A. Janeway, and R. A. Ezekowitz. "Phylogenetic Perspectives in Innate Immunity." *Science* **284** (1999): 1313–1318.

[11] Klein, J. *Immunology: The Science of Self-Nonself Discrimination.* New York: John Wiley, 1982.

[12] Lacroix-Desmazes, S., S. V. Kaveri, L. Mouthon, A. Ayouba, E. Malanchere, A. Countinho, and M. D. Kazatchkine. "Self-Reactive Antibodies (Natural Autoantibodies) in Healthy Individuals." *J. Immunol. Methods* **216** (1998): 117–137.

[13] Lancet, D., E. Sadovsky, and E. Seidemann. "Probability Model for Molecular Recognition in Biological Receptor Repertoires: Significance to the Olfactory System." *Proc. Natl. Acad. Sci. USA* **90** (1993): 3715–3719.

[14] Manna, Z., and A. Pnueli. *The Temporal Logic of Reactive and Concurrent Systems.* Berlin: Springer-Verlag, 1992.

[15] Matzinger, P. "Tolerance, Danger, and the Extended Family." *Ann. Rev. Immunol.* **12** (1994): 991–1045.

[16] Moalem, G., R. Leibowitz-Amit, E. Yoles, F. Mor, I. R. Cohen, and M. Schwartz. "Autoimmune T Cells Protect Neurons from Secondary Degeneration after Central Nervous System Axotomy." *Nat. Med.* **5** (1999): 49–55.

[17] Schwartz, M., and I. R. Cohen. "Autoimmunity and Tissue Maintenance." *Immunol. Today* **21** (2000).

[18] Segel, L. A, and R. L. Bar-Or. "On the Role of Feedback in Promoting Conflicting Goals of the Adaptive Immune System." *J. Immunol.* **163** (1999): 1342–1349.

Diversity in the Immune System

José A. M. Borghans
Rob J. De Boer

1 INTRODUCTION

Diversity is one of the key characteristics of the vertebrate immune system. Lymphocyte repertoires of at least 3×10^7 different clonotypes [2] protect humans against infections, while avoiding unwanted immune responses against self-peptides and innocuous antigens. It is this lymphocyte diversity that forms the main difference between the immune systems of invertebrate and vertebrate species. Invertebrates are protected from pathogenic invasions by broad-spectrum pathogen-associated recognition molecules, recognizing conserved pathogenic structures [29, 33]. On top of these innate responses, which have been preserved in vertebrate species, vertebrates evolved an adaptive immune system, which has the capacity to respond to a virtually infinite variety of antigens. Adaptive immunity evolved when gene rearrangements were employed to generate highly diverse lymphocyte repertoires [1, 20, 34].

Another important source of diversity in the immune system is due to the genes coding for major histocompatibility (MHC) molecules. For a cellular immune response to be induced, the proteins of a pathogen need to be degraded into peptides, which are subsequently bound to MHC molecules on the surface of antigen-presenting cells. The resulting MHC-peptide complexes can be recognized by T-cell receptors. In humans, each individual expresses three classical MHC class I genes (HLA A, B, and C), and three MHC class II gene pairs (coding for the α and β chains of HLA DP, DQ, and DR) [28]. The population diversity of histocompatibility molecules is extremely large, and predates the evolution of vertebrates. For some MHC loci, more than one hundred different alleles have been identified [49, 70]. Due to the high population diversity of MHC molecules, different individuals typically mount immune responses against different subsets of peptides of any pathogen. Pathogens that escape from presentation by the MHC molecules of one particular host, may thus not be able to escape from presentation in another host with different MHC molecules.

The mechanisms underlying the diversity of the adaptive immune system and the MHC complex are very different. The diversity of lymphocyte receptors is due to the evolution of somatic diversification mechanisms [1]. Genes coding for the V, D, and J segments of lymphocyte receptors are somatically rearranged, and imprecise joining of the gene segments, addition of nucleotides, and somatic hypermutation subsequently add to the diversity of lymphocytes [28]. The result is an extremely diverse, semirandomly generated, repertoire of lymphocytes that bind their ligands with great specificity. The diversity of MHC molecules, in contrast, is not due to any special diversification processes. The mutation rate of MHC molecules is similar to that of most other genes [47, 57]. Studies of nucleotide substitutions at MHC loci have revealed that there is Darwinian selection for diversity at the peptide-binding regions of MHC molecules [25, 26, 47, 48]. Contrary to lymphocyte receptors, MHC molecules bind their ligands with great degeneracy [23, 31, 36].

This chapter gives a review of our research on the evolutionary selection pressures underlying the diversity of lymphocytes and MHC molecules. We hypothesize that the adaptive immune system stores the appropriate effector mechanisms against the antigens it encounters (see also Swain et al. [64]). For example, lymphocytes specific for food- and self-antigens switch to a tolerant mode, while lymphocytes recognizing pathogens switch to a particular responsive mode. Once lymphocytes have been instructed as to which type of immune response to mount, they recall their appropriate effector mechanism whenever they recognize their specific epitope [53, 64]. The immune system thereby learns to associate antigens with the appropriate type of immune response against them. Recall responses may be harmful, however, if different antigens requiring different modes of response trigger the same clonotype. The likelihood of such inappropriate responses increases with the degree of crossreactivity of lymphocytes and with the number of peptides per antigen that are presented to the immune system.

In section 2 we show that a somatically learning adaptive immune system requires a high degree of diversity. Repertoire diversity allows the immune system to reconcile specificity (which is required to avoid inappropriate, crossreactive immune reactions) with reactivity to many antigens (see also Borghans and De Boer [9, 10] and Borghans et al. [11]). Interestingly Cohen discusses in this book how the immune system may achieve a high degree of specificity using degenerate receptors. In section 3 we investigate why the number of different MHC molecules expressed per individual is so limited as compared to the large population diversity of MHC molecules. We demonstrate that it is unlikely that the individual MHC diversity is limited due to T-cell repertoire depletion during negative selection, as has been proposed [17, 28, 47, 69] and modeled [18, 44, 66] before. We demonstrate that the selection pressure for more individual MHC diversity vanishes once of the order of ten different MHC molecules per individual have been expressed. Excessive individual MHC diversity has the added disadvantage that it increases the chance to mount inappropriate immune responses, such as autoimmune responses by clones that have escaped tolerance induction. The limited number of MHC molecules per individual may thus reflect a compromise between recognition of many antigens and avoidance of self-reactivity. In section 4, we demonstrate that despite the limited expression of MHC molecules per individual, host-pathogen coevolution can account for a very large *population* diversity of MHC molecules. Using a genetic algorithm, we show that MHC diversity is to be expected in host populations adapting to pathogens with short generation times (see also Beltman [5]).

2 DIVERSITY OF LYMPHOCYTES

During a primary immune reaction, the immune system has to decide which type of immune response is most appropriate [64]. No immune response should be induced against self-antigens and innocuous antigens, while pathogens are eliminated by qualitatively different immune responses, varying from cellular to humoral responses, and varying in, for example, immunoglobulin isotype and cytokine expression [28, 64, 76]. The decision as to which type of immune response to mount is based upon many factors, such as signals from the innate immune system [6, 15, 16, 21, 29, 30, 38, 39, 40, 56], the local tissue environment [77], tissue damage [37], and success-driven feedback mechanisms (see Segel [58, 59]). These signals collectively form the "context" of an antigen.

We hypothesize that, apart from dealing with antigens, one of the main functions of the adaptive immune system is to store the appropriate modes of response against different antigens in differentiated lymphocytes [10, 11]. If effector or memory clones recognize a subset of the epitopes that are expressed by an antigen, they contribute to the antigen context, and provide information on the type of immune response that is to be induced. Being fairly independent of costimulatory signals, such instructed lymphocytes help to eliminate pathogens upon re-encounter even before any tissue damage has been done, and help to induce appropriate immune responses against new antigens that correlate with previously encountered antigens. Instructed lymphocytes can also direct the differentiation of new naïve lymphocytes. Tolerant T cells have been shown to be able to transfer their nonresponsive phenotype to other, naïve cells [51, 72], even if those naïve cells have a different specificity [65].

Analogously, memory lymphocytes of a certain responsive mode may direct the differentiation of new naïve clonotypes [10, 32], for example, *via* cytokine secretion or interactions with dendritic cells [13, 54, 55].

If instructed lymphocytes are too crossreactive, however, they may induce inappropriate responses [3, 45, 46, 75]. Here we show the results of a simulation model that we have developed to study under which circumstances storage of appropriate effector mechanisms can help the induction of new, appropriate immune responses, while avoiding inappropriate, crossreactive responses. We find that lymphocyte specificity is required to avoid inappropriate immune responses, and that repertoire diversity does not hamper the role of instructed lymphocytes in the induction of immune responses to new antigens.

2.1 STORAGE OF APPROPRIATE IMMUNE RESPONSES

We simulate the storage of effector mechanisms against different antigens in an immune system with R_0 different clonotypes. The immune system is sequentially challenged with different antigens, each requiring a certain type of immune response, and each consisting of e different (immunodominant) epitopes. Both the appropriate type of response to an antigen, and the clonotypes recognizing its epitopes, are selected randomly. Each clonotype has a certain mode. Clonotypes specific for the S different tolerance-inducing self-epitopes are initialized in the tolerant mode; all other clonotypes are initially naïve. Due to recognition of an antigen, naïve clonotypes may switch to a particular responsive mode (such as Th1, Th2, IgA, IgE, etc.). Different modes are represented by integer numbers $0, 1, 2, \ldots, m$, where 0 means naïve, 1 means tolerant, and $2, 3, \ldots, m$ identify particular responsive modes. In our simulations, every epitope that the immune system encounters is recognized by precisely one clonotype, which is selected randomly. Repertoire diversity is thus inversely related to the crossreactivity of clonotypes. Depending on the degree of lymphocyte crossreactivity, one clonotype may recognize multiple epitopes.

Whenever epitopes of antigens in our simulations are recognized by previous memory clones, these memory clones determine what type of immune response is induced. The modes of response suggested by different memory clonotypes might not be identical, however. Any conflicts are resolved by treating each signal as a "vote" in the decision-making process. The ultimate decision is the mode for which there is a majority count. In case there is a tie, the decision is chosen randomly from the largest votes. In the absence of cross-reacting memory lymphocytes, we assume that the combination of the innate immune response, the context of the antigen, and possibly feedback mechanisms, ultimately leads to the appropriate type of immune response. This might not be unreasonable, because the innate immune system has learned about different kinds of pathogens and antigenic contexts over evolutionary time.

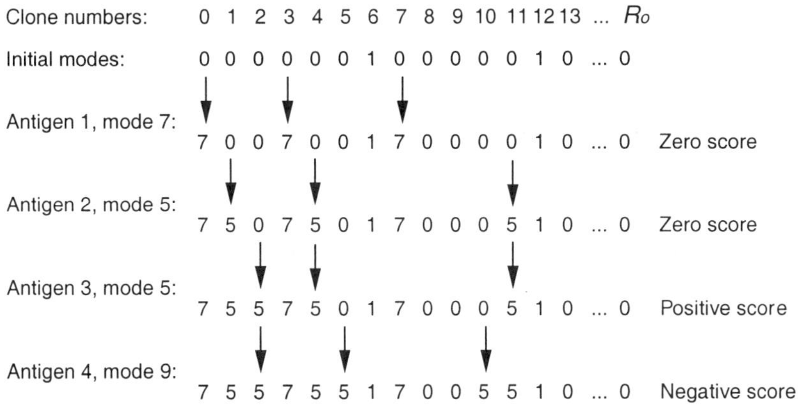

FIGURE 1 A simple example of a simulation with $e = 3$ different epitopes per antigen. After self-tolerance induction most clonotypes are naïve (i.e., mode 0), except clonotypes 6 and 12 which have been initialized in the tolerant mode (i.e., mode 1). The first antigen has to be rejected by an immune response of mode 7, and triggers clonotypes 0, 3, and 7. Since these three clonotypes are naïve in the primary response, the decision as to which type of immune response to mount is made by the innate immune system. Thus, clonotypes 0, 3, and 7 become memory clones of mode 7, antigen 1 is rejected, and no score is obtained. Similarly, antigen 2 triggers three naïve clonotypes, which subsequently switch to memory mode 5. Antigen 3 triggers two memory clones that overlap with antigen 2 (i.e., clones 4 and 11), and triggers the naïve clone 2. Because of the memory votes by clones 4 and 11, an immune response of mode 5 is triggered. This yields a positive score. Clone 2 correctly switches to mode 5. Antigen 4, requiring mode 9, coincidentally triggers a memory clone (2) which is in mode 5. Thus, an inappropriate immune response is induced, yielding a negative score. naïve clonotypes 5 and 10 incorrectly switch to mode 5.

In our simulations, once a decision has been made, all naïve clonotypes involved in a primary immune response switch to the corresponding memory mode. Even if an inappropriate response is triggered, naïve lymphocytes switch (to the incorrect) mode. In accordance with experimental data, memory clonotypes do not switch mode [41, 52]. If previous memory clones have the majority vote and thereby establish the correct mode of response, a positive score is given. All cases in which previous memory clones establish an incorrect mode of response yield a negative score. In the default situation, in which naïve lymphocytes adopt the mode of the innate immune system, no score is added. Figure 1 provides an example of a small simulation.

Obviously, the adaptive immune system will only give a positive contribution to the decision-making process if there are groups of structurally related antigens that require a similar type of immune reaction. To account for such groups of antigens, a fraction P_m of all antigens in our simulations is a mutant of another antigen. Mutant and wild-type antigens always require identical modes of response and share half of their epitopes; the other epitopes are chosen randomly.

2.2 SOMATIC LEARNING REQUIRES LYMPHOCYTE SPECIFICITY

Figure 2 illustrates how the performance of an immune system that has been challenged with one thousand different antigens depends on the diversity of the lymphocyte repertoire. All antigens have been presented to the immune system only once; i.e., we study a "worst case" scenario, ignoring the conventional benefits of immunity obtained when the same antigen rechallenges the immune system. The two panels show the fraction of challenges yielding a positive score, and a negative score, respectively. The different curves in figure 2 depict different levels of correlation between the pathogens, i.e., $P_m = 0$ (solid), $P_m = 0.1$ (dotted), and $P_m = 0.2$ (dashed).

Figure 2(a) shows that memory clones help to make correct decisions whenever (i) there is some correlation between the antigens *and* (ii) the lymphocyte repertoire is sufficiently specific. At a very low repertoire diversity, hardly any positive score is obtained because most lymphocytes have been tolerized by self-epitopes (see also De Boer and Perelson [18]). At an intermediate repertoire diversity, the repertoire is no longer depleted during tolerance induction but the positive scores that are obtained are largely coincidental. Even if there is no correlation between the antigens (see the solid curve), these positive scores occur because of random crossreactions. Above a diversity of $R_0 = 10^5$ clonotypes, this randomness disappears and the positive scores hardly depend on the diversity of the immune system. Whatever the diversity of the system, a recurring epitope always triggers the same clonotype. Increasing the repertoire size R_0, and hence the specificity of the system, therefore does not impair the positive contribution of memory lymphocytes to the decision making during immune reactions.

Figure 2(b) demonstrates that lymphocyte systems of low diversity are prone to make mistakes due to crossreactivity. At a low diversity, previous memory clones specific for epitopes of unrelated antigens tend to induce wrong types of immune responses; on the other hand, clones that have previously been tolerized by self-epitopes hinder the induction of immune responses to subsequent pathogens. Figure 2(b) shows that such mistakes (i) disappear at a large repertoire diversity, and (ii) hardly depend on the correlation between the antigens.

Summarizing, these simulations demonstrate that in immune systems that store the appropriate modes of responsiveness against many different antigens, the avoidance of harmful, inappropriate responses requires a highly specific immune repertoire. High specificity counteracts the demand that all antigens should be recognized, however [35]. Although the current simulation model does not allow for nonrecognition, we know from previous modeling [9, 11] that responsiveness against many antigens can be reconciled with specificity by selecting for a sufficiently diverse immune repertoire.

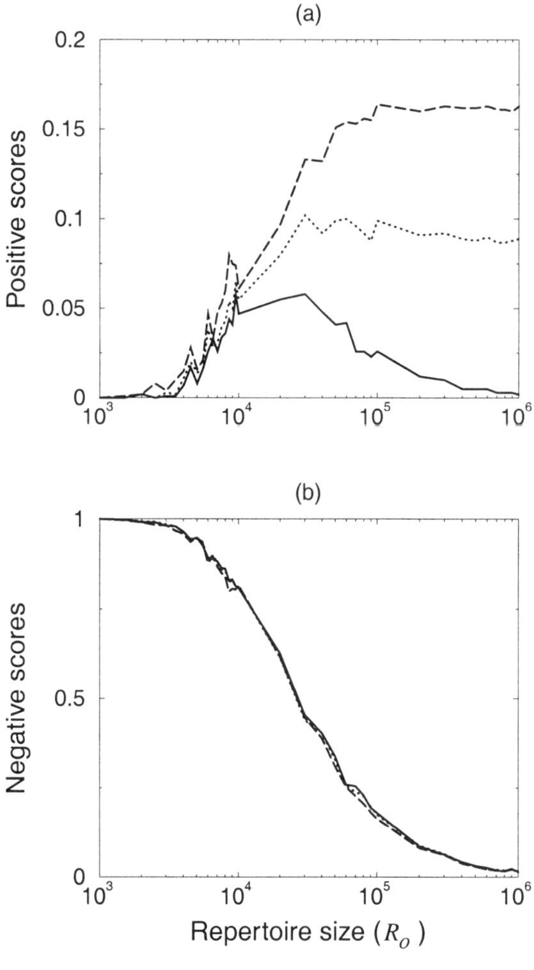

FIGURE 2 The performance of lymphocyte systems of different diversities (R_0) challenged with one thousand different antigens. (a) The fraction of challenges that yield a positive score thanks to previous memory clones making correct decisions. (b) The fraction of challenges yielding a negative score due to inappropriate immune responses induced by previous memory clones or lack of responsiveness due to cross-reactive tolerant clones. The different curves denote different degrees of correlation between the antigens that are encountered: $P_m = 0$ (uncorrelated antigens, solid curves), $P_m = 0.1$ (dotted curves), and $P_m = 0.2$ (dashed curves). Related antigens share 50% of their epitopes. There are $e = 6$ different epitopes per antigen, $S = 10^3$ self-antigens, and ten different modes ($m = 9$).

3 WHAT LIMITS THE INDIVIDUAL MHC DIVERSITY?

In sharp contrast with the highly specific binding between epitopes and lymphocytes, peptide binding to MHC molecules is very degenerate [23, 36]. The chance that a random peptide binds a random human MHC molecule is 0.1% to 10% [31]. Degenerate MHC-peptide binding allows the immune system to present a great variety of peptides and, hence, to mount immune responses against many pathogens. It is generally thought that this selective advantage also explains why individuals tend to be MHC heterozygous (see, e.g., Doherty and Zinkernagel [19], Hughes and Nei [25, 26, 27], and Takahata and Nei [67], and section 4 of this chapter). Indeed, in a study of patients infected with HIV-1, it was shown that the degree of heterozygosity of MHC class I loci correlated positively with a delayed onset of AIDS [14].

Since immunity against pathogens requires the presentation of pathogen peptides on host MHC molecules, the number of MHC genes expressed in vertebrates is, in fact, surprisingly small. Just like favoring MHC heterozygosity, one would expect evolution to favor the expression of many MHC genes per individual. In reality, however, the MHC diversity per individual (i.e., of the order of ten different MHC molecules [28]) pales into insignificance in comparison to the huge diversity of MHC alleles in populations (i.e., up to hundreds of alleles per locus [49, 70]). Using a probabilistic model, we here study which mechanisms may underlie the limited expression of different MHC molecules per individual. In the next section we will investigate the large population diversity of MHC molecules.

It is often quite loosely argued that the number of different MHC molecules per individual is limited due to self-tolerance induction [17, 28, 47, 69]. During negative selection in the thymus, clonotypes that recognize thymic MHC-peptide complexes with too high an affinity are tolerized [43]. Excessive expression of MHC molecules might thus lead to depletion of the T-cell repertoire. Nowak et al. [44] translated this verbal argument into a mathematical model and concluded that self-tolerance induction can indeed account for a realistically low individual MHC diversity. We have criticized this model by a different calculation, leading to the opposite conclusion that negative selection fails to explain the limited MHC diversity observed in nature [12].

Consider an individual with M different MHC molecules and a total lymphocyte repertoire consisting of R_0 different clones. Expression of many different MHC molecules reduces the functional T-cell repertoire due to negative selection. On the other hand, it enlarges the functional repertoire due to positive selection: only T cells that bind thymic MHC-peptide complexes with sufficient affinity enter the functional T-cell repertoire [22, 71]. If a peptide is presented by one of the MHC molecules of a host, the number of clones that can possibly recognize the peptide-MHC complex is the number of clones R_M that is positively selected by the particular MHC molecule, and not negatively selected by any of the M MHC molecules of the host, i.e.,

$$R_M = hR_0(1-t)^M \ . \qquad (1)$$

Here, t is the fraction of thymocytes that is deleted by negative selection per MHC molecule and h is the chance that a T lymphocyte surviving negative selection is positively selected on the particular MHC molecule. Equation (1) reflects the negative effect of expression of many different MHC molecules on the number of lymphocytes surviving negative selection. It has been estimated that approximately 90% of all thymic T cells fail to be positively selected on any of the MHC molecules of a host [68]. At least 50% of all *positively selected* T cells have been shown to undergo negative selection in the thymus [68]. The remaining 5% of all thymic T cells end up in the mature repertoire [61, 71]. Since an individual has typically of the order of ten different MHC molecules, these experimental estimates translate into $h = 0.005$ and $t = 0.005$ per MHC molecule.

If the individual is exposed to an antigen consisting of e different (immunodominant) epitopes, the chance P_i to make an immune response is:

$$P_i = 1 - (1 - q + q(1-p)^{R_M})^{eM} \ . \qquad (2)$$

Here, q is the chance that an MHC molecule presents a randomly chosen peptide and p is the chance that a clonotype that has been positively selected by the MHC molecule in question recognizes a random peptide presented by that MHC molecule. No immune response is induced if, on all MHC molecules, all epitopes are either not presented (with chance $1-q$), or presented but not recognized by any of the R_M clonotypes (with chance $q(1-p)^{R_M}$). Equation (2) reflects the positive effect of expression of many different MHC molecules on both the presentation of antigens and the positive selection of lymphocytes.

The solid curve in figure 3 shows that good protection against pathogens is achieved (i.e., $P_i \simeq 1$) for an individual MHC diversity between 10 and 2000 different molecules. This result thus contradicts the conclusion drawn by Nowak et al. [44] that the individual MHC diversity is limited to avoid repertoire depletion during tolerance induction. Instead we find that repertoire depletion occurs only at an unrealistically high individual MHC diversity. Since different MHC molecules select basically nonoverlapping sets of T-cell clones [7, 22], addition of extra MHC molecules tends to *enlarge* the functional repertoire. The essential difference between the model by Nowak et al. [44] and the current model is that in the previous model, T cells that fail to be positively selected on a particular MHC molecule can nevertheless be negatively selected on that MHC molecule [12]. Thus, the realistically low MHC diversity claimed by Nowak et al. [44] hinges upon an unrealistically stringent negative selection.

Having disputed the commonly accepted argumentation that negative selection limits the individual MHC diversity [17, 18, 28, 44, 47, 66, 69], the question remains which other mechanism can explain the limited number of

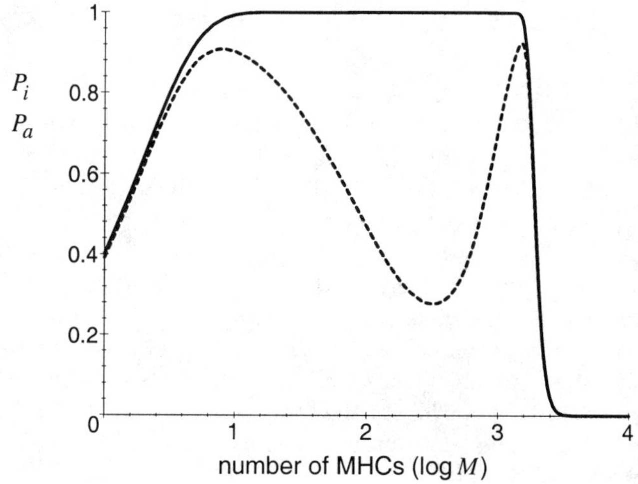

FIGURE 3 The solid curve denotes the chance P_i to mount an immune response as a function of the number of different MHC molecules per individual M. The curve shows that T-cell repertoire depletion occurs only at an unrealistically high individual MHC diversity. The dashed curve denotes the chance P_a to mount an *appropriate* immune response. Once an individual expresses of the order of ten different MHC molecules, additional MHC diversity increases the chance that autoimmune responses are induced. Parameters are: $q = 0.05$ [31], $p = 10^{-8}$, $R_0 = 10^9$, $e = 10$ [18], $h = 0.005$, $t = 0.005$, and $S_i = 2 \times 10^4$.

MHC molecules per individual. The solid curve in figure 3 suggests one possibility. The flat top of the curve demonstrates that the selection pressure for a higher MHC diversity vanishes once about ten different MHC molecules per individual have been expressed. This rather limited individual MHC diversity may thus simply be sufficient to have a good chance to present and respond to antigens.

3.1 AVOIDING INAPPROPRIATE RESPONSES

Elaborating on the theme of section 2, we investigate an alternative explanation for the limited individual MHC diversity, namely the need to avoid inappropriate immune responses. As discussed above, inappropriate responses occur when different antigens requiring different modes of responsiveness trigger the same clonotype. An example of an inappropriate response is when a self-specific clonotype that is ignorant of its self-epitope is triggered by a crossreacting foreign epitope and subsequently induces an autoimmune disease [3, 45, 46, 75]. The likelihood of such inappropriate immune responses increases with the number of epitopes that are presented to the immune system.

Once there are sufficient MHC molecules to ensure presentation of antigens, having a greater diversity of MHC molecules may thus be detrimental.

To study this hypothesis, we extend the above-described model with the chance P_t to stay tolerant to all self-peptides. This is expressed as the chance that during an immune response, on all of the M MHC molecules of a host, foreign epitopes are either not presented (with probability $1-q$), or presented but not recognized by any of the responding, ignorant self-specific clonotypes (with probability $q(1-pa)^{R_M}$):

$$P_t = (1 - q + q(1 - pa)^{R_M})^{eM} \ . \tag{3}$$

The probability a that a clone from the functional repertoire is ignorant and self-specific[1] is given by:

$$a = 1 - (1 - p)^{qS_i M^*} \ , \tag{4}$$

where S_i denotes the number of self-epitopes that fail to induce self-tolerance, and M^* denotes the expected number of MHC molecules that positively select one particular clone from the functional repertoire:

$$M^* = \frac{Mh}{(1 - (1 - h)^M)} \ . \tag{5}$$

Note that the decrease in P_t with increasing M is due to (i) the increasing presentation of foreign epitopes, and (ii) the increasing fraction of ignorant, self-specific lymphocytes a, due to the increasing number of peptide-MHC complexes formed by self-antigens that fail to induce tolerance. The chance P_a to mount an appropriate response to an antigen is the chance P_t to stay tolerant minus the probability that all clones fail to respond:

$$P_a = P_t - (1 - P_i) \ , \tag{6}$$

where P_i is given by eq. (2).

The dashed curve in figure 3 shows that involving the chance to mount an autoimmune response yields a sharply defined, low optimal MHC number, i.e., eight MHC molecules per individual (left-hand top). Yet, the chance P_a to make an appropriate immune response in that optimum remains close to one. Apparently, the system can reconcile the need to respond to many antigens with the need to avoid crossreactive, autoimmune responses, by selecting for a relatively low MHC diversity. At the left-hand top of the P_a curve, adding MHC molecules hardly increases the chance P_i to mount an immune response against an antigen (see the solid curve), while it significantly decreases the chance P_t to stay self-tolerant (see the dashed curve).

[1]Note that eq. (3) may give an underestimation of P_t if clones recognize multiple peptide-MHC complexes coming from one antigen. In our parameter setting, this chance is negligible for $M < 10^5$ since the probability that a particular clone recognizes an MHC-peptide complex during challenge with one antigen is $M^*t < 5 \times 10^{-4}$.

Interestingly, the dashed curve in figure 3 has a second peak at a very high number of different MHC molecules per individual. At the second peak, both self- and foreign epitopes are presented as many different MHC-peptide complexes. The immune system then finds a balance between prevention of autoimmunity due to a severely depleted repertoire, and immunity against foreign antigens thanks to the formation of many different peptide-MHC complexes per epitope.[2] This scenario is extremely wasteful, since at the top only 0.06% of the total lymphocyte repertoire survives thymic selection. If autoimmunity is less of a problem, the P_a curve looses the two sharply defined peaks. For example, if lymphocytes are highly specific (e.g., $p = 10^{-9}$), the risk of autoimmunity by crossreactions becomes negligible, and the P_a curve and the P_i curve become almost indistinguishable. Nevertheless, the dashed curve in figure 3 shows that an increase in autoimmunity due to crossreactions is a possible side-effect of expression of a large individual MHC diversity.

Summarizing, our model suggests that the evolution of a limited number of MHC genes per individual does not result from repertoire depletion during self-tolerance induction in the thymus. Instead, it may either reflect a low requirement of MHC diversity due to degenerate peptide-MHC binding (solid curve, fig. 3), or reflect the need to avoid inappropriate, crossreactive immune responses (dashed curve, fig. 3).

4 POPULATION DIVERSITY OF MHC MOLECULES

Despite the limited expression of different MHC molecules per individual, the MHC diversity of populations is extremely large. A commonly held view is that MHC polymorphism is due to selection favoring MHC heterozygosity. Since MHC molecules are codominantly expressed, and different MHC molecules bind different peptides, MHC heterozygous hosts can defend themselves against a larger variety of pathogens compared to MHC homozygous individuals. This hypothesis is known as the theory of "overdominance" or "heterozygote advantage" [19, 25, 26, 27, 67]. Alternatively, it has been proposed that the large polymorphism of MHC molecules is due to the high speed at which pathogens adapt to their hosts, due to their relatively short generation times. Since evolution will favor pathogens that avoid presentation by the most common MHC molecules in the host population, there will be a permanent selection force favoring hosts that carry rare—e.g., new—MHC molecules. Since hosts with rare MHC alleles have a relatively high fitness, the frequency of rare MHC alleles will increase and common MHC alleles will become less frequent. The result of this "frequency-dependent selection" is a dynamic equilibrium, maintaining a polymorphic population [4, 8, 62, 63].

The mechanisms behind the selection for MHC polymorphism have been debated for over three decades. It has been argued that selection for het-

[2]The position and height of the second peak should be taken with care since our equations may become imprecise at very high values of M.

erozygosity alone cannot explain the high MHC diversity observed in nature [48, 73]. Several models have been developed to study the effects of selection for heterozygosity and frequency-dependent selection on the polymorphism of MHC molecules (see, e.g., Takahata and Nei [67], Wills [73], and Wills and Green [74]). To our knowledge, however, a direct comparison of both hypotheses in one model has never been made. In order to make such a comparison, we have simulated the coevolution of hosts and pathogens using a genetic algorithm [24].

4.1 A SIMULATION MODEL OF MHC DIVERSITY

An extensive description of our model has been published previously [5]. We here confine ourselves to a very brief summary of the model structure. In our model, hosts are diploid and consist of bit strings representing their MHC alleles; pathogens are haploid, and their peptides are also represented by bit strings. Peptide presentation by an MHC molecule may occur at different positions on the MHC molecule, and is modeled by complementary bit matching. If the number of complementary bits at the best matching position on an MHC molecule exceeds a predefined threshold, a peptide is considered to be presented by the particular MHC molecule. In the simulations presented here, the chance that a random MHC molecule presents a randomly chosen peptide is 7.3%. Hosts carrying different MHC molecules will therefore typically present different peptides of pathogens.

At each generation, every host in our simulations interacts with every pathogen. The fitness of a host is proportional to the fraction of pathogens that it can present; the fitness of a pathogen is proportional to the fraction of hosts that it can infect without being presented by the host's MHC molecules. All individuals are replaced by fitness-proportional reproduction at the end of each generation. During reproduction, point mutations can occur. One cycle of fitness determination, reproduction, and mutation defines a generation. To account for the shorter generation time of pathogens, we let pathogens go through several generations per host generation.

4.2 MHC DIVERSITY BY HOST-PATHOGEN COEVOLUTION

To study the origin and maintenance of the MHC polymorphism, all hosts in our simulations initially express one and the same MHC molecule, while the pathogens are initialized randomly. The average fitnesses of the pathogens and the hosts are initially close to 0.5 (see fig. 4). Thanks to their relatively short generation times, the pathogens in our simulations evolve to evade presentation by the MHC molecules of the hosts. Since there is no initial MHC diversity, the average fitness of the pathogens immediately increases (see fig. 4(a)). Any pathogen that is able to infect one host is able to infect all hosts and, hence, rapidly takes over the pathogen population. As a consequence, the average fitness of the hosts initially drops (see fig. 4(b)). Under this selection

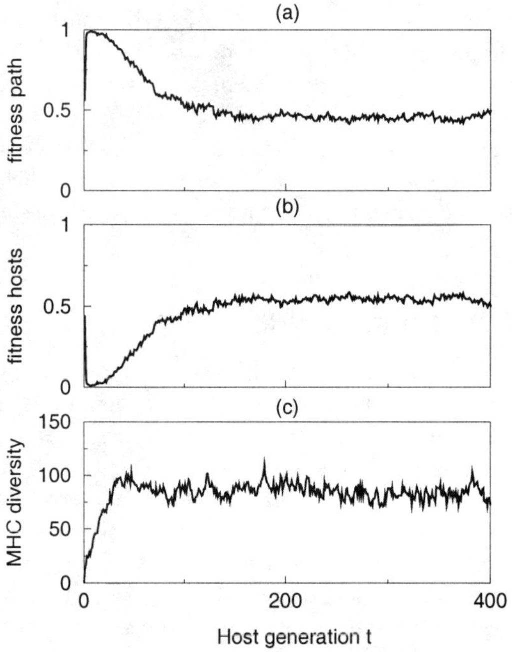

FIGURE 4 The average fitnesses of pathogens (a) and hosts (b), and the average number of different MHC molecules in the population (c), in a coevolutionary simulation in which the pathogens evolve one hundred times faster than the hosts, plotted against the host generation t. The coevolution is initialized with MHC-identical hosts and random pathogens. Results come from a simulation with 200 hosts, each carrying 1 MHC gene with 2 alleles, and 50 different pathogen species, each consisting of maximally 10 different pathogen genotypes, which carry 20 different epitopes each. The epitopes are 12 bits long, while the MHC molecules are 35 bits long. A peptide is presented by an MHC molecule if at least 11 out of 12 bits bind. The probability of mutation, i.e., a bit flip, is $\mu = 0.001$ per bit per generation.

pressure caused by the pathogens, the hosts develop an MHC polymorphism: the number of different MHC molecules in the host population rapidly increases to reach a high equilibrium diversity (see fig. 4(c)). Figure 5 demonstrates that the eventual MHC population diversity that is attained depends on the relative generation time of the pathogens. The faster the pathogens evolve, the larger the resulting MHC polymorphism.

In order to study to what extent the arising MHC diversity is caused by selection for heterozygosity and to what extent by frequency-dependent selection, we performed simulations in which the pathogens do not evolve. Instead, the hosts are exposed to a new, randomly chosen pathogen popula-

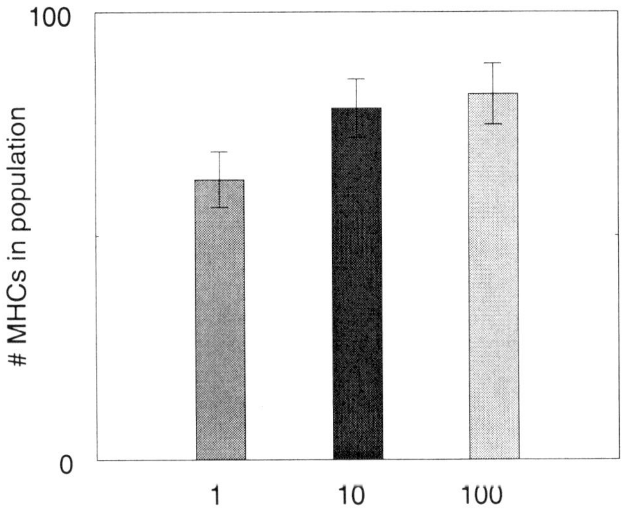

FIGURE 5 MHC molecules become polymorphic. The average number of different MHC molecules arising in the host population increases with the speed at which the pathogens coevolve. Results are shown for three different simulation types: 1: pathogens evolving as fast as the hosts, 10: pathogens evolving ten times faster than the hosts, 100: pathogens evolving one hundred times faster than the hosts. The averages were taken over one hundred generations, between $t = 900$ and $t = 1000$. The error bars denote the standard deviations of the average host and pathogen fitnesses in time. For parameters, see the legend of figure 4.

tion at every host generation (denoted by R). Due to the absence of pathogen evolution, these simulations reflect the MHC diversity that develops under selection for heterozygosity only. Figure 6(a) shows that mere selection for heterozygosity gives rise to an MHC polymorphism that is almost twice as small as the polymorphism arising when hosts and pathogens coevolve. To check if the MHC molecules arising in the host population are really different from each other, and do not differ at a few mutations only, we have also plotted the average Hamming distance between all different MHC molecules in the host population (fig. 6(b)). We find that host-pathogen coevolution increases the genetic distance between MHC molecules. We therefore conclude that rapidly coevolving pathogens provide a considerably larger selection pressure for a functionally diverse set of MHC molecules than mere selection for heterozygosity.

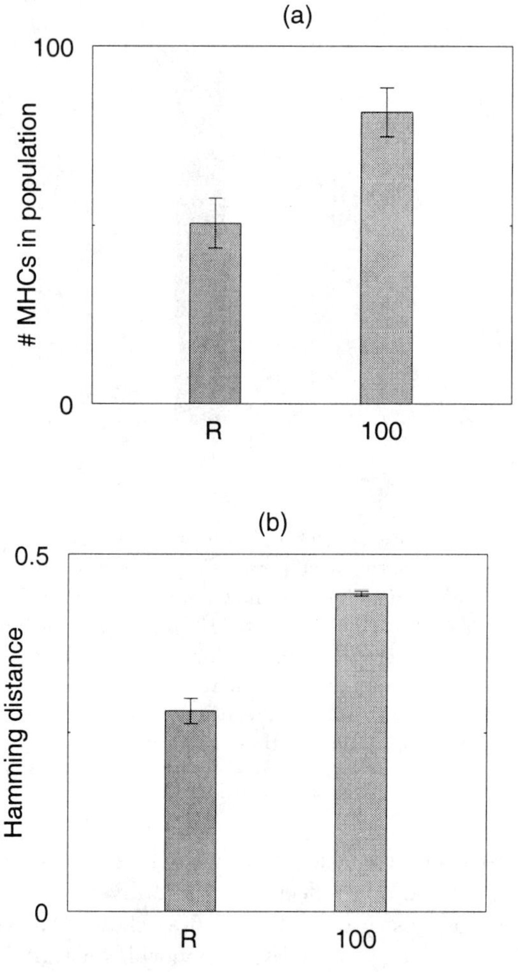

FIGURE 6 Selection for heterozygosity *versus* frequency-dependent selection. (a) The average number of different MHC molecules in the host population, and (b) the average Hamming distance between the different MHC molecules. We have plotted a coevolutionary simulation in which the pathogens evolve one hundred times faster than the hosts (100), and a simulation in which the pathogens do not evolve, but are instead chosen randomly at every host generation (R). The coevolutionary simulation (100) represents the MHC diversity that evolves in the presence of both frequency-dependent selection and selection for heterozygosity, while the simulation with random pathogens (R) represents the MHC diversity that evolves under selection for heterozygosity only. The diversity of pathogens in the R simulation was adjusted to the typical pathogen diversity evolving in the coevolutionary simulations.

5 DISCUSSION

In this chapter, we have studied several sources of diversity in the vertebrate immune system. In particular, we have studied the diversity employed by lymphocytes, which are responsible for the *recognition* of antigens, and the diversity of major histocompatibility (MHC) molecules, which are responsible for the *presentation* of antigens to the immune system. In principle, lymphocytes and MHC molecules are involved in the same task, i.e., to allow immune responses to many foreign antigens, while avoiding inappropriate responses such as autoimmunity. Given the diversity of foreign and self-molecules, it is perhaps not surprising that both MHC molecules and lymphocytes have a high degree of diversity. Nevertheless, they differ fundamentally in the level at which their diversity is expressed. While any vertebrate individual expresses a huge diversity of B and T lymphocytes, the diversity of MHC molecules is mainly evident at the population level. This suggests that MHC and lymphocyte diversity play quite distinct functional roles.

We have proposed that the existence of a diverse lymphocyte system reflects an adaptation of vertebrate hosts to a quickly changing pathogenic world. By storing the appropriate modes of response against different antigens, the vertebrate immune system is able to learn on a somatic time scale. This allows the immune system (i) to respond more promptly and appropriately upon re-encounter of an antigen, even if some of its epitopes have mutated, and (ii) to respond appropriately to whole classes of correlated antigens, even if the immune system has been exposed to only one of their members [60]. We have shown that such a somatically learning system requires sufficient specificity and diversity. If lymphocytes are too crossreactive, inappropriate responses may be induced when unrelated antigens trigger one and the same clone. Immune diversity is required to reconcile reactivity to many antigens with a very specific storage of the appropriate immune responses against them [9, 10, 11].

In sharp contrast with the specificity of lymphocytes, MHC molecules bind their ligands with great degeneracy [23, 31, 36]. This degeneracy, combined with the large degree of heterozygosity of MHC loci, allows the presentation of a large variety of T-cell epitopes to the immune system. Regarding the role of MHC molecules in antigen presentation, it is surprising that the number of different MHC molecules expressed per individual is so limited compared to the large population diversity of MHC molecules. A commonly used argument is that a large individual MHC diversity would impair the T-cell repertoire during self-tolerance induction [17, 18, 28, 44, 47, 66, 69]. As we have shown, however, extra MHC molecules mainly deplete lymphocytes that were not positively selected anyway in the absence of those MHC molecules. As a result, a very wide range of individual MHC diversities—varying from 10 to 2000 different MHC molecules per individual—yields excellent immunity against antigens. The selection pressure for a larger MHC diversity within an individual, however, fades away once there are of the order of ten different MHC molecules per individual. This suggests that the limited individual MHC diversity found in nature reflects a lack of selection for more MHC diversity than what is needed for sufficient presentation of antigens. This is in agreement with the fact that only little correlation has been found between MHC haplotypes and resistance against particular infectious diseases [50, 74].

In contrast to the lack of correlations between MHC molecules and resistance against infectious diseases, strong correlations have been found between certain MHC haplotypes and susceptibility to autoimmune diseases [42, 74]. Such correlations are to be expected if autoimmunity is due to mimicry between foreign-peptide-MHC complexes and self-peptide-MHC complexes. We have extended our model with autoimmunity, by including ignorant self-specific clonotypes that can be triggered by foreign antigens. Our analysis demonstrates that avoidance of crossreactive, autoimmune responses yields a selection pressure for a limited individual MHC diversity.

Despite the fact that different selection pressures may limit an individual's MHC diversity, our genetic algorithm shows that there is selection for a large diversity of MHC molecules at the population level. A large population diversity of MHC molecules allows different individuals to respond differently to identical antigens, thereby giving protection against coevolving pathogens. Just like the individual diversity of lymphocytes, the population diversity of MHC molecules may thus reflect an adaptation of slowly evolving hosts in a rapidly changing world of pathogens.

ACKNOWLEDGMENTS

We are grateful to André Noest and Joost Beltman for their contributions to sections 3 and 4, respectively, and thank Can Keşmir for discussing the concepts and equations of this chapter.

REFERENCES

[1] Agrawal, A., Q. M. Eastman, and D. G. Schatz. "Transposition Mediated by RAG1 and RAG2 and Its Implications for the Evolution of the Immune System." *Nature* **394** (1998): 744–751.

[2] Arstila, T. P., A. Casrouge, V. Baron, J. Even, J. Kanellopoulos, and P. Kourilsky. "A Direct Estimate of the Human $\alpha\beta$ T-Cell Receptor Diversity." *Science* **286** (1999): 958–961.

[3] Bachmaier, K., N. Neu, L. M. De la Maza, S. Pal, A. Hessel, and J. M. Penninger. "*Chlamydia* Infections and Heart Disease Linked through Antigenic Mimicry." *Science* **283** (1999): 1335–1339.

[4] Beck, K. "Coevolution: Mathematical Analysis of Host-Parasite Interactions." *J. Math. Biol.* **19** (1984): 63–77.

[5] Beltman, J. B., J. A. M. Borghans, and R. J. De Boer. "MHC Polymorphism: A Result of Host-Pathogen Coevolution." In *Virulence Management: The Adaptive Dynamics of Pathogen-Host Interactions*, edited by U. Dieckmann, H. Metz, M. Sabelis, and K. Sigmund. Cambridge, MA: Cambridge University Press, in press.

[6] Bendelac, A., and D. T. Fearon. "Innate Pathways that Control Acquired Immunity." *Curr. Opin. Immunol.* **9** (1997): 1–3.

[7] Bevan, M. J. "In Thymic Selection, Peptide Diversity Gives and Takes Away." *Immunity* **7** (1997): 175–178.

[8] Bodmer, W. F. "Evolutionary Significance of the HL-A System." *Nature* **237** (1972): 139–145.

[9] Borghans, J. A. M., and R. J. De Boer. "Crossreactivity of the T-Cell Receptor." *Immunol. Today* **19** (1998): 428–429.

[10] Borghans, J. A. M., and R. J. De Boer. "Adaptive Immunity as a Specific Storage System of Immunological Decisions." (submitted).

[11] Borghans, J. A. M., A. J. Noest, and R. J. De Boer. "How Specific Should Immunological Memory Be?" *J. Immunol.* **163** (1999): 569–575.

[12] Borghans, J. A. M., A. J. Noest, and R. J. De Boer. "What Limits the Individual MHC Diversity?" (submitted).

[13] Bottomly, K. "T Cells and Dendritic Cells Get Intimate." *Science* **283** (1999): 1124–1125.

[14] Carrington, M., G. W. Nelson, M. P. Martin, T. Kissner, D. Vlahov, J. J. Goedert, R. Kaslow, S. Buchbinder, K. Hoots, and S. J. O'Brien. "HLA and HIV-1: Heterozygote Advantage and B*35-Cw*04 Disadvantage." *Science* **283** (1999): 1748–1752.

[15] Cohen, I. R. "The Cognitive Paradigm and the Immunological Homunculus." *Immunol. Today* **13** (1992) 490–494.

[16] Cohen, I. R. "The Cognitive Principle Challenges Clonal Selection." *Immunol. Today* **13** (1992): 441–444.

[17] Cohn, M. "Diversity in the Immune System: 'Preconceived Ideas' or Ideas Preconceived?" *Biochimie* **67** (1985): 9–27.

[18] De Boer, R. J., and A. S. Perelson. "How Diverse Should the Immune System Be? *Proc. Roy. Soc. Lond. B Biol. Sci.* **252** (1993): 171–175.

[19] Doherty, P. C., and R. M. Zinkernagel. "Enhanced Immunological Surveillance in Mice Heterozygous at the H-2 Gene Complex." *Nature* **256** (1975): 50–52.

[20] Du Pasquier, L., and M. Flajnik. "Origin and Evolution of the Vertebrate Immune System." In *Fundamental Immunology*, edited by W. E. Paul, 605–650. New York: Raven Press, 1998.

[21] Fearon, D. T., and R. M. Locksley. "The Instructive Role of Innate Immunity in the Acquired Immune Response." *Science* **272** (1996): 50–53.

[22] Fink, P. J., and M. J. Bevan. "Positive Selection of Thymocytes." *Adv. Immunol.* **59** (1995): 99–133.

[23] Fremont, D. H., M. Matsumura, E. A. Stura, P. A. Peterson, and I. A. Wilson. "Crystal Structures of Two Viral Peptides in Complex with Murine MHC Class I H-2Kb." *Science* **257** (1992): 919–927.

[24] Holland, J. H. *Adaptation in Natural and Artificial Systems.* Ann Arbor, MI: University of Michigan Press, 1975.

[25] Hughes, A. L., and M. Nei. "Pattern of Nucleotide Substitution at Major Histocompatibility Complex Class I Loci Reveals Overdominant Selection." *Nature* **335** (1988): 167–170.

[26] Hughes, A. L., and M. Nei. "Nucleotide Substitution at Major Histocompatibility Complex Class II Loci: Evidence for Overdominant Selection." *Proc. Natl. Acad. Sci. USA* **86** (1989): 958–962.

[27] Hughes, A. L., and M. Nei. "Models of Host-Parasite Interaction and MHC Polymorphism." *Genetics* **132** (1992): 863–864.

[28] Janeway, C. A., and P. Travers. *Immunobiology. The Immune System in Health and Disease.* New York, London: Garland Publications, 1997.

[29] Janeway, C. A. "Approaching the Asymptote? Evolution and Revolution in Immunology." *Cold Spring Harbor Symp. Quant. Biol.* **54** (1989): 1–13.

[30] Janeway, C. A. "The Immune System Evolved to Discriminate Infectious Nonself from Noninfectious Self." *Immunol. Today* **13** (1992): 11–16.

[31] Kast, W. M., R. M. Brandt, J. Sidney, J. W. Drijfhout, R. T. Kubo, H. M. Grey, C. J. Melief, and A. Sette. "Role of HLA-A Motifs in Identification of Potential CTL Epitopes in Human Papillomavirus Type 16 E6 and E7 Proteins." *J. Immunol.* **152** (1994): 3904–3912.

[32] Lehmann, P. V., T. Forsthuber, A. Miller, and E. E. Sercarz. "Spreading of T-Cell Autoimmunity to Cryptic Determinants of an Autoantigen." *Nature* **358** (1992): 155–157.

[33] Loker, E. S. "On Being a Parasite in an Invertebrate Host: A Short Survival Course." *J. Parasitol.* **80** (1994): 728–747.

[34] Marchalonis, J. J., and S. F. Schluter. "Development of an Immune System." In *Primordial Immunity: Foundations for the Vertebrate Immune System*, edited by G. Beck, E. L. Cooper, G. S. Habicht, and J. J. Marchalonis, 1–11. New York: Annals of the New York Academy of Sciences, 1994.

[35] Mason, D. "A Very High Level of Crossreactivity is an Essential Feature of the T-Cell Receptor." *Immunol. Today* **19** (1998): 395–404.

[36] Matsumura, M., D. H. Fremont, P. A. Peterson, and I. A. Wilson. "Emerging Principles for the Recognition of Peptide Antigens by MHC Class I Molecules." *Science* **257** (1992): 927–934.

[37] Matzinger, P. "Tolerance, Danger, and the Extended Family." *Ann. Rev. Immunol.* **12** (1994): 991–1045.

[38] Medzhitov, R., and C. A. Janeway. "On the Semantics of Immune Recognition." *Res. Immunol.* **147** (1996): 208–214.

[39] Medzhitov, R., and C. A. Janeway. "Innate Immunity: Impact on the Adaptive Immune Response." *Curr. Opin. Immunol.* **9** (1997): 4–9.

[40] Medzhitov, R., and C. A. Janeway. "Innate Immunity: The Virtues of a Nonclonal System of Recognition." *Cell* **91** (1997): 295–298.
[41] Murphy, E., K. Shibuya, N. Hosken, P. Openshaw, V. Maino, K. Davis, K. Murphy, and A. O'Garra. "Reversibility of T Helper 1 and 2 Populations is Lost after Long-Term Stimulation." *J. Exp. Med.* **183** (1996): 901–913.
[42] Nepom, G. T., and H. Erlich. "MHC Class-II Molecules and Autoimmunity." *Ann. Rev. Immunol.* **9** (1991): 493–525.
[43] Nossal, G. J. "Negative Selection of Lymphocytes." *Cell* **76** (1994): 229–239.
[44] Nowak, M. A., K. Tarczy-Hornoch, and J. M. Austyn. "The Optimal Number of Major Histocompatibility Complex Molecules in an Individual." *Proc. Natl. Acad. Sci. USA* **89** (1992): 10896–10899.
[45] Ohashi, P. S., S. Oehen, K. Buerki, H. Pircher, C. T. Ohashi, B. Odermatt, B. Malissen, R. M. Zinkernagel, and H. Hengartner. "Ablation of 'Tolerance' and Induction of Diabetes by Virus Infection in Viral Antigen Transgenic Mice." *Cell* **65** (1991): 305–317.
[46] Oldstone, M. B., M. Nerenberg, P. Southern, J. Price, and H. Lewicki. "Virus Infection Triggers Insulin-Dependent Diabetes Mellitus in a Transgenic Model: Role of Anti-Self (Virus) Immune Response." *Cell* **65** (1991): 319–331.
[47] Parham, P., R. J. Benjamin, B. P. Chen, C. Clayberger, P. D. Ennis, A. M. Krensky, D. A. Lawlor, D. R. Littman, A. M. Norment, H. T. Orr, R. D. Salter, and J. Zemmour. "Diversity of Class I HLA Molecules: Functional and Evolutionary Interactions with T Cells." *Cold Spring Harbor Symp. Quant. Biol.* **54** (1989): 529–543.
[48] Parham, P., D. A. Lawlor, C. E. Lomen, and P. D. Ennis. "Diversity and Diversification of HLA-A,B,C Alleles." *J. Immunol.* **142** (1989): 3937–3950.
[49] Parham, P., and T. Ohta. "Population Biology of Antigen Presentation by MHC Class I Molecules." *Science* **272** (1996): 67–74.
[50] Potts, W. K., and E. K. Wakeland. "Evolution of Diversity at the Major Histocompatibility Complex." *TREE* **5** (1990): 181–187.
[51] Qin, S., S. P. Cobbold, H. Pope, J. Elliott, D. Kioussis, J. Davies, and H. Waldmann. "'Infectious' Transplantation Tolerance." *Science* **259** (1993): 974–977.
[52] Reiner, S. L., and R. A. Seder. "Dealing from the Evolutionary Pawnshop: How Lymphocytes Make Decisions." *Immunity* **11** (1999): 1–10.
[53] Richter, A., M. Lohning, and A. Radbruch. "Instruction for Cytokine Expression in T-Helper Lymphocytes in Relation to Proliferation and Cell Cycle Progression." *J. Exp. Med.* **190** (1999): 1439–1450.
[54] Ridge, J. P., F. Di Rosa, and P. Matzinger. "A Conditioned Dendritic Cell Can be a Temporal Bridge between a $CD4^+$ T-Helper and a T-Killer Cell." *Nature* **393** (1998): 474–478.

[55] Rissoan, M. C., V. Soumelis, N. Kadowaki, G. Grouard, F. Briere, R. De Waal Malefyt, and Y. J. Liu. "Reciprocal Control of T-Helper Cell and Dendritic Cell Differentiation." *Science* **283** (1999): 1183–1186.

[56] Romagnani, S. "Induction of Th1 and Th2 Responses: A Key Role for the 'Natural' Immune Response?" *Immunol. Today* **13** (1992): 379–381.

[57] Satta, Y., C. O'huigin, N. Takahata, and J. Klein. "The Synonymous Substitution Rate of the Major Histocompatibility Complex Loci in Primates." *Proc. Natl. Acad. Sci. USA* **90** (1993): 7480–7484.

[58] Segel, L. A., and R. L. Bar-Or. "On the Role of Feedback in Promoting Conflicting Goals of the Adaptive Immune System." *J. Immunol.* **163** (1999): 1342–1349.

[59] Segel, L. A. "Diffuse Feedback from a Diffuse Informational Network: In the Immune System and Other Distributed Autonomous Systems." This volume.

[60] Selin, L. K., S. R. Nahill, and R. M. Welsh. "Cross-Reactivities in Memory Cytotoxic T Lymphocyte Recognition of Heterologous Viruses." *J. Exp. Med.* **179** (1994): 1933–1943.

[61] Shortman, K., D. Vremec, and M. Egerton. "The Kinetics of T Cell Antigen Receptor Expression by Subgroups of $CD4^+8^+$ Thymocytes: Delineation of $CD4^+8^+3^{2^+}$ Thymocytes as Post-Selection Intermediates Leading to Mature T Cells." *J. Exp. Med.* **173** (1991): 323–332.

[62] Slade, R. W., and H. I. McCallum. "Overdominant vs. Frequency-Dependent Selection at MHC Loci." *Genetics* **132** (1992): 861–864.

[63] Snell, G. D. "The H-2 Locus of the Mouse: Observations and Speculations Concerning Its Comparative Genetics and Its Polymorphism." *Folia. Biol. (Praha)* **14** (1968): 335–358.

[64] Swain, S. L., L. M. Bradley, M. Croft, S. Tonkonogy, G. Atkins, A. D. Weinberg, D. D. Duncan, S. M. Hedrick, R. W. Dutton, and G. Huston. "Helper T-Cell Subsets: Phenotype, Function and the Role of Lymphokines in Regulating their Development." *Immunol. Rev.* **123** (1991): 115–144.

[65] Taams, L. S., A. J. M. L. Van Rensen, M. C. P. Poelen, C. A. C. M. Van Els, A. C. Besseling, J. P. A. Wagenaar, W. Van Eden, and M. H. M. Wauben. "Anergic T Cells Actively Suppress T-Cell Responses *via* the Antigen-Presenting Cell." *Eur. J. Immunol.* **28** (1998): 2902–2912.

[66] Takahata, N. "MHC Diversity and Selection." *Immunol. Rev.* **143** (1995): 225–247.

[67] Takahata, N., and M. Nei. "Allelic Genealogy under Overdominant and Frequency-Dependent Selection and Polymorphism of Major Histocompatibility Complex Loci." *Genetics* **124** (1990): 967–978.

[68] Van Meerwijk, J. P., S. Marguerat, R. K. Lees, R. N. Germain, B J. Fowlkes, and H. R. MacDonald. "Quantitative Impact of Thymic Clonal Deletion on the T-Cell Repertoire." *J. Exp. Med.* **185** (1997): 377–383.

[69] Vidović, D., and P. Matzinger. "Unresponsiveness to a Foreign Antigen Can Be Caused by Self-Tolerance." *Nature* **336** (1988): 222–225.

[70] Vogel, T. U., D. T. Evans, J. A. Urvater, D. H. O'Connor, A. L. Hughes, and D. I. Watkins. "Major Histocompatibility Complex Class I Genes in Primates: Coevolution with Pathogens." *Immunol. Rev.* **167** (1999): 327–337.

[71] Von Boehmer, H. "Positive Selection of Lymphocytes." *Cell* **76** (1994): 219–228.

[72] Waldmann, H., S. Qin, and S. Cobbold. "Monoclonal Antibodies as Agents to Reinduce Tolerance in Autoimmunity." *J. Autoimmun.* **5** (1992): 93–102.

[73] Wills, C. "Maintenance of Multiallelic Polymorphism at the MHC Region." *Immunol. Rev.* **124** (1991): 165–220.

[74] Wills, C., and D. R. Green. "A Genetic Herd-Immunity Model for the Maintenance of MHC Polymorphism." *Immunol. Rev.* **143** (1995): 263–292.

[75] Zhao, Z. S., F. Granucci, L. Yeh, P. A. Schaffer, and H. Cantor. "Molecular Mimicry by Herpes Simplex Virus-Type 1: Autoimmune Disease after Viral Infection." *Science* **279** (1998): 1344–1347.

[76] Zinkernagel, R. M., M. F. Bachmann, T. M. Kundig, S. Oehen, H. Pirchet, and H. Hengartner. "On Immunological Memory." *Ann. Rev. Immunol.* **14** (1996): 333–367.

[77] Zinkernagel, R. M., S. Ehl, P. Aichele, S. Oehen, T. Kundig, and H. Hengartner. "Antigen Localisation Regulates Immune Responses in a Dose- and Time-Dependent Fashion: A Geographical View of Immune Reactivity." *Immunol. Rev.* **156** (1997): 199–209.

T Cells Obey the Tenets of Signal Detection Theory

André J. Noest

1 INTRODUCTION: DESIGNING LYMPHOCYTES

As a prerequisite to mounting effective and safe immune responses, lymphocytes must *detect* signals from unknown "intruders," amidst the noise of a harmless background. To clarify the functional importance of lymphocyte properties with respect to this basic but nontrivial task, I ask: What properties would a lymphocyte have if one *designed* it according to the theory of statistically optimal detection, with minimal regard for biological constraints? I will show that all functional properties of such "designer lymphocytes" match surprisingly well with the following properties of real T cells:

- Clonal diversity and antigen specificity, with only one receptor type per cell.
- "Serial triggering" of many T-cell receptor (TCR) molecules per ligand molecule [26].
- Activation if the triggered "TCR-count" [22] crosses a high threshold value, which is adjusted by costimulation [27].
- Transformation of agonist into antagonist ligands by a single mutation [25].
- Positive and negative selection by self-peptides [2, 4, 13], and a much reduced sensitivity ("anergy") of potentially self-reactive cells.
- Response suppression by extremely strong stimuli ("high-zone tolerance" [17]).

- Antigen-specific, clonally independent production of nonspecific responses (cytokines).

Each of these properties follows from a common statistical formalism which describes optimal detection. Thus, by construction, all parts of the design collaborate to solve the detection task. Moreover, the mathematical form in which the design originates can often be used to compute its detection performance. Here, I can only give a largely nonmathematical sketch of basic aspects of the design, which may be enough to illustrate how it matches real T cells. Elsewhere [19], I give a more thorough account of the construction and analysis of the design.

Note that the mathematical formalism generates the design; it is not a rationalization of informal arguments, or a means of "reverse engineering" T cells from experimental observations. This design approach also differs from more usual models, in which one *assumes* a certain functional structure and then derives its behaviour. Here, the structure is derived from basic principles. In deriving the design, I avoid relying on specifically biological or immunological ideas or experimental data. Thus, the results can be compared to real immune systems without inherent bias. If they match, this can't be due simply to a logical circularity, as is often a risk with data-driven modeling. Conversely, mismatches between design and reality (at the functional level) could be very useful for indicating where the detection function of real T cells suffers from mechanistic constraints or historical "accidents" during (co-)evolution. For adaptive immune systems, no clear mismatches actually turn up, but they emerge abundantly when comparing to immune systems of, for example, invertebrates. The large disparity between these two systems seems to hinge on the (in)ability to generate a large diversity of receptors.

To emphasise that a large set of features is implied by the single basic task of detection, I ignore the distinctions between various effector functions, which are often found to correlate with the distinctions between different lymphocyte subtypes. I also ignore the "preprocessing" of intruders (leading to presentation of their peptide fragments as signals for possible detection), which is vitally important but quite distinct in its implications. I simply assume that antigen-presenting cells allow T cells to sample a wide variety of signals (ligands). If pathogens can sabotage this presentation, it cannot be undone at the detection stage, underlining the natural separation between the two tasks.

2 DERIVING A BASIC FUNCTIONAL DESIGN

Any design starts by defining a task. Here I take this to be the near-optimal detection of unpredictable "intruder" signals amidst a wide variety of "background" signals. The system does not know *a priori* which of many possible types of signals exist, nor does it know the source (background or intruder) of any signal. Indeed, both sources may even contribute to the same type of signal. Unless noted otherwise, I assume that the response should be aimed at the same type of signal (epitope) as was detected. The precise nature of this response can be left undefined here. Indeed, the detection stage should be critical to the reliability of stopping an infection by intruders which have evolved to be only weakly "visible," unless the effector stages of the immune response would be very unreliable. I neglect this latter possibility.

This "immune surveillance" task bears a striking resemblance to the well-studied engineering problem of detecting (e.g., electronic) signals of unknown targets among background noise or interference [29]: The goal is to detect even weak intruder signals with high probability, but keeping "false alarms" very rare. The generality of the statistical theory underlying all (near-)optimal detection systems allows the present design to borrow several notions and analytical techniques from engineering, but the general approach and its results should be readable without familiarity with this field. Some readers may find it useful to skip to the Discussion (section 5) before reading the body of the paper. The mathematical aspects of deriving and analysing the design are relegated to another paper [19].

2.1 DETECTION THEORY AS GUIDING PRINCIPLE

To introduce detection theory in its most basic form, assume for a moment that there is only one type of intruder that might occur. Let P_d be the detection probability when the intruder is present, and let P_f be the probability of "false alarm," i.e., of a detection occurring while the intruder is actually absent. The natural time scale on which to define P_d and P_f is the typical contact time between a T cell and an antigen-presenting cell, i.e., several hours [12].

"Optimal" detection performance has to somehow combine a maximal P_d with a minimal P_f, since false alarms due to signals from self-components would risk triggering an autoimmune disease. Perfect detection would be trivial to achieve if background signals were constant. In fact, they are always "noisy." Any detection system then suffers from errors of two types: Besides "false alarms," at rate P_f, there will be "missed detections," at rate $1 - P_d$.

The classical Neyman-Pearson test (e.g., Whalen [29]) specifies the method for optimal detection we seek: at *any* given P_f, the maximal P_d is achieved by a likelihood-ratio test on the set of available signals s_i. With the reasonable assumption that the s_i are statistically independent, it is natural to use the logarithm of the likelihood ratio, yielding the test definition

$$\lambda = \sum_i \left[\log p_i(\rho; s_i) - \log p_i(0; s_i)\right] > \theta \ , \tag{1}$$

where θ is a suitable threshold, ρ is the signal-to-noise ratio (SNR) of the intruder whose presence is being tested, and the $p_i(\rho; s_i)$ and $p_i(0; s_i)$ are the probability distributions of the signals s_i, respectively with $(\rho > 0)$ and without $(\rho = 0)$ the intruder present.

Note that the definition eq. (1) of the "test quantity" λ determines how any available "data" $\{s_i\}$ should be processed before comparison to a threshold θ. Thus, the functional design I seek is fully encoded by the expression in eq. (1). Deriving the design is just a stepwise decoding of this very compact and general prescription. Note also that the result depends on the distributions $p_i(\rho; s_i)$. Thus, the design is dictated by the basic nature of its task and by the statistics of any available data.

Being dependent on the stochastic signals s_i, the test quantity λ will also be stochastic, with a ρ-dependent probability distribution $p_\lambda(\rho;\lambda)$ which can be found from the $p_i(\rho; s_i)$ via eq. (1). By definition, we then have $P_d = \int_\theta^\infty p_\lambda(\rho;\lambda)$, with P_f given by the special case $\rho = 0$. Given some acceptable P_f value, the required threshold θ is thus seen to depend on the distribution $p_\lambda(0;\lambda)$. This already implies the notion that each detector must "adapt" in some way to the background statistics of its test quantity, which cannot be assumed to be known *a priori.*

It is useful to note from eq. (1) that the detection results are not affected by any monotonic nonlinearity applied to the test quantity λ, as long as θ is transformed similarly. This allows a huge (but precisely delimited) freedom in implementing the signal processing which should produce λ from the inputs s_i. Evolution may well have exploited this freedom for improving the detection performance of T cells toward the ideal derived here.

The basic notions of detection which I have just introduced are illustrated in figure 1. Note how P_f, P_d, and θ are related to each other via the cumulative probability of λ, which is determined by the input distributions $p(\rho; s_i)$ via eq. (1).

2.2 DETECTOR DESIGN PROPERTIES: MATCH WITH T CELLS

The task of deriving the detector design consists essentially of spelling out the functional implications of the demand that each cell computes the test quantity λ in real time. Given the definition eq. (1), it is clear that the design will in general depend on $p_i(\rho; s_i)$, i.e., on the statistical properties of the input signals at any intruder level ρ.

2.2.1 "Multiple Hypothesis Testing" Implies a Diversity of Selective Receptors.

Multiplicity: Many unknown pathogen types can appear, and each can produce a few ligands out of a huge set of possible types. Thus, the temporary simplification of a single known intruder is dropped, and the single Neyman-Pearson test generalizes to "multiple hypothesis" testing—each hypothesis is associated with the presence of one intruder-related ligand type. To cover all possible ligands, a very large set of independent detectors (cells) is then needed. Since each of their test quantities is still of the same mathematical form as eq. (1), we may work out most of the design while focusing on just a single detector from the set.

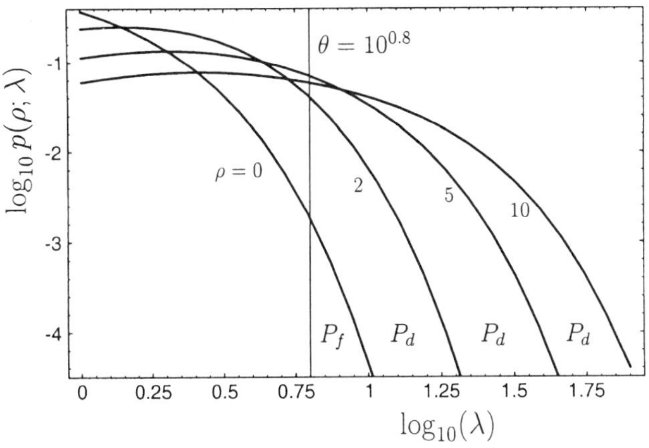

FIGURE 1 A basic illustration of statistical detection: a test quantity λ, derived from noisy signals s_i, is compared to a threshold θ. Without an intruder (SNR $\rho = 0$), the statistics of λ is described by the background distribution $p(0;\lambda)$. Its upper-tail area causes "false alarms," at a (small) rate $P_f = \int_\theta^\infty p_\lambda(0;\lambda)d\lambda$ set by θ. When an intruder appears, ρ increases, causing the distribution $p_\lambda(\rho;\lambda)$ to shift to the right and to broaden. This increases the detection rate $P_d = \int_\theta^\infty p_\lambda(\rho;\lambda)d\lambda$. Useful detection performance (say, $P_d > 0.5$, while P_f is many orders smaller) usually requires a considerable SNR ρ.

Selectivity and specificity: For each signal i in eq. (1), the termwise subtraction s $\log p_i(\rho; s_i) - \log p_i(0; s_i)$ implies that each detector must apply a selective "filter" to the set of available signals $\{s_i\}$ before further processing: its test quantity λ must not depend on signals s_i whose statistics are unaffected by the presence of the specific intruder that the test is designed to detect. Indeed, all irrelevant signals will have $p_i(\rho; s_i) = p_i(0; s_i)$, so that eq. (1) demands excluding them from λ. This fits the intuitive notion that "interference" must be filtered out before applying the detection threshold.

The immunological implementation is obvious: ligand-specific receptors are needed. Note that the demand is only for enough specificity to avoid significant interference from other ligands that a cell is likely to encounter; potential "cross reactivity" [14] with other members of the astronomically larger set of *possible* ligands is irrelevant.

Diversity: The minimal diversity of the set of receptor types should scale with the required degree of specificity, since coverage of all possible intruder-related ligands must be guaranteed. More quantitative conclusions about the required specificity and diversity can be derived by putting more detailed optimization demands on the system [5, 6, 7].

In any case, the general feature of a large diversity of specific T cells is seen to follow from basic detection theory when the type of signal produced by an intruder is essentially unpredictable.

2.2.2 Early Amplification and Integration Implies "Serial TCR Triggering."

The need to evaluate eq. (1) can also be shown [19] to imply that each detector should amplify and integrate its specific set of signals at the earliest possible stage. The interpretation of this result is that the test quantity λ should be made less noisy than the raw input signals s_i, which will often have strong Poisson sampling noise due to the low mean rate with which rare ligand molecules can be available to the detectors (T cells). Mere signal transduction (essentially just "copying") of the input signals before using them to trigger a "large" response would only add more sampling noise, which degrades the information about possible intruders which is carried in the original s_i.

"Amplification" of the input signals means that each available ligand molecule must generate a larger number of "messenger" molecules; that is, the ligand must act as a *catalyst.* The "substrate" of this catalysis can then only be the actual receptor which binds the ligand. The simplest kinetic scheme is

$$\mathrm{L} + \mathrm{T} \underset{k_d}{\overset{k_a}{\rightleftharpoons}} (\mathrm{LT}) \overset{k_c}{\rightarrow} (\mathrm{LU}) \overset{k_d}{\rightarrow} \mathrm{L} + \mathrm{U}\,. \tag{2}$$

Thus, ligand (L) and TCR (T) associate at rate k_a to a complex (LT); its TCR part undergoes some irreversible conversion(s) to a product U, say, at a rate k_c. Dissociation of the complex, at a rate k_d (irrespective of the conversion state of the TCR) then frees the ligand L, allowing it to convert more T's to U's. The typical time for an L molecule to go through a cycle of association and dissociation is $\tau_c = (k_a T)^{-1} + k_d^{-1}$. Per cycle, a T is converted to a U with probability $k_c/(k_c + k_d)$. The accumulated U's can be processed further to mount a cell-level response. Apart from the need to apply a threshold, eq. (1) does not specify such later stages, which are indeed no longer critical to the sensitivity of the system. Hence, I ignore the later stages here.

Note that this simplest amplification scheme also performs temporal integration, by accumulating the converted TCRs (U). Just such an accumulation and thresholding of converted TCRs has been quantified experimentally as the basis of T-cell activation ("TCR counting") [22, 27].

The combined effect of amplification and integration can be expressed in terms of a net "gain" factor, i.e., the total output $U(\tau) - U(0)$ produced by a (bound plus free) amount of ligand L^*, within an integration time τ. In the linear regime, the result is most simply put in the form of a normalized gain $\tilde{\alpha}$, which corresponds to a formal $\tau = 1/k_c$. Under a quasi-steady state approximation (slow T depletion, and $\tau \gg \tau_c$), the $\tilde{\alpha}$ then depends only on two rescaled rates $\tilde{k_a} = k_a/k_c$, and $\tilde{k_d} = k_d/k_c$

$$\tilde{\alpha} = \left(\frac{\tilde{k_d}}{1+\tilde{k_d}}\right)\left(\frac{\tilde{k_a}T}{\tilde{k_d}+\tilde{k_a}T}\right) . \tag{3}$$

Note that the system should avoid reaching its maximal gain ($\tilde{\alpha} \to 1$), which occurs in the regime $\tilde{k_a}T \gg \tilde{k_d} \gg 1$. In this regime, the antigen specificity of the receptor is lost, since many ligands with distinct k_a, k_d would be amplified equally instead of selectively. Avoiding this regime robustly is possible because k_a has an upper bound, set by diffusion. Indeed, this limit appears to be reached by many ligands "similar" to the best matching one [2]. Keeping the TCR level T below a fixed maximum such that $\max(\tilde{k_a}T) < 1$, at given k_c, maintains the receptor-specific differences between the antigen gains $\tilde{\alpha}$ set by the rate-constant pairs (k_a, k_d). Clearly, the standard notion that receptors rank ligands in terms of their affinity $K = k_a/k_d$ is inadequate for describing the function of this scheme.

The loss in normalized gain $\tilde{\alpha}$ due to staying in the "selective" regime can be compensated easily by the scale factor τk_c in the full signal gain $\alpha = \tilde{\alpha}\tau k_c$. Given the observed $\tau = \mathcal{O}(10^4)$ sec and a plausible conversion rate $k_c = \mathcal{O}(0.1)$/sec (of the same order as the typical k_d for "reasonable" ligands [2]), we find a net gain $\alpha = \mathcal{O}(10^2)$. This fits well with observations [26] on the number of TCRs per ligand molecule which disappear from the cell surface during the course of T-cell activation.

Note also that adding a second conversion step [21] to the scheme, say, (LU)$\to$(LV) with the same rate k_c, makes $\tilde{k_d}$ independently available as the ratio U/V. As in the many-step scheme proposed before [16], this allows a T cell to *discriminate* between ligands that differ in their (k_a, k_d)-pair, even if they produce the same total signal $(U + V)$. Kinetic discrimination will not be analysed in more detail here, since the focus in this paper is on *detection*.

2.2.3 Adaptive Background Scaling for False-Alarm Rate Control Implies "Tolerance."

To complete a basic working design, we need to ask what is required for keeping the false alarm rate P_f of each detector very low. An explicit value for P_f need not be chosen here, but its order of magnitude should be well below the inverse of the number of cells performing the detection task.

I now write the test quantities as u_j, to denote the numbers of type j TCRs T_j converted to U_j in a time τ. As was illustrated in figure 1, the choice of P_f fixes the required θ, given background distributions $p_j(0; u_j)$, by $P_f = \int_\theta^\infty p_j(0; u_j)du_j$. (In case the θ are j dependent, one simply redefines u_j in units of θ_j.) Thus, each cell type j must tune its parameters somehow to the specific $p_j(0; u_j)$, which will differ for each j. The basic need for adaptation has been proposed before [9], but the present approach differs by being derived from the optimal detection formalism.

Let $\hat{u}_j$ denote the U_j count which would occur without adaptation. To a good approximation, one expects all background distributions $p_j(0; \hat{u}_j)$ to be merely scaled copies of each other, since the $\hat{u}_j$ fluctuations should be due to the same sampling mechanisms (see section 3) acting on different mean ligand concentrations. The scale factors are then measured most simply by the expected values $b_j = \int_0^\infty \hat{u}_j p_j(0; \hat{u}_j) d\hat{u}_j$. The scaling assumption means that $p_j(0; \hat{u}_j) = b_j^{-1} p(0; \hat{u}_j/b_j)$, with $p(0; \hat{u}_j/b_j)$ being the j-independent distribution of *proportional* fluctuations due to T cells finding their ligands presented by various types of antigen presenting cells, at various times and places. The $p(0; \hat{u}_j/b_j)$ should also cover residual Poisson noise remaining after the integration stage.

The functional implications are clear: the proper test quantity to use is $u_j = \hat{u}_j/b_j$. With this rescaling, setting θ to achieve some allowed P_f needs to be done only once, and for all clone types j, for example, during cell maturation. Scaling $\hat{u}_j$ by b_j also fits well into the serial triggering scheme: the simplest approach is to scale the gains $\tilde{\alpha}_j$ per cell type. The type j TCR levels T_j are the natural parameters to use for the required "gain control." Indeed, in the proper regime, $\tilde{\alpha}_j$ depends almost linearly on T_j.

What remains then is the need to make the TCR levels T_j be proportional to $1/b_j$. This too fits well into the existing scheme. In fact, the required result can be shown [19] to arise naturally by down regulation of T_j during prolonged exposure to the background. This may occur before the cell is allowed to function as a detector (i.e., in thymus), or even throughout its active life.

The effect of the described adaptive rescaling of the background constitutes a form of "tolerization"—The gain of each cell is calibrated so that it has a stable false-alarm rate P_f. Still, the cell has the maximal P_d possible given its background. Downregulated self-reactive T cells would require very strong stimuli to get activated. Given more moderate stimuli, they would appear to be "anergic." As a slight extension of this scheme, one could delete cells in which adaptation drives the gain so low that their u_j could never reach θ for relevant intruder levels. This approach is actually thymic "negative selection." When the adaptation is allowed to occur continually, but on a time scale much slower than τ, the result would be a form of "peripheral tolerance." Note that in all such schemes, negative selection, TCR down regulation and anergy should be driven by ligands which act as *agonists* peripherally.

3 DETECTION PERFORMANCE OF THE BASIC DESIGN

Quantifying the detection performance means determining the relation between the detection rate P_d, the SNR ρ, and the false alarm rate P_f. The receptor index j can now be dropped, since all detections occur independently, and the adaptive gain control scales all u_j backgrounds to have the same probability distribution $p(0; u)$. Detection performance is then fully determined by the probability distribution $p(\rho; u)$ for general SNR ρ, since both P_d and P_f are upper-tail integrals of $p(\rho; u)$. Several sources of fluctuation contribute to $p(\rho; u)$, as follows.

After each integration time τ, u is Poisson distributed with an expected value h which is a sum over ligand levels, weighted by the ligand-specific gain factors $\tilde{\alpha} k_c \tau$. However, h itself also fluctuates from one detection attempt to another, since T cells sample ligand levels at various times and places, and on various antigen-presenting cells. Background ligands and intruder ligands can contribute to h. The distributions of the two terms, which are so far experimentally unknown, can be chosen via a parsimony (maximum entropy) principle [24] (see Noest [19] for details). One finds that both terms must be exponentially distributed, with the ratio of the two widths equal to the SNR ρ. Assuming the two terms to be additive and independent, one can then compute the detection rate $P_d = Q(\rho; \theta)$ in terms of an explicit function $Q(\rho; \theta)$ (see the full paper by Noest [19]).

The required threshold θ is found by solving $P_f \equiv Q(0; \theta)$. Neglecting the residual Poisson noise, one has simply $Q(0; \theta) = e^{-\theta}$, yielding $\theta = \ln(P_f)$, a very slow dependence. Note that any practical false-alarm rate requires $\theta = 15 - 30$ (in units of the mean number of TCRs triggered by background). This high θ justifies neglecting the Poisson noise. It also fits the situation in real T cells [27]: several thousand TCRs must be converted to activate the cell.

The detection rate $P_d = Q(\rho; \theta)$ of the most basic design is now fully determined as a function of the SNR ρ, at any given P_f. A graph of P_d vs. $\log_{10} \rho$ is shown in figure 2, as the dashed line on the left. The other, fully drawn graphs are for more realistic variants, in which the integration time τ is also a random variable. Indeed, one expects fluctuations in the interval that a T cell stays in functional contact with an antigen-presenting cell (APC), unless very tight control mechanisms exist. Fluctuation in τ contributes multiplicative noise to the test quantity u, and will thus reduce performance, showing P_d-graphs shifted to the right. The probability distribution $p(\tau)$ must be exponential when bound T-cell/APC pairs dissociate with first-order kinetics, for example, via one rate-limiting step. More generally, one may consider an n-step process, causing τ to become gamma distributed, $p(\tau) = n^n \tau^{n-1} e^{-n\tau} / \Gamma(n)$, where the mean of τ has been scaled to one.

Representative graphs of $P_d = Q(\rho; \theta)$ for this family of models are plotted in figure 2. Note the roughly ten-fold sensitivity loss when τ fluctuates exponentially ($n = 1$), relative to the $n \to \infty$ limit of a constant $\tau = 1$. Even moderate control of τ fluctuations, for example by n-step kinetics, yields a good improvement over the uncontrolled $n = 1$ case, but the effect saturates quickly for about $n > 4$.

In figure 3, I show how very large variations in P_f, which induce only moderate variations in θ, affect the "detection limit," which I define as the SNR ρ^* at which the detection rate P_d reaches a reasonably large value P_d^*. In all examples shown, I choose $P_d^* = 0.5$. The overall chance of failing to detect an intruder must clearly be much smaller than $1 - P_d^* = 0.5$; perhaps 10^{-3} is a more realistic upper bound. At $P_d^* = 0.5$, only ten independent T-cell

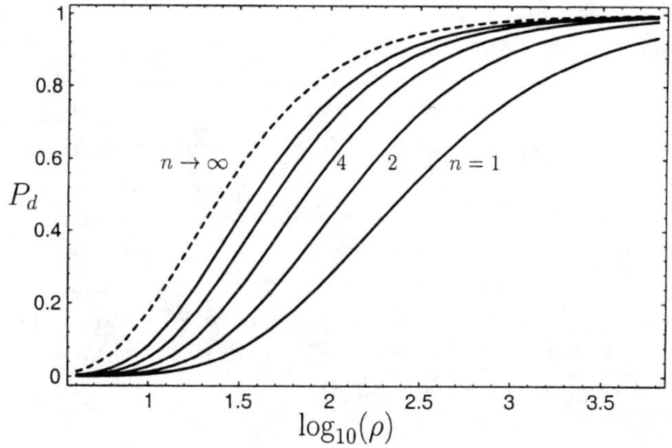

FIGURE 2 P_d vs. SNR ρ, at $P_f = 10^{-8}$, for a stochastic contact time τ, controlled by an n-step process with $n = 1, 2, 4, 8, 16$. The dashed ($n \to \infty$) graph is for fixed $\tau = 1$.

activation attempts are required to achieve this goal. Indeed, it is not unreasonable to assume that, for each clone, ten T cells meet the relevant APCs, or that any single T cell could undergo ten sequential contacts with such APCs. Note also that with an ill-controlled contact time (small n), the detection limit ρ^* is slightly more dependent on P_f. This reflects the longer upper tail of the background distribution for small n.

4 EXTENDING THE BASIC DESIGN

4.1 EXPLOITING NONSPECIFIC SIGNALS: COSTIMULATION

The basic design so far has only exploited signals which are specific, in the sense of (ideally) contributing to only one of the many hypotheses being tested. However, it is too restrictive to assume that all available signals have such specificity. How should the design use nonspecific signals? As before, the Neyman-Pearson formalism dictates the optimal solution.

As a simple example, consider just one extra signal a which is fully nonspecific, while the usual signals $\{s_i\}$ are now explicitly assumed to be fully specific. Note that the relevant meaning of a signal being "specific" is independent of its being generated by one type of pathogen or by many. Thus, the same signal, say, lipopolysaccharide (LPS) which occurs in many bacteria, may even play both roles: it is a "specific" signal when triggering an anti-LPS clonal response, and a "nonspecific" signal when it aids other responses.

Under a few realistic assumptions, it can be shown [19] that the new required test quantity is merely a simple extension of the log-likelihood form eq. (1). It requires only one extra sum term containing a in the formal test quantity λ_j for each clone j:

$$\lambda_j = \log p(\rho_j; a) - \log p(0; a) + \sum_i \log p_i(\rho_j; s_i) - \log p_i(0; s_i) \,. \tag{4}$$

The new a contribution is, as usual, nonzero whenever the SNR ρ_j affects the distribution $p(\rho_j; a)$, i.e., if a contains any information related to the presence of intruder j.

Immunologically, the fact that the a term contributes to all j identifies this term as a specificity-independent input, i.e., as "costimulation" or "signal-2," which T cells receive from activated APCs via, for example, the B7-CD28 interaction. The functional effect of costimulation on T cells fits what the design demands: it lowers the T-cell threshold [27] and probably also boosts the net signal amplification via the conversion rate k_c, since it recruits the TCR phosphorylation machinery [28, 30]. Given the freedom in applying any monotonicity to the test quantity, both effects fit the theoretical demands.

4.2 SPIKY BACKGROUND IMPLIES "HIGH-ZONE TOLERANCE"

Throughout the design discussed so far, the optimal test quantity was found to be a nondecreasing function of the input, but this needs not be so. Indeed, a nonmonotonic function can be shown [19] to be required if one allows the background to be much more "spiky" than hitherto assumed. "Spiky" means that occasionally, and for a short time, the background signals show a very large peak. The required test quantity is then found to be a one-hump nonlinear function $f(u)$ of the previously prescribed quantity u. Intuition may

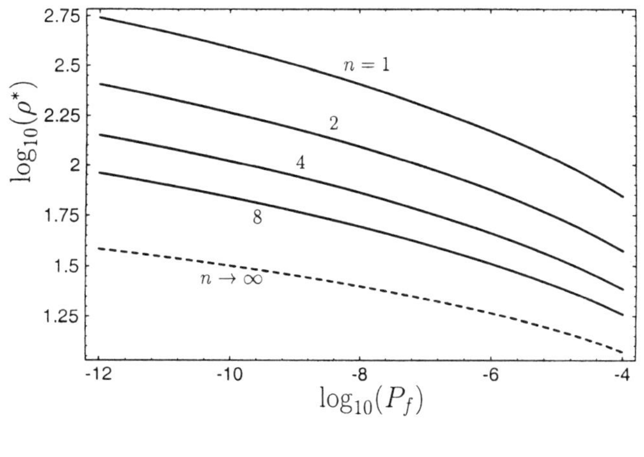

FIGURE 3 The detection limit, i.e., the SNR ρ^* at which $P_d = 0.5$, plotted against the allowed false-alarm rate P_f. Drawn graphs are for stochastic τ, controlled by an n-step process with $n = 1, 2, 4, 8$. The dashed ($n \to \infty$) graph is for fixed $\tau = 1$.

already lead one to expect that rare large "outliers" are best "clipped" to prevent false alarms, but the formal derivation shows how this is to be done precisely, by fixing the required function $f(u)$.

A reduced or suppressed response for very strong stimulation is in fact a classical feature of real T-cell activation, known as "high-zone tolerance" [15, 17]. Here, I do not explore what may be a whole range of different processes to which this term is often applied. For now, it is sufficient to note that a "humped" type of response follows from optimal detection in a spiky background.

4.3 ANTIGEN-SPECIFIC ACTIVATION OF NONSPECIFIC RESPONSES

So far, I assumed that the response to a detection would be directed against the same object (epitope) as was detected. Now, it is useful to allow responses that act against a broad class of intruders. For example, T cells produce cytokines such as interferon-γ which hinders replication of all viruses. Note that such broad-spectrum responses are triggered as part of the antigen-specific activation. Why? At first sight, one might think that a broad-spectrum response should be triggered only by nonspecific signals.

The correct strategy is again found by applying the general test formalism, but this time the required test quantity generalises to a sum over conditional likelihood ratios. Written in terms of the previously found quantities u_j, the required test quantity is $H = \sum_j p_i(\rho_j; u_j)/p_j(0; u_j)$. Analysis [19] then proves that threshold detection on this optimal H can be approximated very well by taking the logical OR of many independent detections based on the individual u_j. Thus, one may simply continue to use the same detector design that was found without any consideration of nonspecific responses.

The only extension that is probably required is in setting the threshold θ for the new task, since specific and nonspecific responses will generally allow different false alarm rates. Thus, I predict that different responses can be triggered in the same cell at different thresholds, while still sharing most of the common machinery for signal selection, amplification, and integration.

4.4 CORRECTION FOR EARLY SIGNAL BLURRING IMPLIES "ANTAGONISM"

Before signals are actually detected, the distinction between "similar" but independent signals will always be blurred to some extent: Protein processing may excise ligands with slightly varying lengths, or a nominally unique ligand may be presented in slightly varying modified forms, for example, by cobinding to MHC and TCR in distinct conformations, which also fluctuate by thermal motion. Left uncorrected, this would reduce the overall specificity of the response, even if the nominal TCR specificity is very high. Furthermore, TCR down regulation would reduce the signal amplification of clones whose cognate ligand shape is "near" that of a self-ligand.

One can design a (partial) remedy for this problem, and this turns out to fit well to experiments showing antagonistic effects on T-cell activation when slightly mutated ligands are used. To ease analyzing the problem, one may associate each (nominal) signal type (shape) with a specific point in a "shape space" [23], having say D-dimensional coordinates $x = (x_i)_{i=1}^{D}$. "Blurring" then means that the nominal signal spikes $s(x)$ become finite-width peaks $r(x)$. With physical and biochemical limitations preventing further reduction of the blur at its source (preprocessing and presentation), one is forced to try reducing the blur at the level of T cells.

Mathematically, blurring is described by convolution $r(x) = B(x) \star s(x)$, where $B(x)$ is the blur "kernel." For the sake of computing a simple explicit example (see fig. 4), I take $B(x) = \exp(-\sum_{i=1}^{D} |x_i|)$. One then seeks a kernel $C(x)$, representing signal processing by the T cell, such that the overall system specificity kernel $S(x) = C(x) \star B(x)$ acquires shorter tails than $B(x)$. Using regularisation theory, I have derived [19] a robust form of $C(x)$ that turns $S(x)$ into a Gaussian of width σ, which has the wanted fast decay of tails for shape distances beyond σ.

The striking feature of this $C(x)$ kernel, for any reasonable σ, is its "Mexican hat" form: a positive ("agonist") response near the optimal ligand, a negative ("antagonist") response for ligands at distances of a few times σ, and a negligible ("null") response for more distant ligands. An example with $\sigma = 1/2$ is shown in figure 4. A minimal σ value is set by the need for complete coverage of shape space by the attainable T-cell diversity.

It may be worth noting that the designed antagonism should only be visible experimentally when probing with slightly mutated ligands. Mutations at the level of unprocessed pathogens should not "see" the antagonism, at least on average. Indeed, the overall effect of blurred processing and detector antagonism is designed to produce a single, short-tailed peak $S(x)$ in shape space, as shown in figure 4. Pathogen mutants could only exploit imperfections in the implementation, or attempt to sabotage its machinery.

In reality, antagonistic T-cell responses are indeed often found when using single-mutant forms of (presumably) optimal ligands [25]. So far, the function of this antagonism has been quite puzzling; in fact, it has been viewed as a dangerous flaw ("Achilles heel" [2]) of T cells, which might allow mutant pathogens an easy escape from detection. The present derivation provides a clear functional role for antagonism, and denies the putative danger as long as the system fits the specifications derived here.

A natural implementation of antagonism would be via the multistep phosphorylation of triggered TCRs [18], as predicted by the "kinetic proofreading" model [16]. Earlier in this paper, I noted how the two-step simplification of this approach [21] emerges as the natural extension of the design for selective signal amplification and integration.

4.4.1 Adaptation Revisited: Positive Selection on Self-Peptides.

Independent of the implementation, another prediction follows if one allows the deblurring

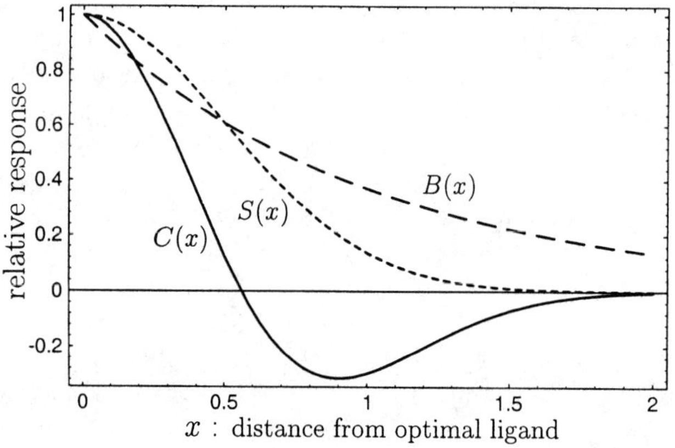

FIGURE 4 Blurring and deblurring of the response to signals, at three stages of antigen processing. All graphs show blur kernels, scaled to unit maximum, sectioned along one of the shape-space dimensions (see text). $B(x)$: Blurring due to antigen preprocessing and presentation. $C(x)$: Deblurring kernel (with $\sigma = 0.5$), showing antagonism for ligands in a halo around the TCR's best match. $S(x)$: Overall result, being the convolution of $B(x)$ and $C(x)$. Note the much smaller tails then those of $B(x)$.

scheme to be already in effect when the T-cell gain is adaptively calibrated by its background signals. Consider T cells that do not find their cognate ligand among the background during thymic maturation, and assume that the preprocessing and presentation of self-antigen in thymus is less blurred than in the periphery. If the T cells do see ligands which lie in the halo around the perfect match, their gain-control T will *increase.* It would be natural to use this as a "survival" signal, consistent with (but opposite in effect to) the earlier assumption that strong T depletion would lead to cell death. This prediction fits at least qualitatively with thymic "positive selection." The negative selection process noted earlier would continue, driven by ligands with a near perfect fit, lying in the central positive lobe of $C(x)$.

This design predicts that T cells would also be *positively* selected by self-ligands. Recent data [4] supports this. The apparent paradox of two opposite effects driven by the same set of ligands is now resolved by pointing to the role of the mismatch (distance x) between the cognate and the actual ligands. Moreover, the design shows how the need for both processes emerges naturally from the task of handling blurred signals.

5 DISCUSSION

Applying detection theory to the task of immune surveillance, I have derived a substantial list of functional properties that are required for achieving near-optimal detection performance. This approach shows how the functional design of an idealized T cell is dictated by its assumed task and the statistics of the available signals. It also guarantees functional coherence of all parts of the machinery. Moreover, it specifies the design mathematically, and thus enables explicit calculation of the detection performance.

The most basic assumptions are that many distinct, unpredictable signals exist, each consisting of a noisy background and a possible intruder signal. This dictates the use of a large, diverse set of detector units (cells), each of which must apply a specific filter (receptor) to the pool of available signals (ligands), before threshold detection. It also shows the need for adaptation to the cell-specific background level. Adding that the formal signals are time series of discrete events occurring at a low rate (ligand binding events) dictates early signal amplification and temporal integration. The simplest way of implementing these demands biochemically turns out to be "serial receptor triggering" [12, 26, 27]. All other design features follow from the formalism similarly, being dictated by a few added or modified assumptions about the statistics of the available signals.

Because the design ignored both specific data on T cells as well as general biological considerations, it is a nontrivial outcome that each of the design features matches well with the major functional properties of real T cells.

The most basic features (e.g., cellularity, specificity, diversity) are very well known experimentally, but deriving them from simple statistics quantifies their fundamental role in solving a highly parallel detection problem. Similar conclusions apply for many of the other features I derived, but some features seem to be not yet as well characterised experimentally as befits their functional importance, as identified by the design. For example, the substantial dependence of the detection limit on the relative size of fluctuations in the contact time τ suggests that some relevant control mechanisms should exist. In other cases, for example, the antagonistic effects of mutated ligands, and in high-zone tolerance, the phenomenon as such is well known, but the present theory provides a clear functional reason for the effect, independently of the underlying mechanism(s).

Given the systematic neglect of real biological constraints, it should be expected that many clear mismatches between design and reality occur. Of course, many vital aspects of real immunity (e.g., presentation, memory) were simply kept outside the scope of the design. Nevertheless, the list of design features is sufficiently large to allow useful comparison with real immune system architecture. As argued below, the outcome of this comparison changes drastically depending on whether one compares the design to "adaptive" or "innate" immune systems.

In "adaptive" immune systems [11], which use a huge diversity of clones, the T cells appear remarkably free of clear functional mismatches with the design. Note that one actually expects striking mismatches if real T cells would reflect an evolutionary history dominated by biochemical constraints, accidentally fixed arbitrary choices, or adaptations that only counteract a few specific subversion strategies by pathogens. Although such historical events are likely to have occurred, they have apparently not left large scars on the basic detection functionality. In fact, coevolution may even have driven immune systems toward the present near-optimal detection strategy: many immune evasion schemes used by viruses [20] subvert or inhibit the degradation of pathogens and presentation of their antigens. Reduced antigen presentation only sharpens the need for higher sensitivity of the detection stage, which may accelerate evolution toward functional equivalence with the optimal design I have sketched here.

Fundamental mismatches are found when comparing the design to the simpler "innate" (germline specified) immune systems such as occur, for example, in invertebrates [10] or even plants [3]. The most striking distinction is the much smaller receptor diversity. Only a few dozen signals seem to be specifically detected, each of which can betray the presence of a broad class of pathogens. This mismatch with the design sharpens the notion that the evolution of adaptive, high-diversity immune systems constituted a fundamental change of strategy in the defence against pathogens, after a transposon insertion enabled recombinase activity to generate the huge diversity of receptors found in T and B cells of adaptive immune systems [1]. Without this, the detection strategy I derived here would be essentially impossible. It will be interesting to apply the design approach to a task definition which leads to nonadaptive immunity, in either of the two roles it is known to play, i.e., as an independent system in lower organisms, or as a subsumed part of the dual-tier vertebrate immune system [8].

ACKNOWLEDGMENTS

I am very grateful to José Borghans, Paulien Hogeweg, Can Keşmir, Rob de Boer, Lee Segel, and Jorge Carneiro for providing important information, criticism, and encouragement. This work was sponsored by NWO/SLW and by Dutch AIDS Fund grant PccO-1317.

REFERENCES

[1] Agrawal, A., Q. M. Eastman, and D. G. Schatz. "Transposition Mediated by RAG1 and RAG2 and Its Implications for the Evolution of the Immune System." *Nature* **394** (1998): 744–751.

[2] Alam, S. M., P. J. Travers, J. L. Wung, W. Nasholds, S. Redpath, S. C. Jameson, and N. R. J. Gascoigne. "T-Cell-Receptor Affinity and Thymocyte Positive Selection." *Nature* **381** (1996): 616–620.

[3] Baker, B., P. Zambryski, B. Staskawicz, and S. P. Dinesh-Kumar. "Signalling in Plant-Microbe Interactions." *Science* **276** (1997): 726–733.

[4] Barton, G. M., and A. Y. Rudensky. "Requirement for Diverse, Low-Abundance Peptides in Positive Selection of T Cells." *Science* **283** (1999): 67–70.

[5] Borghans, J. A. M., and R. J. de Boer. "Crossreactivity of the T-Cell Receptor." *Immunol. Today* **19** (1998): 428–429.

[6] Borghans, J. A. M., A. J. Noest, and R. J. de Boer. "How Specific Should Immunological Memory Be?" *J. Immunol.* **163** (1999): 569–575.

[7] de Boer, R. J., and A. S. Perelson. "How Diverse Should the Immune System Be?" *Proc. Roy. Soc. Lond. B* **252** (1993): 171–175.

[8] Fearon, D. T., and R. M. Locksley. "The Instructive Role of Innate Immunity in the Acquired Immune Response." *Science* **272** (1996): 50–53.

[9] Grossman, Z., and W. E. Paul. "Adaptive Cellular Interactions in the Immune System: The Tunable Activation Threshold and the Significance of Subthreshold Responses." *Proc. Natl. Acad. Sci. USA* **89** (1992): 10365–10369.

[10] Hoffmann, J. A., F. C. Kafatos, C. A. Janeway Jr., and R. A. B. Ezekowitz. "Phylogenetic Perspectives in Innate Immunity." *Science* **284** (1999): 1313–1318.

[11] Janeway, C. A., and P. Travers. *Immunobiology.* London: Current Biology Ltd., 1996.

[12] Lanzavecchia, A., G. Iezzi, and A. Viola. "From TCR Engagement to T-Cell Activation: A Kinetic View of T-Cell Behaviour." *Cell* **96** (1999): 1–4.

[13] Margulies, D. H. "An Affinity for Learning. *Nature* **381** (1996): 558–559.

[14] Mason, D. "A Very High Level of Crossreactivity is an Essential Feature of the T-Cell Receptor." *Immunol. Today* **19** (1998): 395–404.

[15] Matis, L. A., L. H. Glimcher, W. E. Paul, and R. H. Schwartz. "Magnitude of Response of Histocompatibility-Restricted T-Cell Clones is a Function of the Product of the Concentrations of Antigen and Ia Molecules." *Proc. Natl. Acad. Sci. USA* **80** (1983): 6019–6023.

[16] McKeithan, T. W. "Kinetic Proofreading in T-Cell Receptor Signal Transduction." *Proc. Natl. Acad. Sci. USA* **92** (1995): 5042–5046.

[17] Mitchison, N. A. "Induction of Immunological Paralysis in Two Zones of Dosage." *Proc. Roy. Soc. Lond. B* **161** (1964): 275–292.

[18] Neumeister Kersh, E., A. S. Shaw, and P. M. Allen. "Fidelity of T-Cell Activation through Multistep T-Cell Receptor ζ Phosphorylation." *Science* **281** (1998): 572–575.

[19] Noest, A. J. "Designing Lymphocyte Functional Structure for Optimal Signal Detection: *Voilà*, T-Cells." Submitted.

[20] Ploegh, H. L. "Viral Strategies of Immune Evasion." *Science* **280** (1998): 248–253.

[21] Rabinowitz, J. D., C. Beeson, D. S. Lyons, M. M. Davis, and H. M. McConnell. "Kinetic Discrimination in T-Cell Activation." *Proc. Natl. Acad. Sci. USA* **93** (1996): 1401–1405.

[22] Rothenberg, E. V. "How T Cells Count." *Science* **273** (1996): 78–79.

[23] Segel, L. A., and A. S. Perelson. "Computations in Shape-Space: A New Approach to Immune Network Theory." In *Theoretical Immunology II*, edited by A. S. Perelson and S. A. Kauffman, 321–343. Santa Fe Institute Studies in the Sciences of Complexity, Proc. Vol. III. Redwood City, CA: Addison-Wesley, 1988.

[24] Shannon, C. "The Mathematical Theory of Communication." *Bell System Tech. J.* **27** (1948): 379–423 & 623–656.

[25] Sloan-Lancaster, J., and P. M. Allan. "Altered Peptide Ligand-Induced Partial T-Cell Activation: Molecular Mechanisms and Role in T-Cell Biology." *Ann. Rev.* **14** (1996): 1–27.

[26] Valitutti, S., S. Muller, M. Cella, E. Padovan, and A. Lanzavecchia. "Serial Triggering of Many T-Cell Receptors by a Few Peptide-MHC Complexes." *Nature* **375** (1995): 148–151.

[27] Viola, A., and A. Lanzavecchia. "T-Cell Activation Determined by T-Cell Receptor Number and Tunable Thresholds." *Science* **273** (1996): 104–106.

[28] Viola, A., S. Schroeder, Y. Sakakibara, and A. Lanzavecchia. "T Lymphocyte Costimulation Mediated by Reorganisation of Membrane Microdomains." *Science* **283** (1999): 680–682.

[29] Whalen, A. D. *Detection of Signals in Noise.* New York: Academic Press, 1971.

[30] Wülfing, C., and M. M. Davis. "A Receptor/Cytoskeletal Movement Triggered by Costimulation during T-Cell Activation." *Science* **282** (1988): 2266–2269.

Diffuse Feedback from a Diffuse Informational Network: In the Immune System and Other Distributed Autonomous Systems

Lee A. Segel

1 INTRODUCTION

The immune system is composed of trillions of cells, of tens of types, linked by hundreds of signaling chemicals and functioning entirely democratically, without a "leader." Evolutionarily ancient innate immunity gives a first, relatively indiscriminate response to pathogens, followed later by the adaptive response—which features antigen-specific cells. A majority of contemporary immunologists probably agree with Kaufmann [30] who, citing Fearon and Locksley [17], asserts that "It is now clear that the innate immune system, by promptly reacting to microbial components, determines the generation of the appropriate immune response."

The premise of this essay, however, is that the initial reactive response—both due directly to the innate response and instructed by it—is modified during the days or weeks that follow the microbial challenge, by information that the immune system collects about how well it is performing. This premise is not really heretical. Indeed, after the bald statement quoted above, Kaufmann later adds the qualification, "Although direct recognition of bacteria-specific patterns appears to be the central mechanism that instructs the innate immune system, additional mechanisms may contribute" [28]. It is the "additional mechanisms" beyond the first response that will be discussed here, for both the innate and the adaptive immune system. My main goal is to provide a unifying conceptual framework for what I believe to be an important set of "additional mechanisms," based on a variation of the notion of feedback. In the concluding section, consideration will be given to application of this framework to other distributed autonomous systems, both natural and artificial.

In its pure form, then, the classical view of the immune response treats it as a preprogrammed reflex. The innate system senses the stimulus of a pathogenic challenge and responds appropriately, both directly with means to kill pathogens and indirectly, via cytokines, to marshall appropriate arms of the adaptive response. For example, the innate system selects between the two classes of T-helper cells, Th1 and Th2). A closely related characterization of the classical view is that it regards the immune system as organized from the top down: evolution has shaped an immune organization that upon sensing a particular pathogen challenge can automatically select an appropriate response from among a number of prearranged possibilities.

I think that the stimulus-response view of the immune system requires modification. In particular, there are several reasons for believing that the immune system cannot be organized in an exclusively top-down fashion. (i) The immune system was not fashioned *de novo* to solve a problem. The system evolved and presumably gradually improved its performance in some sense, but did so primarily by low-level tinkering with the means that it inherited from the past. Moreover, since biological complexity leads to biological variability, there are, for example, a wide distribution of cytokine production patterns with concomitant different responses to the same stimulus in different individuals [37]. (ii) In large part because of (i), the relevant biological organization is so complex and so sensitive (in some respects) to small "accidents" that a purely top-down organization inevitably will suffer frequent failures. Antigen presentation can serve as another example of the immune system's complexity. In this volume Sercarz [44] describes various baroque aspects of the presentation subsystem that lead him to the conclusion that an aleatoric (probabilistic) view of this system is appropriate. (iii) The immune system not only acts to combat pathogens, but it has several other functions such as assisting in the wound healing process and even promoting the development of the physical structure immune system organs [20]. It is unlikely, for example, that a macrophage can be unerringly controlled by top-down commands given that this cell must not only contend with pathogens but also must participate in reconstruction of damaged tissue. (iv) Since the generation time of pathogens is so much shorter than that of their hosts, pathogens have had the opportunity to evolve many ways to evade the immune response (see, for example, Ahmed and Biron [2]) and will continue to do so. The unceasing development of novel countermeasures by pathogens argues against exclusive reliance on a rigid program of response regulation by the host.

It can be asserted that the "helter skelter" and aleatoric aspects of the immune system do not excessively mar its efficiency, because evolutionary tinkering resulted in the establishment of a variety of different approaches for dealing with any given pathogen—each of which is fairly effective. Moreoever, it would seem advantageous if a population contains alternative approaches for a given pathogen. Thus, sabotage of one of these approaches may have little effect. There is certainly a measure of truth in these assertions. But be that as it may, I shall argue that the immune system employs relatively simple and likely devices to turn tendencies toward redundancy and indeterminism from inevitable and reasonably functional consequences of evolution to strong advantage.

I suggest that the organization of the immune system is a networked combination of top-down and bottom-up. The bottom-up aspect employs a type of "diffuse feedback," which is the mechanism for exploiting the overlapping and stochastic aspects of the immune response. I shall argue that during the course of its reply to each incident of pathogen invasion the immune system tests the various responses available to it and selects the more promising among them. Essential in guiding the development of the immune response is the signaling provided by the system's highly complex network of signal chemicals (cytokines), their receptors, and the intracellular chemicals that combine and process "messages" from the receptors in order to modify gene action. To help make sense of this complexity, I will define and explain the concept of a "diffuse informational network."

2 HOW THE IMMUNE SYSTEM RESPONDS TO CHALLENGE

The body faces various disturbances to its homeostasis. Among these, invasions of dangerous pathogens are arguably of central interest to the immune system. Yet, as we have mentioned, there are other homeostatic problems with which the immune system deals. I propose that the following four principles characterize the immune system's efforts to restore homeostasis.

Principle H1. *In meeting its challenges, the immune system can be regarded as possessing various overlapping and even contradictory goals.* Two important, conflicting, goals are "killing dangerous pathogens" and "avoiding harm to self." (We leave to the future the problem of formulating potential goals for functions of the immune system that are not connected with repelling invasion of dangerous pathogens.)

Principle H2. *The immune system's initial response is top-down in organizational structure and is preprogrammed to be dominantly in a direction that has been selected by evolution as appropriate to the immune goals. Nonetheless, the initial response is broad spectrum in that it contains a variety of different alternatives, in addition to the dominant response.*

Principle H3. *Sensors continually collect information on the state of the body and in particular on the extent of progress toward the immune system's various goals.* Given the goals mentioned in H1, it is to be expected that among the sensors are receptors that respectively bind molecules whose concentrations are increasing functions of pathogen death and of self-harm. We will soon see that indeed such receptors exist.

Principle H4. *The information at its disposal is fed back to the immune system in "real time" to modify its initial response by improving the action of individual effectors and by selecting classes of more appropriate effectors.*

Discussion Here are comments on the four principles that have just been put forward, including a few examples of their operation.

H1: Goals. The term "goals" has a teleological flavor, but bear in mind that when I refer to the goal, say, of killing pathogens, I, of course, do not mean that the immune system "wants" to kill pathogens. Postulating this goal is merely a brief and thought-provoking way to say "scientific observation and experiment provide strong evidence that evolution has produced a system that is likely to destroy invading pathogens."

Each goal is better formulated by amending it with the phrase "all other things being equal." All other things being equal, the immune system attempts to kill dangerous pathogens and to avoid harm to self. But, in general, all other things are not equal. Killing dangerous pathogens is often not compatible with avoiding harm to self. An example of this incompatibility occurs during attacks by noncytolytic viruses. These viruses can produce infectious progeny without killing the cell that they have invaded. As one would expect, cell killing is the dominant mechanism for ridding the body of noncytolytic viruses [2] even though this mechanism harms self.

Orosz [37] offers another perspective on the fact that immune goals may conflict with other goals when he writes of competing agendas and a clash of interests between the immune network's efforts to limit pathogen damage and the physiological network's efforts to maintain homeostasis.

H2: Initial Broad Spectrum Preprogrammed Response. As has been mentioned, the top-down built-in response is now regarded as primarily associated with the innate immune system. The innate system not only acts rapidly to begin dealing with pathogens but also provides important information to the adaptive system. Thus macrophages bind and engulf a class of bacteria, called "Gram negative" because of their staining characteristics, that bear lipopolsaccharide (LPS) molecules on their surface. This is done via an LPS receptor, whose engagement induces the macrophage to secrete a certain panel of cytokines (IL-1, IL-6, IL-8, IL-12, TNF-α). These cytokines, in turn, activate further responses such as fever and various inflammatory processes [29].

An example of a broad spectrum response is the finding of Coffman et al. [8] that LPS stimulation of B cells from the spleen led to antibody secretion that was predominantly IgM but contained five other isotypes.

Reiner and Seder [39] stress the autonomy of lymphocyte response, by which they mean the potential ability of lymphocytes to make decisions seemingly without much regard for extracellular signals. "There is a stochastic distribution of diverse fates after a broad range of signals." Consonant with the point of view advanced here, they suggest that this autonomy leads to "intrinsic diversification of many effector choices, which can then be secondarily shaped to the cues provided by pathogens and their interactions with innate immunity." How this "secondary shaping" works is a major concern of this chapter.

H3: Progress Testing via Information Chemicals. Progress toward the goal of killing pathogens would be most directly indicated by the presence of pathogen "corpses." This is not necessary, however. **Scalps** will do, i.e., irrefutable evidence that corpses exist. A molecule can, in principle, serve as a pathogen scalp, and hence as a "kill-indicator chemical K" (see below) if it usually occurs in pathogens, not other cell types, and if it is an *interior molecule*—whose presence thus indicates pathogen destruction. Further evidence that a molecule plays the role of a pathogen scalp is the existence of "scalp receptors" whose engagement alters the immune response. As I have discussed elsewhere [42], possible bacterial scalp molecules, which fulfill these criteria, include heat shock proteins, mycolic acid (a constituent of the inner cell wall of bacteria), N-formyl methionine peptides, unmethylated CpG sequences, and also endotoxins that are released when bacteria are damaged.

The goal of avoiding harm to self can be assessed with the aid of a "harm chemical" H that is a measure of self damage. One good candidate for H is a trisulfated disaccharide fragment F that results from the cleavage of the extracellular matrix by the enzyme heparanase [32]. Heparanase is secreted by activated T cells to help enable them reach the site of infection. Fragment F damps inflammation, as would be the expected reaction of the immune system when it "realizes" that it is harming the host.

Other candidates for the role of harm chemicals H are molecules peculiar to the exterior of apoptotic bodies—membrane-bound cell fragments that are characteristic of cells that are induced to commit suicide (apoptosis). In analogy with the corresponding requirements for kill molecules K, H molecules should bind receptors that influence the immune response. See Fadok et al. [16] for a summary of recent efforts to identify changes in apoptotic cells that induce recognition by macrophages, and to discover receptors through which such recognition might be mediated.

Perception of harm to the host is anticipated to have a two-edged effect [43]. Down regulation of the immune response is expected to be generated by harm that stems from the immune system itself, such as that signaled by fragment F. By contrast, upregulation of the immune response should stem from evidence that pathogens are harmful. The following observations, brought to my attention by C. Orosz, are in line with this "prediction."

Experiments by Fadok et al. [16] indicate that when macrophages ingest ultraviolet-irradiated apoptotic neutrophils then the macrophages secrete a variety of cytokines that actively downregulate inflammation. No such suppression is observed when neutrophils opsonized by IgG are ingested. By contrast, participation of apoptotic cells in upregulating the immune response was observed by Albert et al. [3]. These investigators produced apoptotic monocytes by infecting them with influenza virus. Via specific receptors, the apoptotic bodies were efficiently internalized by immature dendritic cells (which are specialized for phagocytosis), but not by mature dendritic cells (which are specialized for T-cell activation). Then the immature dendritic cells matured and presented their associated viral antigens to influenza-specific cytotoxic T cells (CTL), for these CTL lysed the infected monocytes. Macrophages more actively ingested the apoptotic cells but did not present the viral antigen. Based on the evidence in their paper, Albert et al. [3] suggested that dendritic cells in peripheral tissue have the important role of ingesting apoptotic cells and then migrating to the draining lymph nodes where antigen presentation can stimulate CTL if pathogenic epitopes are presented. It is reasonable to assume that CTL would be not be stimulated, and might well be tolerized, if the apoptosis were the result of normal cell turnover; indeed, there is evidence that antigen-presenting cells can induce tolerance to self-antigens that are only expressed in peripheral tissues [25].

The results of these experiments are consistent with the idea that molecules peculiar to the exterior of apoptotic bodies can be regarded as harm chemicals. It seems that the abundant and widely distributed macrophages provide a default interpretation of these harm signals, namely that they arise from natural circumstances or from harm due to the immune system and thus should downregulate the immune response. However, immature dendritic cells in peripheral tissue can upregulate the immune response in the presence of harm, provided that they present evidence of the association of harm and pathogens, i.e., evidence that harmful pathogens are present.

H4: Use of Information to Improve Individual and Collective Response. Examples of improving individual effectors are affinity maturation (selecting "better" antibodies) and appropriately arranging switches in B-cell states from inactivity to proliferating to secretory. One example of response selection, the central way of improving collective response, is choice of antibody class (IgG, IgM,...). Another example concerns selection of Th1 vs. Th2. A third example concerns "choices" that the immune system can make to repel a viral attack. These choices include secreting antibodies to block virus entry into cells, employing antibody to opsonize virus particles and, thus, mark them for destruction by macrophages, killing virus-infected cells, and secreting cytokines such as INF-γ and TNF that downregulate intracellular gene expression by viruses [2]. Often more than one choice of possible effectors is appropriate, for example, choice of *both* Th1 and Th2 [5], but it is certainly not a general rule that "more is better." For example, it has been found that nonprotective antibodies reduce the effectiveness of protective antibodies [36].

Given these ways of improving response, the key question is, how (if at all) does the immune system generate such improvement by means of information concerning progress towards its goals. This brings us to an examination of what Orosz [37] has nicely termed Immuno-informatics ("how the immune system generates, posts, processes, and stores information"). Let us turn to some considerations in this area.

3 IMMUNO-INFORMATICS: THE DIFFUSE INFORMATIONAL NETWORK

The following three observations form an essential part of the experimental foundation of Immuno-informatics. (See the chapter by Denny in this volume [14].)

1. The ligation of appropriate receptors typically induces the secretion of not one but several cytokines; many cytokines bind to several different receptors.
2. Each cytokine typically affects several functions.
3. Each function is affected by several cytokines.

In 1998 alone, more than 12,000 papers were published on cytokines, but "practically nothing is known about the behavior of the network as a whole" [6]. We concentrate here on the cytokine network, but there are other signaling systems that are relevant to the operation of the immune system. One such system is centered on "integrins." These are cell surface receptors whose binding to proteins of the extracellular matrix leads to various types of signal. The same kind of pleiotropy as was just mentioned in observation 3 occurs; most integrins recognize several proteins and the major matrix proteins bind to several integrins [18]. Another complex signaling network, inside the cell, links the genes to integrins and to receptors for cytokines and other signal molecules.

The classical way to view cytokines is as a **command network**. For example, it has been found that the command for B cells to switch to IgG secretion is somehow given by a combination of IL-2, IL-4, IL-6, and INF-γ [1]. An enormous amount of useful information of this nature has been collected.

We have proposed an alternative point of view [43], that the cytokines form a **diffuse informational network** or DIN. Cardinal principles of the DIN are these.

Principle DIN1. The DIN provides information about the state of the immune system, the pathogens, and the host. In particular, when effector cells perform a function, they "advertise" this fact.

Principle DIN2. Information is often coded not in the form of a single "tone" (such as the concentration of a single cytokine) but as a chemical "chord" (concentration of several cytokines).

Principle DIN3. Information generated at some point P typically is effective only near P, and only for a limited period of time. (This will be the case for a pulse of diffusing cytokine, which has a limited half-life and which disperses as it diffuses. Thus super-threshold concentrations are confined to a neighborhood of P.) Yet some information carriers, such as cytokines that reach the blood stream, can be rapidly and broadly circulated.

Principle DIN4. The association of two pieces of information generates new information.

A direct consequence of principle DIN1, that cytokines and other signals provide information, is the expectation that the ligation of different receptors will trigger the production of different cytokines. Consider the macrophage in this light. Depending on circumstances, LPS receptors, Fc-γ receptors, or receptors for apoptotic bodies (among many others) are bound. The corresponding implications are "there is potential danger from gram negative bacteria," "opsonized pathogens are being internalized," and "cells are undergoing apoptosis." Just as the spirit of DIN would lead one to expect, ligation of the different receptors induces "individualized" cytokine chords that can broadcast these variegated implications. Accordingly, I observed [42] that different cytokine chords are stimulated by LPS (IL-1, IL-6, IL-8, IL-12, TNF-α; see above) and by Fc-γ receptors (IL-10). At the time that I called attention to these observations, I was not aware of any results concerning cytokine secretions triggered by receptors for apoptotic bodies, but I wrote that "the prediction is that there would be something novel induced by ligation of these receptors." Indeed, the ligation inhibits inflammation via secretion of TGF-β, PGE-2, and PAF [16].

The assertion that cells "advertise" what they are doing (DIN1) is related to a principle that was formulated by Hogeweg and Hesper [26] in trying to understand how relatively stupid social insects can collectively accomplish complex tasks such as the building of elaborate nests. Hogeweg and Hesper suggested that as the insects wander around, if they come across any task that they can do, then they do it. They called this the **to-do principle**. I propose that if an agent in a complex system carries out one of the tasks that it can do, then the agent gives rise to a specific signal; it "moos" to inform the rest of the collective. This I call the **do-moo principle**.

A model for isotype selection that will be presented shortly is based on the premise that macrophages "moo" in a characteristic fashion when they destroy bacteria. One possibility for accomplishing this is the display of specific bacterial epitopes from the heat shock or "stress" proteins that bacteria manufacture when they are subject to threatening circumstances. This display by the macrophages could activate "homunculus" cells [10], probably specialized to the specific epitopes, which provide a "signal 2" to help upregulate local B cells.

Why might it be that information is represented by a "chord" of several cytokines rather than a single "note" of one cytokine (principle DIN2). Three possibilities that might explain this observation are these. (i) Chords add combinatorial richness to the information repertoire. Here is an example. The number of items of information that can be represented by ten "notes" of chemical concentrations, under the assumption that a tone has different significance if its concentration is respectively high or low, is 20. Suppose instead that the information is represented by a three-note chord. There are $10!/(3!)(7!) = 120$ different three-note chords. Each three-note chord has eight variants of high and low concentrations, giving the possibility of representing 960 items of information—which is two orders of magnitude larger than the previous alternative. (ii) Interpretation of information contained in a single chemical "tone" requires sensitivity to absolute concentration. Interpretation of a chord can be carried out if the cell can gauge ratios of concentrations. Gauging ratios allows the system to compensate to some extent for the attenuation of signal strength with distance. (iii) Chords facilitate redundancy, which is a weapon against pathogen subversion of the signaling system. For example, to establish a piece of information about the system, such as "B cells are secreting antibody," it might be sufficient if two out of three of the following conditions hold: cytokine 4 is high, cytokine 7 is high, or cytokine 9 is low.

We have mentioned that locality of information (see principle DIN3) is engendered when signaling molecules diffuse and decay. Another way that locality is obtained is when information is represented by the display of a special molecule on a cell surface; this information can be "read" by other cells only when they contact the first cell. Locality is also obtained when signaling molecules bind to the extracellular matrix. Such a fixed message is available only to cells that contact the matrix suitably. In his discussion of the possible role of immunologically active molecules that are attached to the matrix, Orosz [37] suggests the attractive metaphor that the matrix "serves as a biochemical bulletin board."

The **associative principle** of DIN4 can add enormous power to the expressiveness of the immune cytokine language. To help illustrate this, recall the discussion in section 2 of the utility that would be generated if the immune system could distinguish between harm from pathogens, which should upregulate response, and harm from the immune system, which should downregulate the response. I mentioned that dendritic cells interpret the harm that

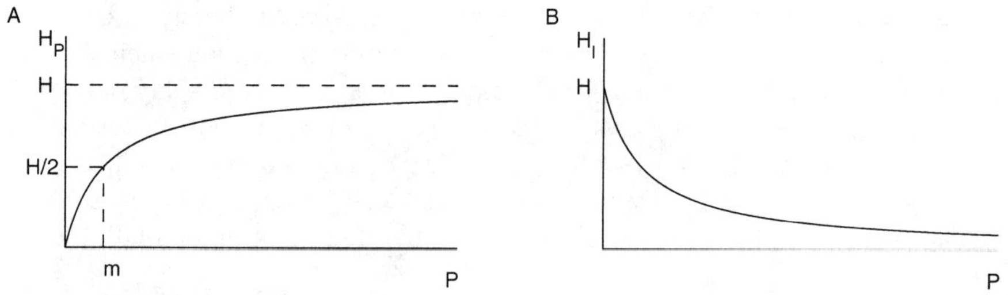

FIGURE 1 A. Graph of harm due to pathogens, H_P, as a function of the pathogen population level P—according to definition (1). B. Harm due to the immune system, according to definition (2).

they sense as due to pathogens if activation of the dendritic cells by apoptotic bodies is accompanied by a display in lymph nodes of pathogen epitopes. This is an association of signals for "harm" and "pathogens" that yields the information "harmful pathogens are in the vicinity." The usefulness of association was illustrated more formally [43] by suggesting that a way to combine information on harm H and pathogen concentration P to yield a measure of pathogen-induced harm H_P is for the immune system to implement the expression

$$H_P = \frac{HP}{m+P}, \tag{1}$$

where m is a half-saturation coefficient (fig. 1A). H_P is approximately equal to the total harm H when P is sufficiently large (much larger than m)—corresponding to the presumption that observed harm must primarily be due to pathogens if the pathogen level is high enough. In other words, strong association between harm H and pathogen level P (high levels of both H and P) signal harm due to pathogens. The remaining harm, $H - H_P$ would be harm due to the immune system, H_I (fig. 1B):

$$H_I = H - H_P = \frac{mH}{m+P}. \tag{2}$$

Similarly the killing of dangerous pathogens can be signaled by simultaneous high values of the killing chemical K and pathogen harm H_P, i.e., a strong association between K and H_P. Thus the product KH_P can be a measure of the extent to which dangerous pathogens are being killed. Of course, it is the spirit of the suggested formulas that matters, not their details. For example, the product $K^2 H_P$ is also a measure of dangerous pathogen killing, with more emphasis (when $K > 1$) on pathogen killing than on whether the killed pathogens are dangerous.

4 REMARKS ON THE EVOLUTION OF SIGNALING

How immune signaling evolved is obviously a central topic in immuno-informatics. Here are some initial thoughts on this matter.

Suppose that at some point in evolution, mutation causes a new capability to arise. It is likely that the new capability, call it function F_1, will generate at least one novel molecule M. (Returning to an example that has already been discussed, cleavage of the extracellular matrix will generate cleavage fragments.) It also seems likely that eventually a mutation will arise that allows one of these trace molecules M to influence a second cellular function F_2. This might occur, for example, if slightly faulty gene duplication leads to the generation of a receptor that binds M, in addition to its previous ligands, and that the new binding effects some action of the genes (perhaps after further mutations). Now that M modifies function, it becomes an informational molecule. In fact, it is a "do moo" molecule, because its presence indicates that cellular function F_1 is operative.

As evolution proceeds, M might come to influence more cellular functions, $F_3, F_4, \ldots$. Even if this does not occur, further developments are likely. This is because, when M influences function F_2, then the role of M changes. Originally M denoted the performance of F_1; now it denotes the performance of both F_1 and F_2. This might, for example, lead to the evolution of an alteration in the original influence of M on F_1.

How might a special signaling molecule evolve? One possibility concerns events that could occur after M comes to have various influences on various processes. It might now happen that function F_1, of which M is a hallmark, ceases to be of relevance except in so far as it gives rise to M. Thus aspects of the original function F_1 would disappear, leaving the secretion of M as the sole residual of the evolutionary history. Once M is secreted, if M has important functions, then evolution should lead to a more sophisticated and responsive secretory apparatus.

Since the immune system has evolved, each new feature must be helpful when mutation first allows it to appear. Because they are limited by a qualifier "all other things being equal," the goals that we have proposed satisfy this condition. For example, a new way to slightly decrease self-damage ought to make an organism more fit. Once this way is incorporated into the genome, it can be partially counteracted by consequences of goals that require some self-damage.

The evolutionary scenarios sketched in our brief examination of "evolutionary Immuno-informatics" lend plausibility to the idea of viewing signaling chemicals primarily as information bearing. But the command view of informational chemicals can also be supported; it is perfectly reasonable to assert that a mutation gave M a beneficial influence on function F_2, say, upregulating F_2, so that the command to "upregulate F_2" was incorporated into the informational repertoire.

5 THE DIFFUSE INFORMATION NETWORK PROVIDES DIFFUSE FEEDBACK

How does the diffuse information network act on the immune system? Modification of behavior is the hallmark of information (see Cohen [10] §66). The diffuse goals that characterize the immune system (homeostatic principle H1) and that are monitored by a variety of sensors (homeostatic principle H3) are what drives what I call a **diffuse feedback**, which tunes the behavior of immune system components in order to achieve better performance. The principles of diffuse feedback are these.

Principle DF1. Diffuse information from many sensors modifies the actions of many cells to improve the situation with respect to the variegated goals.

Principle DF2. Different cells of the immune system are affected differently by the same information.

Principle DF3. Different types of information, and different weightings of the same set of information elements, are employed by a given cell when it decides whether to alter the intensities of its different possible actions—such as proliferation, migration, signaling, performing an effector function, or leaving the field of combat by deactivation, anergy, or death.

The various different cellular responses that are triggered by LPS ligation can serve as a good illustration of how the same piece of information can engender different responses from different cells (principle DF2). One set of responses involves B cells. I have already noted that IL-6 is among the cytokines that switches antibody isotype to IgG. IL-6 also increases antibody production [29]. These commands to B cells are an understandable response to the information that LPS receptors are ligated, for abundant IgG can lead to the removal of Gram negative bacteria by opsonization. The production of acute phase proteins is another appropriate response induced by information that LPS receptors are ligated, for the bacteria that are responsible for the ligation can be destroyed when liver cells are signaled by IL-1 and IL-6 to produce acute phase proteins such as C-reactive protein and the mannose-binding protein [29]. The same information (via the same cytokines) induces the appropriate response in hypothalamus cells of raising body temperature. The same information mobilizes neutrophils and increases vascular permeability. And the same information, via IL-12, tends to encourage Th1 cells that typically promote the cellular killing that is appropriate for intracellular bacteria [15].

The following are some considerations that would lead to the conclusion that different types of information would be required for appropriate decision making concerning different possible activities of the **same** cell (as in Principle DF3), taking B cells as an example. B cells have to make several fundamental decisions—when to become activated to proliferation, when to terminally differentiate into antibody producers, and when to switch isotypes. Perelson et al. [38] have demonstrated for a simple model immune system that under most circumstances the optimum combination of proliferation and antibody secretion is initially to devote resources completely to proliferation and then for all cells to differentiate into antibody secretors at a certain time t_{diff}. At least for simple mathematical models of antibody action on pathogens, a time t_{diff} can be calculated that will lead to the minimal time for pathogen eradication. But the calculation gives no clue how the transition time might be implemented in practice. One simple way to ensure that the transition would occur more or less appropriately is for the immune system to monitor the ratio B/P between the number of B cells and the numbers of pathogens P and to switch to antibody production when B/P reaches some critical value. There must be such a critical value. If the switch value of B/P is set too low, then proliferation should continue, for a small amount of antibody will have little effect. If the switch value of B/P is set too high, then time and resources would be wasted in "overkill."

Let us consider in some detail how signals localized in space could lead to the choice of a suitable class of effectors, in this case, of antibodies. To fix ideas, let us focus on invasion by pathogenic bacteria that are efficiently disposed of by IgG opsonization, which leads to macrophage ingestion and destruction of the bacteria. Suppose that a cluster of B cells near some point X_a employs the isotype IgA and another (successful) cluster near point X_g employs IgG. (Such clusters might occur in germinal centers because, even if they move, daughter cells tend to remain together for a period of time.) Suppose further that macrophages that are destroying bacteria somehow signal this to the immune system (do-moo) and that this "kill signal" is interpreted by every B cell as a command to proliferate more rapidly and to secrete more antibody (of the isotype that is already being secreted). If the macrophage signal is a chemical secretion, the chemical will disperse and become degraded and thus lose its potency as it moves further and further from X_g. If the signal is the appearance of a special molecule on the macrophage, then too the effect of the signal will be confined to B cells near X_g. Thus the "successful" IgG-secreting B cells near point X_g will have their effects amplified, but not the "unsuccessful" IgA secretors near X_a.

See Segel and Lev Bar-Or [43] for more details of the argument just given, including mathematical models that lend it more precision as well as suggestions for its biological implementation. Support for the idea that selection, not just switching, is significant in isotype deployment comes from (in vitro) evidence that at best there is a small probability per cell division of a cytokine-induced isotype switch [13]. Thus it appears that selective proliferation of already-existing plasma cells of a suitable isotype must be at least as important as directed switching to new isotypes.

To summarize, I have just discussed two examples of decision making in B cells—when to switch from proliferation to antibody secretion and when and how to secrete an appropriate antibody isotype. I suggested that the first decision might be based on the B cells' "perception" of the B-cell/pathogen ratio and the second on the concentration of a kill chemical. This illustrates the assertion (in Principle DF2) that an immune cell is likely to use different types of information as aids in making the different "decisions" that it faces.

The model for isotype selection that was just presented illustrates one solution of a fundamental problem that distributed autonomous systems face when they try to improve their performance—**credit assignment**. Which of the different elements should take credit for "good work" and therefore have their contributions magnified? The solution suggested above assumes spatial heterogeneity of elements and assigns credit for a job well done at point X to the elements that are near X. The idea is that if something good happens, reward your neighbors. This idea relies on the locality of information (DIN3).

It is odd that the idea of testing "how are you doing," and altering response according to the results of the tests (feedback) is not prevalent in writings about the immune system. After all, quality control is a major concept in cellular biology [27].

6 IMPLEMENTATION OF DIFFUSE FEEDBACK

Here is a simple concrete illustration of how diffuse feedback may work [43]. Information that positive goals are being achieved upregulates the immune response and information concerning the achievement of negative goals downregulates response. Thus, to implement the "evolutionary discoveries" that killing dangerous pathogens is desirable and harm to the host by the immune system is undesirable, the intensity of an effector action of the immune response, E, can, for example, be regulated such that

$$E = \frac{mKH_P}{1 + rH_I + qKH_P} . \tag{3}$$

Graphs of eq. (3) are presented in figure 2.

According to eq. (3), when all other things are equal, the intensity E of effector action indeed increases when processed sensor output indicates that a higher intensity of response will "do more good" in the sense of increasing the measure KH_P of dangerous pathogen killing. And indeed, when all other things are equal, according to eq. (3) E will be decreased by evidence from the sensors that E is doing harm, in the sense of increasing the measure H_I of harm done to the host by the immune system. The coefficient r in eq. (3) measures the relative importance of the good and the harm; the bigger the r the bigger the relative influence of the harm in downregulating the response, compared to the influence of the "good" in upregulating the response. In nature, the size of the coefficient r would be set by evolutionary competition.

Bear in mind that there is not just one effector of the immune response, there are many. Let us denote these by E_i. To make each effector respond

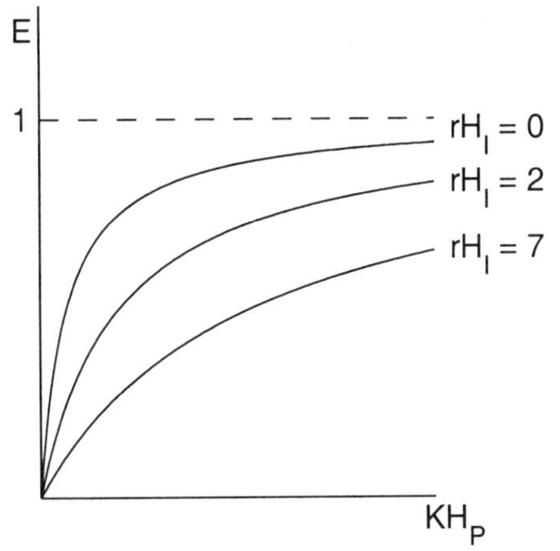

FIGURE 2 Graphs of the intensity E of immune response as a function of the measure KH_P of the desirable feature, dangerous pathogen killing. As is shown, E is inhibited by increasing levels of the undesirable feature, measured by H_I, of harm to the host due to the immune response. Equation (3) gives the assumed functional form of E.

differently to the same information, eq. (3) can be generalized to

$$E_i = \frac{m_i K H_P}{1 + r_i H_I + q_i K H_P} \tag{4}$$

where $i = 1, 2, 3, \ldots, N$. Here N is the number of effectors. According to eq. (4) each effector, in general, weights the same information differently, via different coefficients m_i, r_i, and q_i. Also, the form of eqs. (3) and (4) can be altered in various ways to provide variations in the response. For example, response to the information represented by KH_P and H_I can be made less gradual and more switchlike if these quantities are squared or raised to some higher power. Demonstrated possibilities for intracellular kinetics can, in principle, implement arbitrarily steep switches [19].

Cohen [11] discusses two types of immune decisions: *transformational* (generating a sequential response to a chain of discrete signals) and *reactive* (employing a dynamic representation of the environment and responding in parallel to many signals at once). As it stands, eq. (4) well represents a system component that responds reactively. The possibility of modifying eq. (4) to make it more switchlike illustrates the fuzzy boundary between transformational and reactive systems.

It was mentioned earlier that monitoring of the ratio B/P of B cells to pathogens might provide an adequate algorithm for B cells to "decide" when to switch from proliferation to antibody secretion with the goal of minimizing the time to pathogen eradication. A preferable goal is to minimize pathogen damage, which varies from pathogen to pathogen depending on factors such as reproduction rate and destructive potential. A way to diminish pathogen damage is for switching to depend on the ratio B/H_P of B cells to pathogen harm.

R. Callard mentioned to me that there is a problem with employing the concentration of a chemical K as a measure of successful bacterial killing by the immune system: 10^6 bacteria that are being killed by the immune system might release the same amount of K as 10^9 bacteria that are dying natural deaths. This problem would be solved if K/P were used instead of K, where P is the concentration of pathogens. If the amount of kill chemical per pathogen is high, then it seems clear that pathogens are being efficiently killed.

Let us examine how the formation of the ratio K/P might be implemented physiologically—which is a case study for the implementation of other combinations of sensed quantities. Consider the expression

$$V = \frac{K}{1 + aP + bK} \tag{5}$$

where a and b are constants. Equation (5) has the standard form for the reaction velocity of an enzyme subject to an inhibitor. Thus, eq. (5) could be implemented if P induced an inhibitor of an enzyme that was induced by K, assuming that the concentrations of the enzyme and its inhibitor were respectively proportional to K and P (first-order kinetics). If a were large compared to 1 and b, then V would be approximately proportional to K/P (unless P were very small). This gives a way to display an approximation to the ratio of two effects.

How might pathogen concentrations P be sensed? One possibility is via the LPS receptor, whose ligation by a certain class of bacteria induces a characteristic spectrum of cytokines. Another possible sign of pathogen presence is the set of cytokines secreted by activated B cells.

7 FURTHER EXAMINATION OF THE THEORY

An argument in favor of viewing cytokine concentrations primarily as representing information concerning the system is the fact that this view can explain two of the three central observations concerning the nature of immune signaling that were listed at the beginning of section 3. Why does the same cytokine affect so many different functions? Because a given piece of information should indeed effect the behavior of many different functional entities in the system, upregulating some and downregulating others. Why is a given response affected by so many different cytokines? One reason is that information is given by a "chord" of several cytokines. Another reason is that the necessity to harmonize multiple goals requires that a given response be regulated by multiple inputs. (See eq. (4), for example, where response is modified by both the informational chemicals H and K considered in the simple model, together with the pathogen level.)

Orosz [37] cites findings that production of many cytokines is accompanied by simultaneous production of their antagonists. He suggests that a function of this superficially paradoxical behavior is to force cytokine signals to be intermittent and hence for each new signal to be new evidence of the desirability of the behavior that it fosters. Orosz also mentions the possibility that the immune system is constantly tuning its connectivity. An additional explanation (all explanations could well be true) is suggested by my assertion that immune goals can be contradictory. Inflammation is desirable in so far as it serves a need such as helping to destroy dangerous pathogens, yet inflammation is destructive of tissue. Thus it is not surprising that inflammation-promoting cleavage of the extracellular matrix induces liberation of an inflammation-damping fragment F (see section 2). I suggest that if and only if the cleavage results in the destruction of dangerous pathogens will the evidence that this is the case override the evidence of fragment F that undesirable inflammation is present. Without this override, the inflammation should indeed be damped.

One way to view the balance between the reflexive top-down and feedback-driven reactive bottom-up aspects of the immune response is this. The reflexive response is especially important in dealing with the "tough guys." By contrast the reactive response is especially preoccupied with the "wise guys." The "tough guy" pathogens are old-timers. They have evolved a strategy that has allowed them to copersist with their hosts. A variety of reflexive attacks play an important role in the copersistence of the host in the face of the evolved threat of the tough guys. By contrast, in every pathogen generation there arise mutant strains that threaten to nullify traditional host tactics. These "wise guys" can be countered effectively by reactive defenses that change deployment of weapons systems depending on real-time information concerning which of the available systems best meets the novel attack. Hosts coexist both with the traditional tough guys and the changing spectrum of wise guys. Thus it seems that evolution indeed has selected a suitable combination of both reflexive and reactive immune responses.

Just a word about antecedent work. The ideas expressed here are part of a family of views that stress interconnection and context in the immune system. Grossman was a pioneer in this area [22, 23, 24]. Langman's book [31] mentions ideas of Cohn and Langman concerning the importance of effector choice and how to achieve it. Cohen [9] challenged clonal selection with the cognitive principle. A major novelty in my own approach is the generalization of classical sensor-based feedback to a situation with multiple sensors of multiple progress-indicating signals that can beneficially enhance the attainment of multiple conflicting goals.

How can our speculations be tested further? Christy Warrender, Stephanie Forrest, and I are working on this matter. Our approach is to model true evolution, which alone determines how biology "turns out," by a surrogate: optimization via a genetic algorithm. True evolution has no goals, but it seems an illuminating approximation to postulate that the immune system has evolved toward the overall goal of keeping to a low level the combined damage to the host from pathogen invasion, both direct harm from the pathogen and indirect harm from the immune system's attempts to combat the pathogen. We are constructing simple mathematical models that suitably represent immune systems under various circumstances and using the "biologylike" genetic algorithm method [34] to choose system parameters so that successive iterations drive the system closer to an overall goal such as the one that I just mentioned. We can thus test some of the ideas suggested above by asking, for example, is it true for simple models that if and only if pathogen subversion of the immune system is allowed, then a "chordal" informational system "evolves?" Is it true that a given piece of information will differently affect different functional arms of a model immune system? Does our surrogate evolution typically produce systems where a given function is influenced by many cytokines, in the presence of multiple sensors that test the accomplishment of conflicting goals?

Which is "correct," regarding cytokines as forming a command network or an informational network? In biology, such multiple-choice questions often are correctly answered "all of the above." I believe that this is the case here. For example, returning to a point discussed in section 5, it is permissible to say both that IL-6 promotes switching to IgG and that IL-6 forms part of the message "the LPS receptor has been ligated." But matters may be even more complex. For example, I conjecture that both the possibilities for interpreting IL-6 may be significantly incomplete: in vivo, IL-6 action is modulated so extensively by so many additional factors that an accurate and understandable description of what IL-6 does may remain forever obscure. Support for this conjecture comes from the fact that it often seems virtually impossible to ascertain the function of any given "neuron" in artificial neural networks that have been trained to do a certain task [7, 12].

If the physiological function of a given cytokine were to be essentially unknowable, what hope would there be to attain a primary goal of cytokine studies, to bring cytokine manipulation into the clinical arsenal? One answer is that physicians are not limited to physiological doses of their drugs, and laboratory experiments show that unphysiologically strong doses of certain cytokines can have quite a well-defined effect on a given cell action. Here is where the concept of a command network might usefully be replaced by that of an informational network. Using a cytokine to "command" a cell to perform some function may generate a variety of unwanted side effects, owing to the pleiotropy of cytokine action. It may be preferable to provide a suite of cytokines to "deceive" the organism into "believing" that some suitable situation exists and to hope that all the different responses of the various cell types will be appropriate to the imposed information. For example, if curing an allergy requires shifting from a Th2 response to a Th1 response, this might be accomplished by injecting a panel of cytokines that make the organism believe that it is undergoing an attack by intracellular pathogens.

This essay has been concerned with how diffuse feedbacks can improve the immune response. Allergies and autoimmune diseases are two examples of inappropriate immune responses. It is no surprise that a huge and highly complex autonomous system can go wrong. Understanding of the flaws that might be induced by diffuse feedback must await future study.

8 OTHER DISTRIBUTED AUTONOMOUS SYSTEMS

I have developed several ideas that, in my opinion, help understand the operation of the immune system. I now wish briefly to examine how these ideas might be helpful in understanding other distributed autonomous systems. To that end, it will be helpful to recapitulate some of my major points—but now divorced from any specific reference to the immune system.

I suggest that distributed autonomous systems "would do well" to adopt a fourfold way of responding to challenge—formulating an overlapping and even conflicting set of goals, providing an initial preprogrammed response that is diversified but skewed to handle likely challenges, testing for progress with respect to the goals, and altering the response in accord with test results. A diffuse informational network is an efficient way to provide tests for progress and to communicate the results of the tests. Such a network deploys a variety of sensors to provide information about the state of both the environment and of the system itself. The "do moo" principle furnishes an example of the type of information that should be provided. This principle asserts that when an effector unit carries out a task, it "moos" a signal that conveys this information.

How can information concerning progress in fulfilling system goals be effectively fed back into the system? Simple systems can be controlled in the face of uncertainty by classical feedback. If there is a single well-defined goal, as time goes on the feedback alters the output of a "plant" (in the industrial sense) that can achieve the goal, with the aid of a measuring instrument that can quantify progress toward the goal. A standard example is a thermostat, which appropriately regulates a heater depending on the difference between the actual temperature and a predetermined desirable temperature.

A generalization of classical feedback is required in the face of the variety of overlapping and sometimes contradictory goals that characterize complex systems. Classical feedback operates amidst uncertainty to sense the present state of a particular quantity and thereby to maintain the level of that quantity within narrow bounds. In diffuse feedback, information is deployed among many effectors to improve the situation with respect to a multitude of goals. Diffuse feedback employs a variety of sensors, each of which reports on measurements that can indicate progress toward one or more of the goals. Results of these measurements differently influence the multiple "plants." The influence is such that it simultaneously promotes positive goals and diminishes negative trends. In biological realizations of diffuse feedback the relative weightings of the various influences are selected by evolution. Whether the field of application is biological or not, no optimization need be attained by diffuse feedback. It is sufficient that the feedback improves the response in some sense.

The feedback is "diffuse" in several respects. The sensors give information that is typically blurred, whether because of inaccuracies in measurement or because a few measurements cannot completely capture a complex situation. The feedback does not influence just one plant, but several. There is not one goal but many, and the feedback typically proceeds by messages that diffuse from their source, which weakens their impact at larger distances from the source.

At first sight, diffuse communication may seem a weakness that must be accepted as what evolution has provided in many biological contexts. By contrast, in distributed robotics systems, for example, brainlike "wired" communication would seem to be superior. However, such interrobotic message transfer is profligate of equipment and power. Furthermore, local communication allows a simple attack on the credit assignment problem. Information testifying to a successful system action is distributed locally; in the absence of action at a distance, it is indeed among the local agents that those responsible for the success are to be found. (See Segel [41] for additional comparisons of biological and robotic distributed autonomous systems.)

In biology, what other distributed autonomous systems might provide arenas for the operation of ideas gleaned from the study of the immune system? The metabolic system presents itself as one natural candidate. The metabolic system is relatively simple, compared to the immune system. There are no perpetually threatening enemies, "only" the task of supplying the organism's shifting needs for substances and energy. The complexities of this task are tangibly illustrated by the fact that enzymes often have rather a large number of regulatory sites. Phosphofructokinase, for example, has at least six [4]. This fact alone is virtually sufficient to make clear that the metabolic system applies diffuse feedbacks to mediate conflicting demands for its attention. "The medium is the message" [33] in the metabolic system. The various molecules that are produced themselves act as signals to regulate their own production and that of other molecules. There is no "chord" of signals, which is in line with the conjecture in section 3 that robustness to sabotage may be the principal reason for the chordal structure. In studies of the metabolic system, genetic algorithms have already been used to mimic evolution [40].

Gordon [21] provides enlightening comparisons between immune systems and ant colonies. In particular she proposes a hypothesis that can be regarded as an interesting possible example of the do-moo principle. Observations indicate that the effort devoted by certain types of ants to maintence work is a decreasing function of the effort devoted to foraging. Gordon suggests that this might occur because a nest maintenance worker's probability of leaving the nest might decrease if it meets a forager without seeds, for "just inside the nest entrance, a forager with a seed is a returning forager." To be precise, the principle illustrated is the "did-moo" principle—"I did forage, as you can see by my seed."

As chapters in this book illustrate, ideas gleaned from a study of the immune system can find application in fields that are of major concern in computer science—computer security, artificial intelligence, and adaptive knowledge management. Segel and Bar-Or [41] discuss application of ideas from immunology to distributed robotics.

There is a temptation to make analogies between immune system behavior and high-level human behavior. For is not the immune system exhibiting self-awareness and even self-consciousness when it monitors how it is doing in its task of combating pathogens? I would answer yes. There is nothing mysterious here. Self-awareness and self-consciousness seem to be natural (and hence "evolved") efforts, not only by the body and the brain, but by a variety of distributed autonomous systems, to take into account their own actions and their own internal states as part of the information that can feed back to modify future thoughts and actions (see Mitchell [35]).

I have discussed the relative advantages of viewing a collection of signals as providing information on the system state and on the environment, or, alternatively, as representing commands to system components. I mentioned that physicians might find it advantageous to learn the language of physiological information representation so that they can manipulate what a system perceives and leave it to the various system components to react appropriately to this perception. For large-scale artificial systems the spirit of this approach seems especially fruitful. The systems are typically so complex that it is difficult for a designer to provide commands that will suit his or her purposes. It might well be easier to design a system for adequate information representation and then leave it to some type of artificial evolution, for example by genetic algorithms, to decide how to achieve these purposes in a relatively efficient manner.

ACKNOWLEDGMENTS

Thanks to C. Orosz and C. Warrender for valuable comments on a preliminary version of the manuscript.

REFERENCES

[1] Abbas, A. K., A. H. Lichtman, and J. S. Pober. *Cellular and Molecular Immunology*, 2d ed. Philadelphia, PA: W. B. Saunders, 1994.

[2] Ahmed, R., and C. A. Biron. "Immunity to Viruses." In *Fundamental Immunology*, edited by W. E. Paul, 1295–1334. Philadelphia, PA: Lippincott-Raven, 1999.

[3] Albert, M. L., S. Frieda, A. Pearce, L. M. Francisco, B. Sauter, P. Roy, R. L. Silverstein, and N. Bhardwaj. "Immature Dendritic Cells Phagocytose Apoptotic Cells via $\alpha_v\beta_5$ and CD36, and Cross-present Antigens to Cytotoxic T Lymphocytes." *J. Exp. Med.* **188** (1998): 1359–1368.
[4] Alberts, B., D. Bray, J. Hemes, M. Raff, and K. Roberts. *The Cell.* New York: Garland Publishers, 1983.
[5] Allen, J. E., and R. Maizels. "Th1–Th2: Reliable Paradigm or Dangerous Dogma?" *Immunol. Today* **18** (1997): 387–393.
[6] Callard, R., A. J. T. George, and J. Stark. "Cytokines, Chaos, and Complexity." *Immunity* **11** (1999): 507–513.
[7] Carpenter, G. A., and A.-H. Tan. "Rule Extraction: From Neural Architecture to Symbolic Representation." *Connections Sci.* **7** (1995): 3–28.
[8] Coffman, R. L., B. W. P. Seymour, D. A. Lebman, D. B. Hiraki, J. A. Christiansen, B. Shrader, H. M. Cherwinski, H. F. J. Savelkoul, F. D. Finkelman, M. W. Bond, and T. R. Mosmann. "The Role of Helper T-Cell Products in Mouse B-Cell Differentiation and Isotype Regulation." *Immunol. Rev.* **102** (1988): 5–28.
[9] Cohen, I. R. "The Cognitive Principle Challenges Clonal Selection." *Immunol. Today* **13** (1994): 441–444.
[10] Cohen, I. R. *Tending Adam's Garden: Evolving the Cognitive Immune Self.* San Diego, CA: Academic Press, 2000.
[11] Cohen, I. R. "The Creation of Immune Specificity." This volume.
[12] D'Alche-Buc, F., V. Andres, and J.-P. Nadel. "Rule Extraction with Fuzzy Neural Network." *Intern. J. Neural Systems* **5** (1994): 1–11.
[13] Deenick, E. K., J. Hasbold, and P. D. Hodgkin. "Switching to IgG3, IgG2b, and IgA is Division Linked and Independent, Revealing a Stochastic Framework for Describing Differentiation." *J. Immunology* **163** (1999): 4707–4714.
[14] Denny, T. N. "Cytokines: A Common Signaling System for Cell Growth, Inflammation, Immunity, and Differentiation." This volume.
[15] Diefenbach, A., H. Schindler, M. Röllinghoff, Y. M. Yokoyama, and C. Bogdan. "Requirement for Type 2 NO Synthase for IL-12 Signaling in Innate Immunity." *Science* **284** (1999) 951–955.
[16] Fadok, V. A., D. L. Bratton, A. Konowal, P. W. Freed, J. Y. Westcott, and P. M. Henson. "Macrophages that Have Ingested Apoptotic Cells in vitro Inhibit Proinflammatory Cytokine Production through Autocrine/Paracrine Mechanisms Involving TGF-β, PGE2, and PAF." *J. Clin. Invest.* **101** (1998): 890–898.
[17] Fearon, D. T., and R. M. Locksley. "The Instructive Role of Innate Immunity in the Acquired Immune Response." *Science* **272** (1996): 50–54.
[18] Giancotti, F. G., and E. Ruoslahti. "Integrin Signaling." *Science* **285** (1999): 1028–1032.
[19] Goldbeter, A., and D. E. Koshland, Jr. "Sensitivity Amplification in Biochemical Systems." *Qtr. Rev. Biophys.* **15** (1982): 555–591.

[20] Golovkina, T. V., M. Shlomchik, L. Hannum, and A. Chervonsky. "Organogenic Role of B Lymphycytes in Mucosal Immunity." *Science* **286** (1999): 1965–1968.

[21] Gordon, D. "Task Allocation in Ant Colonies." This volume.

[22] Grossman, Z. "Recognition of Self and Regulation of Specificity at the Level of Cell Populations." *Immunol. Rev.* **79** (1984): 119–138.

[23] Grossman, Z. "Contextual Discrimination of Antigens by the Immune System: Towards a Unifying Hypothesis." In *Theoretical and Experimental Insights into Immunology*, edited by A. S. Perelson and G. Weisbuch, 71–89. NATO ASI Series, vol. H66. Berlin, Heidelberg: Springer-Verlag, 1992.

[24] Grossman, Z., and W. E. Paul. "Adaptive Cellular Interactions in the Immune System: The Tunable Activation Threshold and the Signficance of Subthreshold Responses." *Proc. Natl. Acad. Sci. USA* **89** (1992): 10365–10369.

[25] Heath, W. R., C. Kurts, J. F. A. P. Miller, and F. Carbone. "Cross-Tolerance: A Pathway for Inducing Tolerance to Peripheral Tissue Antigens." *J. Exp. Med.* **187** (1998): 1549–1553.

[26] Hogeweg, P., and B. Hesper. "Evolution as Pattern Processing: TODO as a Substrate for Evolution." In *From Animals to Animats*, edited by J. A. Meyer and S. A. Wilson. Cambridge, MA: MIT Press, 1991.

[27] Hurtley, S. M. "Frontiers in Cell Biology: Quality Control." *Science* **286** (1999): 1881.

[28] Janeway, C. A. "The Immune System Evolved to Discriminate Infectious Nonself from Noninfectious Self." *Immunol. Today* **13** (1992): 11–16.

[29] Janeway, C. A., Jr., and P. Travers. *Immunobiology.* Oxford: Blackwell Scientific, 1994.

[30] Kaufmann, S. H. E. "Immunity to Intracellular Bacteria." In *Fundamental Immunology*, edited by W. E. Paul, 1335–1372. Philadelphia, PA: Lippincott-Raven, 1999.

[31] Langman, R. E. *The Immune System.* San Diego, CA: Academic Press, 1989.

[32] Lider, O., L. Cahalon, D. Gilat, R. Hershkoviz, D. Siegel, R. Margalit, O. Shoseyov, and I. R. Cohen. "A Disaccharide that Inhibits Tumor Necrosis Factor α is Formed from the Extracellular Matrix by the Enzyme Heparanase." *Proc. Natl. Acad. Sci. USA* **92** (1995): 5037–5041.

[33] McLuhan, M. *Understanding Media: The Extensions of Man.* Cambridge, MA: MIT Press, 1994.

[34] Mitchell, M. *An Introduction to Genetic Algorithms.* Cambridge, MA: MIT Press, 1996.

[35] Mitchell, M. "Analogy Making as a Complex Adaptive System." This volume.

[36] Nussbaum, G., R. R. Yuan, A. Casadevall, and M. D. Scharff. "Immunoglobulin G3 Blocking Antibodies to the Fungal Pathogen *Cryptococcus neoformans*." *J. Exp. Med.* **183** (1996): 1905–1909.

[37] Orosz, C. G. "An Introduction to Immuno-ecology and Immunoinformatics." This volume.

[38] Perelson, A. S., M. Mirmirani, and G. F. Oster. "Optimal Strategies in Immunology. I. B-Cell Differentiation and Proliferation." *J. Math. Biol.* **3** (1976): 325.

[39] Reiner, S. L., and R. A. Seder. "Dealing from the Evolutionary Pawnshop: How Lymphocytes Make Decisions." *Immunity* **11** (1999): 1–10.

[40] Ross, J., and M. O. Vlad. "New Approaches to Complex Chemical Reaction Mechanisms." This volume.

[41] Segel, L. A., and R. Lev Bar-Or. "Immunology Viewed as the Study of an Autonomous Decentralized System." In *Artificial Immune Systems and Their Applications*, edited by D. Dasgupta, 65–88. Berlin: Springer-Verlag, 1998.

[42] Segel, L. A. "How Can Perception of Context Improve the Immune Response." In *Irun Cohen Festschrift*, edited by L. Steinman. Jerusalem: Bialik Institute, 2000.

[43] Segel, L. A., and R. Lev Bar-Or. "On the Role of Feedback in Promoting Conflicting Goals of the Adaptive Immune System." *J. Immunol.* **163** (1999): 1342–1349.

[44] Sercarz, E. "Distributed, Anarchic Immune Organization: Semi-Autonomous Golems at Work." This volume.

Multistep Navigation and the Combinatorial Control of Cell Positioning: A General Model for Generation of Living Structure Based on Studies of Immune Cell Trafficking

Eugene C. Butcher
Ellen F. Foxman
Junliang Pan
Eric J. Kunkel

1 SUMMARY AND INTRODUCTION

As a model for tissue morphogenesis, we have studied the mechanisms that target leukocytes into specific tissues and microenvironments in vivo. We have found that both leukocyte extravasation from the blood and leukocyte migration in the context of chemoattractant arrays, such as those known to exist within tissues, are directed and targeted by multistep processes. In these targeting processes, cellular homing can be controlled at any one of multiple steps and, therefore, targeting is determined combinatorially—each sequential step can be mediated by interchangeable combinations of receptor-ligand pairs.

Combinatorial targeting processes share several desirable features necessary for the evolutionary development and maintenance of specific cell targeting events. First, they allow the evolutionary generation of novel targeting events by use of preexisting receptor-ligand pairs in novel combinations. Second, they allow for the specificity of the overall process—the ultimate segregation of different cells to desired locations—to exceed that of the individual steps involved. This lack of a requirement for complete selectivity at each step, in turn, permits overlap between sequential steps, overlap in receptor usage, and multiple receptor usage at individual steps. This overlap then provides the potential for extraordinary robustness and resilience in cell targeting in the face of mutation and regulatory variation. These features, well-illustrated by the physiology of leukocyte trafficking and the "morphogenesis" of the immune response, also provide a paradigm for understanding the diversity, resilience, and specificity of developmental morphogenetic events in the face of evolutionary change.

2 MULTISTEP PARADIGM FOR LEUKOCYTE-ENDOTHELIAL CELL RECOGNITION AND RECRUITMENT FROM THE BLOOD

The migration of circulating leukocytes from the blood into tissues is controlled by a multistep process of leukocyte-endothelial cell recognition. Blood-borne leukocytes have microvillous processes that bear specialized adhesion molecules that initiate contact with the endothelial cell lining of postcapillary venules (called capture or tethering), and support reversible rolling on the vessel wall under blood-flow shear forces. Rolling allows the white blood cell to sample the endothelial surface for activating factors (acting through G-protein-linked serpentine receptors of the chemoattractant family) that can trigger activation of the cell's surface integrins. Integrins are heterodimeric adhesion molecules that, upon activation either by affinity or avidity regulation, can arrest rolling leukocytes on endothelial cells expressing appropriate immunoglobulin superfamily ligands. Such activation and arrest can occur within less than a second in model systems. The cell then has several minutes to "decide" whether to migrate across the endothelium into the surrounding tissue, a decision again thought to be regulated by chemoattractant molecules. In essence, this process consists of a multistep algorithm in which each step represents a "yes" or "no" decision point [3, 4].

Because the ultimate success of extravasation and recruitment depends on successful completion of each step, and several different receptor-ligand pairs can substitute for each other at each of these steps, the process provides for combinatorial generation of both specificity and diversity in leukocyte homing from the blood. This process has not only been modeled in vitro but, more importantly, has now been confirmed to be the basis of leukocyte extravasation and lymphocyte homing in several physiologic situations, including naïve T-cell and B-cell homing into lymph nodes [14] and Peyer's patches [1, 15], memory lymphocyte homing to the skin [2, 5] and gut [8, 16], and neutrophil extravasation into sites of inflammation [10, 13].

3 MULTISTEP NAVIGATION OF LEUKOCYTES THROUGH CHEMOATTRACTANT ARRAYS

We have tried to elucidate the mechanisms that allow chemoattractants to direct leukocyte homing into defined microenvironments, or to locate target cells or sites within a tissue. Many different cell surface receptors participate in this process as well, but we have focused on describing the potential contributions of chemoattractants and their receptors on leukocytes. Recent discoveries have revealed that all leukocytes (and indeed, in all likelihood, all migratory cells) display multiple chemoattractant receptors on their surface [17]. Conversely, resident tissue cells secrete multiple chemoattractants, resulting in overlapping chemoattractant arrays [11]. Therefore, a leukocyte entering a recruiting tissue likely encounters many different chemotactic signals to which it can respond. Leukocytes, therefore, must navigate through complex chemoattractant arrays, and in so doing they must migrate from one chemoattractant source to another until they reach their effector targets.

By modeling cell migration in the context of multiple chemoattractant gradients, presented in different spatial configurations, we have found that leukocytes (in this case neutrophils) can navigate to their target by migrating in sequence towards one chemoattractant source, then another [6, 7]. By evaluating directional persistence and chemotaxis during neutrophil migration under agarose, we showed that cells migrating away from a local chemoattractant, against a gradient, display true chemotaxis to distant agonists, often behaving as if the local gradient were without effect. This behavior allows cells to find their way through complex chemoattractant environments, navigating in a step-by-step fashion through attractant arrays.

We described two interrelated properties of migrating cells that allow this to occur. First, migrating leukocytes can integrate competing chemoattractant signals, responding as if to the vector sum of the orienting signals present. Second, migrating cells display "memory" of their recent environment: a cell's perception of the relative strength of orienting signals is influenced by its history, so that cells prioritize newly arising or encountered attractants. We propose that this cellular memory, by promoting sequential chemotaxis to one attractant after another, is in fact responsible for the integration of competitive orienting signals over time, and allows combinations of chemoattractants to guide leukocytes in a step-by-step fashion to their destinations within tissues. Just as for leukocyte-endothelial cell recognition and recruitment from the blood, this behavior permits combinations of chemoattractants to act in series (combinatorially) to guide cells to specific destinations.

Multistep processes exhibit amazing specificity and diversity while maintaining a great deal of robustness and reliability. Below, we discuss simple models for how multistep processes, such as leukocyte-endothelial cell interactions, can exhibit these complex properties.

4 COMBINATORIAL PROCESSES DISPLAY SPECIFICITY AND DIVERSITY

Because several different receptor-ligand pairs can operate interchangeably at each step in leukocyte-endothelial cell interactions, or during chemotactic navigation, a remarkable diversity of specific targeting events can be generated from a limited number of receptor-ligand pairs. In a simplistic formulation, let S equal the number of sequential steps involved in a process and let N equal the number of unique, but interchangeable, receptor-ligand pairs at each step. If we assume that each receptor-ligand pair is used in only one step (so that K, the total number of receptor-ligand pairs, is equal to NS), that the number of interchangeable pairs is the same for each step (N is constant), and that only one of the N receptor-ligand pairs at each step is used at a time, then the number of specific independently determinable "homing events," Y, is equal to N^S. In this simple model, the complexity that can be specified increases exponentially with the number of steps in the targeting process. For instance, if there were $S = 5$ steps in the process, and $N = 2$ receptor-ligand pairs are available for interchangeable use at each step, then $Y = 2^5$ or 32 different combinations of the five different receptor-ligand pairs could be used to generate independent hypothetical "homing events," each being a unique combination of receptor-ligand pairs that are able to mediate the given multistep process. Let us consider another simple example. With $K = 6$ receptor-ligand pairs (A through F) and $S = 2$ steps, we have $N = 3$ pairs per step. Let A, B, and C be available for step one and D, E, and F be available for step two. Unique combinations ($Y = 3^2 = 9$ of them) would then be AD, AE, AF, BD, BE, BF, CD, CE, and CF. These combinations satisfy the assumptions that (1) each receptor is assigned to one step only, and (2) only one of the assigned receptors is used in a combinatorial event. Note that the interchangeability requirement means that A, B, and C at step one could each mediate the step equally well, thus allowing the combinations shown.

Evolution of a truly novel receptor-ligand pair is likely unusual and, of course, the human body only has about 100,000 genes available, which is a remarkably small number of units of information considering how complex we are as biological systems, as individuals, and as an evolving species. It is therefore of interest to apply the model described above ($Y = N^S$) to a situation in which the number of receptors available is limiting and determine the number of steps, S, which optimizes the number of combinations, Y, or "homing events." Since $N = K/S$ (see above), Y can be written as a function of S only, giving $Y = (K/S)^S$.

We can now calculate the optimum number of steps which maximizes the number of "homing events" or the complexity of the homing process. For example, if we limit ourselves to 10 receptor-ligand pairs, Y is optimal for $\sim$4 steps, giving the potential of $\sim$40 independent targeting combinations. At 100 pairs, the optimal number of steps is $\sim$37, potentially encoding 10^{16} targeting combinations. And at 100,000 receptors (roughly the number of genes in our genome), the optimum number of steps is (surprisingly) still relatively small at $\sim$36,700, but nonetheless offers the potential for independently specifying an astronomically large number of different targeting combinations ($\sim 10^{15,966}$)! At this level of information content, one could easily specify the three-dimensional position of every macromolecule in our bodies (of which there are approximately 10^{23} in a 90-kg human assuming the average protein is about 25 kDa) to within fractions of an angstrom and even track their positions over time!

We observe that a relatively large K is necessary to allow multiple combinations. If $K = S$, then there is only one combination possible which severely limits unique targeting events. If $K = 2S$, then the number of combinations is already increased by 2^S (up to the maximum combinations as dictated by the equation $Y = (K/S)^S$). In real life, for instance, leukocyte recruitment into a tissue may only consist of, say, rolling, activation, adhesion, diapedesis, and sequential migration to two serial chemoattractant sources (~6 steps). Therefore, with even 2 different receptor-ligand pairs available for use at each step, there are already 64 possible combinations, or enough to uniquely target 64 different cell types. Interestingly, recent studies suggest that there are indeed this many different leukocyte subtypes, and probably even more when all the subsets of memory and effector lymphocytes are considered. As we will discuss below, a large K is also required to make the system robust, by allowing more than one receptor-ligand pair to operate in parallel at each step. Moreover, our assumption that each molecule is only used once in a multistep event is a limitation that is not true in biology; the ability to use receptors at more than one (nonsequential) step can greatly increase the potential number of independent targeting events that can be specified.

5 COMBINATORIAL PROCESSES DISPLAY ROBUSTNESS AND RELIABILITY

In the above simplistic discussion, we assumed that only one receptor-ligand pair out of N available at each step mediates the interaction during that step. However, while this situation would generate the maximum number of combinatorial possibilities, multistep mechanisms in this case would be highly susceptible to any genetic (or other) molecular defect, regulatory variation in receptor or ligand expression, or pathogen-mediated disregulation of targeting processes (such as expression of nonfunctional viral chemokine mimics). Multistep processes can combine high specificity with robustness because the specificity at the end of the process is the product of the specificity at each of the individual steps. Thus, highly specific targeting events can be specified by a sequence of events even though each step of the process displays a limited selectivity. As an example, if memory lymphocytes were to interact only twice as well as naïve lymphocytes at each of the 4 steps involved in recruitment from the blood, 16 times more memory cells would enter the tissue despite the relatively small difference in selectively at each step. Moreover, the robustness, or reliability, of the multistep process can be greatly increased by having even slight redundancies at each step in the process, thus converting a purely serial process into a serial process with parallel redundancy.

Multistep selectivity and reliability arises during leukocyte-endothelial cell interactions in vivo, and can be modeled by our simple analysis. Of particular interest is the increase in the process robustness when more than one receptor-ligand pair is utilized at each step and each pair is capable of mediating the specific interaction independently. Say, for example, that two receptor-ligand pairs (four individual molecules) participate coordinately at each step. If we demand that, as before, each receptor-ligand pair out of K pairs can only be used once in the process, then the total number of independently determinable "homing events" Y is now equal to $((K/2)/S)^S$, so that the diversity is reduced by $(1/2)^S$. For 100 receptors, the optimum number of steps is now reduced from ~37 to ~18, and the total number of independently determinable, completely specific, "homing events" Y is reduced to $\sim 10^8$. This loss in diversity, however, is compensated for by a remarkable increase in robustness.

Let P_m be the fixed probability that at least one receptor in the repertoire is nonfunctional due to mutation or aberrant expression. If each step in a given "homing event" is mediated by a single receptor ligand pair (out of N pairs at each step), only one molecule of any pair involved in the process must fail to cause the particular multistep process to fail (the hallmark of a serial process). The probability, P_e, that a given multistep targeting event will fail is therefore $P_m P_i$ or the probability that at least one receptor is nonfunctional times the probability, P_i, that a mutated receptor is one of the $(2S)$ individual molecules involved in the specific "homing event" of interest, where $P_i = (2S)/(2K) = S/K$. This is simply the ratio of the number of molecules involved in the particular multistep process of S steps to the total number of molecules susceptible to mutation in the repertoire. Thus, $P_e = P_m S/K$.

In the simple limiting case where all receptors are involved, $P_i = 1$ and the probability of process failure is equal to the probability that any one or more receptors are dysfunctional: $P_e = P_m$. However, if each step involves two parallel receptor-ligand pairs, both sets of receptor-ligand pairs at some individual step would have to fail to make the whole process fail. Recalling that P_m is the probability that some receptor among the $2K$ molecules in the repertoire will be nonfunctional, we see that the probability that one of the $4S$ molecules involved in a given multistep process will be nonfunctional is again $P_e = P_m P_i$, where $P_i = 4S/2K$. Thus, $P_e = P_m(2S/K)$. However, for the multistep process to fail totally, one of the two molecules in the redundant receptor-ligand pair at the same step would also have to fail with a probability of $P_m(2/2K) = P_m/K$. So the probability of total process failure given two pairs of molecules acting in parallel at each step becomes $P_m(2S/K)P_m/K$ or $2SP_m^2/K^2$. Again, in the simplest case where all receptors are involved in the process of interest, $2S/K = 1$, and the probability of process failure is simply P_m^2/K.

It is instructive to use a ratio to compare the probabilities of failure when the number of receptor-ligand pairs used at each step increases from one (with

S_i steps and K_i receptor-ligand pairs in the repertoire) to two, with S_{ii} steps and K_{ii} pairs. This ratio is $(2S_{ii}P_m^2/K_{ii}^2)/(P_m S_i/K_i)$. In the limiting case when all molecules are involved, and assuming $K_i = K_{ii}$, then $S_i = 2S_{ii}$ and the ratio reduces to $(P_m^2/K)/P_m = P_m/K$.

The relative increase in robustness of the multistep process as a whole can now be seen using a simple example. If the probability, P_m, of at least one receptor in the repertoire failing is 0.9 in a nonredundant (only one molecular pair per step) "homing event" with $K = 50$ molecules and $S = 10$ steps, then the probability of homing failure, P_e, would be $0.9S/K = 0.18$. With a redundant system and the same values for P_m, K, and S, P_e would be $2SP_m^2/K^2 \approx 0.006$. Thus, redundancy increases robustness by a factor of ~30 in this example. If P_m were smaller and K were greater (so that the probability of a given receptor being mutated is reduced), the difference becomes even larger. The difference in robustness, or relative probability of failure, of a twofold redundant versus a nonredundant process of ten steps when $P_m = 0.1$ and $K = 1000$ is ~1:5,000!

6 OVERLAP BETWEEN SEQUENTIAL STEPS AS A MECHANISM FOR THE ROBUSTNESS OF MULTISTEP PROCESSES

Although it is easiest to think of sequential steps as independent, in the two types of multistep processes we have explored (leukocyte-endothelial cell interactions and migration in chemoattractant arrays), there is significant functional overlap. This is particularly true for navigation through chemoattractant arrays, in which the gradient fields may overlap extensively and several chemoattractants may signal to the migrating cell through a single receptor [11]. It is also true for the adhesive steps in leukocyte-endothelial cell interactions, in which (for example) selectins are most specialized for initial tethering, but α_4 integrins can also play a role [1], and indeed can substitute to a degree if selectins are not available (as in homing to the gut lamina propria). The β_2 integrins play a dominant role in activation-dependent adhesion and arrest, but here α_4 integrins can also participate [1]. The result is that inhibition or deficiency of any particular player reduces the reliability of trafficking substantially, but that, in most cases, residual trafficking sufficient for immune responses is retained. Such functional overlap between steps, like the involvement of more than one receptor at each step, thus dramatically enhances the resilience of the system to defects in molecules or expression patterns. While this is at the expense of specificity at each step, the specificity of the overall process is retained as described above.

7 RELIABILITY ANALYSIS AS A GENERALIZED DESCRIPTION OF MULTISTEP PROCESSES

We can generalize the above discussion in a relatively simple manner by describing the specificity, diversity, and robustness of a multistep process using a simplified reliability diagram (fig. 1). The reliability, R_p, or probability that a multistep process will succeed, can be defined as $R_p = R_1R_2 \ldots R_i$ where R_i is the reliability of each individual step in the process. A reliability of 1 would mean that the probability of any step in the process succeeding would be 100%. For instance, if there were a 50% probability that a rolling lymphocyte would become activated by a chemoattractant presented on the venular endothelium, then the reliability of that step in the leukocyte-endothelial cell cascade would be 0.5. Clearly, because the reliability of the process is the product of the individual reliabilities of each step of the process, a step which has a low probability of success can severely reduce the success of the overall process even if the other steps are assured of occurring $(R_i = 1) : R_p = (1)(1)(0.1)(1) = 0.1$ or a 10% probability of success. As shown in figure 1, the reliability of a leukocyte becoming firmly adherent on the endothelium during leukocyte-endothelial cell interactions can be represented as $R_p = (R_{\text{capture/rolling}})(R_{\text{activation}})(R_{\text{firm adhesion}})$.

Specificity can be generalized as a difference in the probability of the success of an individual step (or in many steps of the process) for one cell type versus another, or one situation versus another. As an example, consider homeostatic lymphocyte (but not neutrophil) recruitment into a lymphoid tissue in vivo shown in figure 1 (grey boxes). In Peyer's patches, a secondary lymphoid tissue in the gut, lymphocytes must tether to and roll on the endothelium predominantly through a selectin, become activated by a chemoattractant, and adhere to the endothelium through an integrin [1, 15]. Interestingly, both neutrophils and lymphocytes express the major selectin (L-selectin) and integrin ($\alpha_L\beta_2$ integrin) necessary for rolling and adhesion (half-grey/half-white boxes). The only difference is that lymphocytes have a specific chemoattractant receptor, CCR7, that allows them to become activated in response to the chemoattractant, SLC, presented on Peyer's patch endothelium (the box for CCR7/SLC is only grey). Therefore, the reliability of the process can be described as R_p =(probability rolling)(probability activation)(probability adhesion). Based on the values in figure 1, a lymphocyte has an $R_p = (0.92)(0.5)(0.65) = 0.3$ or a 30% chance of becoming adherent whereas for a neutrophil, $R_p = (0.9)(0)(0.3) = 0.0$ or no chance of success because of the inability to become activated by any chemoattractant in a Peyer's patch. Thus, this type of analysis can adequately describe the specificity of a multistep process.

A reliability analysis can also describe the robustness of a multistep process including receptor redundancy and receptor overlap. For a process with one step, the probability of success is the probability of successful engage-

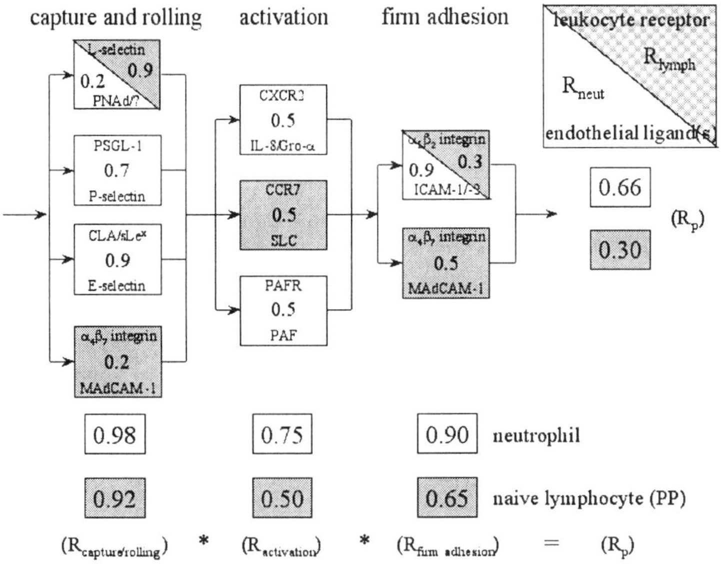

FIGURE 1 Hypothetical reliability diagram for the three-step process of lymphocyte adhesion to the endothelium in a Peyer's patch and neutrophil adhesion to the endothelium in an inflammatory site. Adhesion receptors involved in lymphocyte adhesion in Peyer's patches (PP), a lymphoid tissue in the gut wall, are shown in grey and those involved in neutrophil adhesion at an inflammatory site are shown in white. Boxes with both colors indicate receptors used by both cell types. The reliabilities (probabilities of successfully mediating the given step for each receptor-ligand pair) are given in each box. The parallel reliability of each step (capture/rolling, activation, and firm adhesion) is shown below each step for each cell type and represents the probability that several redundant molecules working together at each step can successfully mediate the step (parallel reliability is greater than any individual reliability for a step with redundant receptor-ligand pairs). The serial reliability, R_p, is shown at the right of the flow chart for each cell type and represents the probability of success of the three-step process for each cell type using its specific receptors. Reliabilities shown are based on experimental data from many laboratories. The process itself is necessarily approximated and simplified (for example, $\alpha_4\beta_7$ plays a very important role in slowing and stabilizing lymphocyte rolling on the endothelium in vivo, even though its reliability of participation in capture or tethering is less than that of L-selectin). Nonetheless, R_p calculated using equations in the text is close to the observed probabilities of lymphocyte or neutrophil arrest under experimental conditions in vivo. Note that as stated in the text, R_p for neutrophil adhesion in Peyer's patches is ~0 because neutrophils lack both CCR7 and $\alpha_4\beta_7$, and Peyer's patch endothelium lacks ligands which can activate neutrophils through their chemoattractant receptors. Similarly, naïve lymphocytes have a very poor reliability of interaction with inflamed endothelium outside of lymphoid tissues because they lack both functional ligands for P- and E-selectin and chemoattractant receptors for neutrophil activators present in these sites.

ment of the single receptor-ligand pair involved, $R_p = R_1$. If R_1 is small, say 0.1, then the process has a 10% likelihood of success. However, if even one redundant receptor-ligand pair with the same probability of success is included in parallel at this step with R_1, the probability of process success becomes $R_p = 1 - (1 - R_1)(1 - R_1) = 0.19$; there is twice the likelihood of success. In fact, because of the parallel process, either individual receptor pair could fail ($R_i = 0$) and the reliability of the process would still be 10%. Thus, as described above, redundancy prevents failure (increases robustness) of the whole multistep process and, at the same time, increases the reliability of any individual step. In such a parallel system, the individual receptor-ligand pairs could have different reliabilities and, therefore, lacking one of the redundant pairs could decrease the whole process reliability to a greater extent than lack of another.

Receptor overlap and redundancy (as described above) occur during lymphocyte homing to Peyer's patches (fig. 1) when α_4 integrins can substitute for selectins in the lymphocyte rolling process, but at a reduced reliability, and α_4 integrins work together with $\alpha_L\beta_2$ integrin to help mediate lymphocyte adhesion. In the hypothetical model of figure 1, then, the success of lymphocyte adhesion increases from about 14% without the contribution of the α_4 integrins, to about 30% including this integrin. As shown in figure 1 (white boxes), neutrophil recruitment during inflammation is also a multistep process with various degrees of redundancy in different steps. The presence of several steps leads to a good deal of reliability in the process as a whole, with about 60% of neutrophils becoming adherent during cytokine-induced inflammation [9]. Because inflammatory neutrophil recruitment is dependent on the regulation of expression of various endothelial adhesion molecules and chemoattractants, redundancy helps to prevent poor recruitment in the face of regulatory variability.

How is the reliability of a given step in a multistep process determined? Defining the reliability of a given step is complicated because it is generally a function of multiple parameters, many which are difficult to quantify. For instance, the reliability of the rolling step of the leukocyte-endothelial cell cascade is a complex function of receptor expression on the leukocyte, blood hemodynamics, ligand expression on the endothelium, morphology of the vasculature in a particular site, etc. These factors can furthermore be influenced by the activation state of the leukocyte and/or endothelium. Another extremely complicating factor is that the reliability of a given step can change with time, e.g., as endothelial ligand expression varies during an inflammatory response in response to feedback mechanisms (as explained below). Of course, the probability of success of a particular step, such as rolling, can be determined empirically for one experimental condition, but the variability of physiologic processes precludes determining the reliability of a given step for multiple situations. However, reliability analysis is valuable for examining how a complex multistep system responds to changes in individual steps in the process, and provides a framework for quantitatively understanding why a multistep process can exhibit complex behavior such as specificity and diversity while maintaining robustness and reliability.

8 FEEDBACK IN MULTISTEP PROCESSES

The reliability diagram in figure 1 illustrates that the reliability of a given step contributes to the probability of success of the multistep process. Although the reliability of any step is a function of many factors, it has become clear that the most dynamic factor predicting the success of a particular step is perhaps the endothelium, which is highly regulated by inflammatory cytokines. Interestingly, many of the cytokines that can regulate endothelial cell function are produced by previously "homed" or resident leukocytes. Different subsets of leukocytes secrete distinct cytokines, and these cytokines can enhance or depress endothelial expression of adhesion molecules and chemoattractants, thus altering the character of the endothelial cells and the probability and specificity of recruitment of different leukocyte subsets as inflammation develops and eventually resolves. Thus, another aspect of the success of a multistep process is the enhancement of the reliability of certain steps by feedback.

During inflammatory leukocyte recruitment there are many conceivable ways in which the first "homed" leukocyte can provide feedback to endothelial cells. For example, the "homed" leukocytes can signal endothelial cells to express more proteins critical for recruiting leukocytes of its own class while suppressing expression of proteins involved in recruitment of other classes of leukocytes. We have found that an endothelial chemokine involved in recruitment of skin-homing memory T cells can be induced by a unique combination of cytokines derived from activated, but not resting, skin-homing memory T cells or naïve T cells [12]. In this scenario, the recruitment of skin-homing memory T cells is amplified when the first homed T cells recognize their antigen(s) and become activated. This type of feedback mechanism could be particularly relevant in chronic inflammation, because, in many cases, only a subset of leukocytes predominate as inflammation progresses. Intriguingly, the feedback from the "homed" leukocytes to endothelial cells bears a striking similarity to that from the end products to the rate-limiting enzymes in a metabolic pathway, although feedback to endothelial cells can be both positive and negative. Indeed, such positive feedback loops in which recruited lymphocytes produce cytokines that act on endothelial cells to upregulate molecules that favor recruitment of the same lymphocyte type, lead to further recruitment and further production of the selectivity-supporting cytokines, etc. This type of feedback loop may parallel the feedback regulation of T_H1 versus T_H2 cytokine production in lymphocytes, in essence leading to quasi-stable, self-supporting homing specificities.

Other feedback mechanisms could include stimulation of endothelial cells by activated antigen-presenting cells (as occurs in lymph nodes), negative feedback mechanisms such as growth-factor release to stimulate wound healing and endothelial adhesion protein downregulation, and more complex feedback networks such as activation of endothelial cells by cytokines secreted from activated macrophages which themselves have recently been activated by recruited T cells. The specificity of the process can therefore be even more highly controlled through cell-to-cell feedback. Feedback mechanisms as a control point in the success of a multistep process can be complex and tightly regulated, and are just being fully appreciated.

9 CONCLUSIONS

We have demonstrated experimentally that both leukocyte extravasation from the blood and leukocyte migration in the context of chemoattractant arrays are directed and targeted by combinatorial multistep processes. Here we have attempted to demonstrate mathematically, using simple models, that such combinatorial multistep processes can display amazing diversity with a limited repertoire of receptors, specificity in the overall process even with limited selectivity at each individual step, extreme robustness in the face of genetic mutation or other functional alteration, and reliability, even with regulatory variation, due to feedback regulation. Further understanding of complex systems such as these will require more involved mathematical analysis coupled with a clearer understanding of how such processes are regulated.

ACKNOWLEDGMENTS

We would like to thank Dr. D. J. Campbell, Dr. R. Alexander, and Dr. L. A. Segel for informative discussions about the models in this chapter.

REFERENCES

[1] Bargatze, R. F., M. A. Jutila, and E. C. Butcher. "Distinct Roles of L-Selectin and Integrins Alpha 4 Beta 7 and LFA-1 in Lymphocyte Homing to Peyer's Patch-HEV in situ: The Multistep Model Confirmed and Refined." *Immunity* **3(1)** (1995): 99–108.

[2] Berg, E. L., T. Yoshino, L. S. Rott, M. K. Robinson, R. A. Warnock, T. K. Kishimoto, L. J. Picker, and E. C. Butcher. "The Cutaneous Lymphocyte Antigen is a Skin Lymphocyte Homing Receptor for the Vascular Lectin Endothelial Cell-Leukocyte Adhesion Molecule 1." *J. Exp. Med.* **174(6)** (1991): 1461–1466.

[3] Butcher, E. C. "Leukocyte-Endothelial Cell Recognition: Three (or More) Steps to Specificity and Diversity." *Cell* **67(6)** (1991): 1033–1036.

[4] Butcher, E. C., M. Williams, K. Youngman, L. Rott, and M. Briskin. "Lymphocyte Trafficking and Regional Immunity." *Adv. Immunol.* **72** (1999): 209–253.

[5] Campbell, J. J., G. Haraldsen, J. Pan, J. Rottman, S. Qin, P. Ponath, D. P. Andrew, R. Warnke, N. Ruffing, N. Kassam, L. Wu, and E. C. Butcher. "The Chemokine Receptor CCR4 in Vascular Recognition by Cutaneous but not Intestinal Memory T Cells." *Nature* **400(6746)** (1999): 776–780.

[6] Foxman, E. F., J. J. Campbell, and E. C. Butcher. "Multistep Navigation and the Combinatorial Control of Leukocyte Chemotaxis." *J. Cell Biol.* **139(5)** (1997): 1349–1360.

[7] Foxman, E. F., E. J. Kunkel, and E. C. Butcher. "Integrating Conflicting Chemotactic Signals. The Role of Memory in Leukocyte Navigation." *J. Cell Biol.* **147(3)** (1999): 577–588.

[8] Hamann, A., D. P. Andrew, D. Jablonski-Westrich, B. Holzmann, and E. C. Butcher. "Role of Alpha 4-Integrins in Lymphocyte Homing to Mucosal Tissues in vivo." *J. Immunol.* **152(7)** (1994): 3282–3293.

[9] Kunkel, E. J., J. L. Dunne, and K. Ley. "Leukocyte Arrest during Cytokine-Dependent Inflammation in vivo." *J. Immunol.* **164(6)** (2000): 3301–3308.

[10] Ley, K. "Molecular Mechanisms of Leukocyte Recruitment in the Inflammatory Process." *Cardiovasc. Resh.* **32(4)** (1996): 733–742.

[11] Mantovani, A. "The Chemokine System: Redundancy for Robust Outputs." *Immunol. Today* **20(6)** (1999): 254–257.

[12] Pan, J., J. J. Campbell, G. Haraldsen, B. Xu, and E. C. Butcher. "Memory T Cells Signal Vascular Expression of the Chemokine TARC: A Positive Feedback Mechanism in Lymphocyte Homing to Skin." In preparation.

[13] von Andrian, U. H., J. D. Chambers, L. M. McEvoy, R. F. Bargatze, K. E. Arfors, and E. C. Butcher. "Two-Step Model of Leukocyte-Endothelial Cell Interaction in Inflammation: Distinct Roles for LECAM-1 and the Leukocyte Beta 2 Integrins in vivo." *Proc. Natl. Acad. Sci. USA* **88(17)** (1991): 7538–7542.

[14] Warnock, R. A., S. Askari, E. C. Butcher, and U. H. von Andrian. "Molecular Mechanisms of Lymphocyte Homing to Peripheral Lymph Nodes." *J. Exp. Med.* **187(2)** (1998): 205–216.

[15] Warnock, R. A., J. J. Campbell, M. E. Dorf, A. Matsuzawa, L. M. McEvoy, and E. C. Butcher. "The Role of Chemokines in the Microenvironmental Control of T- versus B-Cell Arrest in Peyer's Patch High Endothelial Venules." *J. Exp. Med.* **191(1)** (2000): 77–88.

[16] Zabel, B. A., W. W. Agace, J. J. Campbell, H. M. Heath, D. Parent, A. I. Roberts, E. C. Ebert, N. Kassam, S. Qin, M. Zovko, G. J. LaRosa, L. L. Yang, D. Soler, E. C. Butcher, P. D. Ponath, C. M. Parker, and D. P. Andrew. "Human G Protein-Coupled Receptor GPR-9-6/CC Chemokine Receptor 9 is Selectively Expressed on Intestinal Homing T Lymphocytes, Mucosal Lymphocytes, and Thymocytes and is Required for Thymus-Expressed Chemokine-Mediated Chemotaxis." *J. Exp. Med.* **190(9)** (1999): 1241–1256.

[17] Zlotnik, A., J. Morales, and J. A. Hedrick. "Recent Advances in Chemokines and Chemokine Receptors." *Crit. Rev. Immunol.* **19(1)** (1999): 1–47.

Distributed, Anarchic Immune Organization: Semi-autonomous Golems at Work

Eli E. Sercarz

Can a distributed autonomous immune system manage to accomplish its necessary work of insuring survival of each individual and providing a homeostatic safeguard from exogenous and endogenous incursions? The analogies between the nervous and the immune systems have often been pointed out, but it can be argued that the immune system has a more varied input and the capacity to construct a more varied output than the nervous system. The immune state of affairs might be expected to depend on a highly ordered system run by a set of dependable rules and devices that are subject to little random fluctuation. The need for order is overriding, since the life of each individual is at stake, and protection could have been predicted to evolve following a standard course, invariable for each microbe in the environment or for each antigenic determinant constituting the microbe. In fact, this is not at all the case. Even in the creation of the tools of specific recognition, the antibodies and the T-cell receptors, the rearrangement of particular genes constituting these molecules appears to be left up to chance, and the assemblage is unique in each individual.

1 SOFT-WIRED VS. HARD-WIRED SYSTEMS

The hypothesis presented here is that, to a very large extent, the immune system operates with maximal flexibility, in a distributed, modular mode in which collaboration between modules is soft-wired and rather haphazard. Because the very construction of the system depends on the interdigitation of fixed, common elements and variable, individualized elements, there is an inherent wobble which reduces the predictability of the outcome of any confrontation.

Another modus operandi is found in hard-wired systems such as the automobile, and its assembly line. To the basic frame components (e.g., chassis) are added several separate assemblies (e.g., engine, steering system) each of which is comprised of subassemblies (e.g., the piston subassembly includes piston, rings, connecting rod, bearings). The final sleek product comprised of these components operating in synchrony, can be controlled, albeit indirectly, by the driver (see fig. 1). In this hard-wired system, the operator need not be aware of, nor perform adjustments to, the cooperative interaction of the subassembly components. Although the components operate inflexibly within the whole, there are regulatory components (e.g., timing, air intake, carburetion) which can adjust the output of the total assemblage. In comparison, the product of the superassemblies of the immune system (IS) need to emerge from the system with many variable qualities, yet function in a reproducible fashion. The heterogeneity of these qualities is of two general types: first are elements of the receptor repertoires of different lymphocyte subpopulations, which must accommodate to the diversity of ligands in the surrounding universe; second is the lesser diversity of effector structures which collaborate in performing the necessary functions of the IS. Mixed into the equation is a set of extrinsic and intrinsic environmental signalling agents such as antigens, hormones, cytokines, chemokines, receptors for these agents, and molecules which control cell trafficking, synthesis of effectors in various physiologic circuits such as cell death pathways, etc.; these environmental influences lead to an essential unpredictability of the outcome of response. Scattered throughout the IS, however, are mechanisms of regulation which act as major schemes for interaction of components; e.g., regulatory T cells acting in part as components of hard-wired circuitry; members of cell death pathways. In the descriptions to follow, I will nevertheless focus on the aleatory essence of the components of the IS.

In this chapter, I will address the particular qualities of determinants on self and foreign antigens which modulate the nature of the repertoire that is selected and expressed in the immune response. Three specific areas will be explored, united by their relationship to predictability in IS operation: (1) the vagaries of antigen processing, (2) the nature of immunodominance, and (3) the outcome of intercellular regulation. The context of these explorations concerns the expectation among the lay public and also among many immunologists that it should be possible to predict the outcome of any particular immune scenario. In fact, this may be true, provided that we attain a

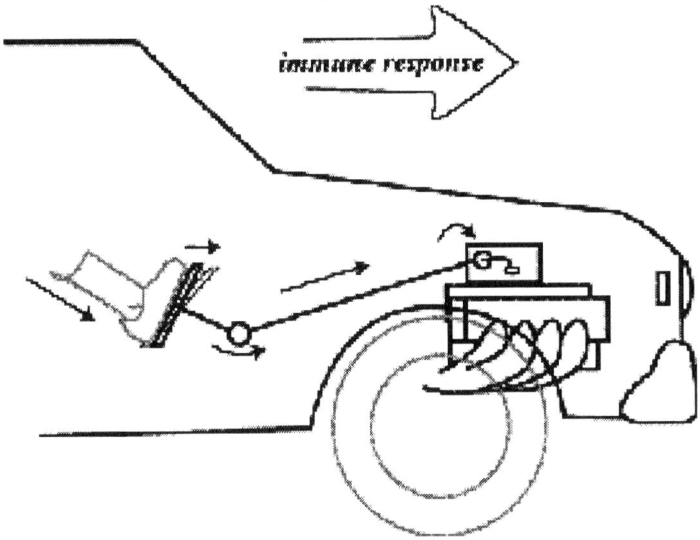

FIGURE 1 Hard wiring. An initiating event leads to a set of events in a fixed order or pattern, often linearly.

level of molecular understanding that at present appears to be unachievable. As an approach to formulating a coherent view about the limits of predictability in immune affairs, I have chosen to consider the Golem, and a "golemic" perspective, to the analysis of this most complicated, distributed autonomous system.

2 THE GOLEMIC ESSENCE

2.1 CHARACTERISTICS OF GOLEMS

Golems are humanoid creatures, created from dust and the earth according to complex rabbinical prescriptions, very carefully recorded over many centuries in the mystical Jewish folklore. The most famous such individual was the Golem of Prague, whose creator was the historical Rabbi Loewy (ca. 1520–1609). Rabbi Loewy was driven to create "Yossef" Golem to guard Jews from evils and persecutions, and other Golems were created down through the years for particular and precise conditions to accomplish prescribed functions, for example, guarding the rabbi, helping the rabbi's wife, performing heroic deeds, etc. [12]. Yossef and other Golems were powerful, had limited capacities to reason, and were often granted magic capabilities, such as invisibility. However, they were never imbued with a soul or the ability to speak, aspects reserved for human beings, and were said to lack sexual drive. One amusing debate concerned whether Golems could help to constitute a minyan, the quorum of ten males necessary for reciting prayers, or whether if one destroyed a Golem, had murder been committed? [12].

However, the central golemic quality from our perspective was their unpredictability, and lack of flexibility. For example, Yossef was commanded to catch a basketful of fish for the Rabbi, but later when fish became available and Yossef was acquitted of this responsibility and asked to return home, he dumped out the valuable fish he had already caught before he returned. Likewise, sometimes Golems turned against their masters, and others were even accused of lecherous intent with the rabbi's wife, (although there was mention about a reciprocal interest of the wife). The Mickey Mouse character in the Sorcerer's apprentice movie version in Fantasia 1940 (repeated in Fantasia 2000), possessed some typical golemic features, and given the task of carrying water, became obsessional, ending in disaster. Thus, Golems were created to possess a limited "workscope," and were endowed with limited adaptability.

2.2 THE GOLEMIC IMMUNE SYSTEM

The immune system can be considered as a conglomerate of subcellular and intracellular golemic networks. The characteristic modular assemblies and subassemblies previously described in the automobile example still hold true in the IS, as will be described in detail below. Figure 2 shows several Golems actively at work within the cell. Each of the subassemblies will run quasi-autonomously with little knowledge about parallel subassemblies. I have termed the components of the subassemblies as "kits," some of which can be removed from the system (e.g., in "knockout" deletional mutant mice) without leaving a completely nonfunctional residue, but which play a unique role in the subassembly. Despite the fact that a kit may add a feature to the system which makes the IS more flexible and responsive, there is considerable redundancy in the system as a whole (see Cohen [2]) so that certain kits may be dispensed with. Just because there is redundancy does not imply that the complete system would not be more fit for the environment; thus, the redundancy is not total, but rather conditional—there are other ways of accomplishing the task, although perhaps not as elegant.

3 THE ALEATORY NATURE OF ANTIGENIC PROCESSING

3.1 THE ANTIGEN PROCESSING SUBASSEMBLY

In the straightforward, hard-wired automobile analogy, pressure on the accelerator pedal induces a complex but totally predictable and analyzable series of events which results finally in forward movement, at a speed decided by the operator (with an assist to the automatic transmission). I will use the system of antigen processing and presentation to illustrate the purported golemic organization of the IS. This is a major site which leads to an aleatory outcome

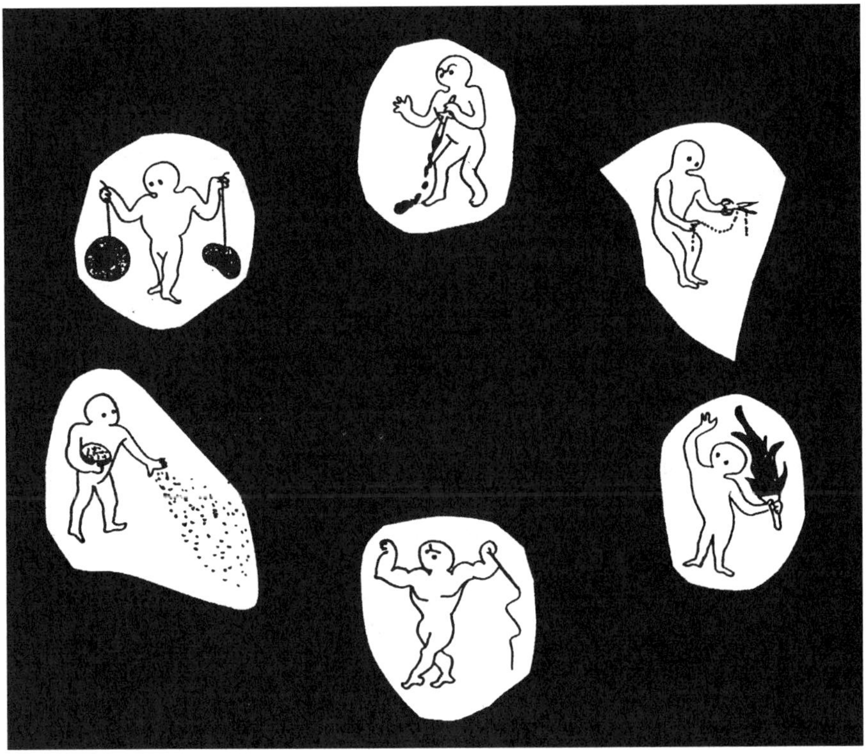

FIGURE 2 An artist's view (Rabyn Blake) of the activities of various Golems within the cell. Starting from the upper left and proceeding clockwise: the Th1/Th2 deviation Golem; the killer Golem; the antigen processing Golem; the inflammatory Golem; the dominance Golem; and the determinant spreading Golem.

of the issue of immunodominance (e.g., Which is the dominant determinant? Which T cell is selected? How does regulation impinge on these choices?).

3.1.1 Determinant Spreading. As an example, we can consider the phenomenon of determinant spreading [8] in self-reactive systems, and some of the superassemblies and subassemblies involved. Very often in autoimmune responsiveness, a single initiating event such as a viral infection occurs, where a dominant antigenic determinant within the virus induces a response in a T cell which cross-reacts with a cryptic self-component. If this response is propagated, an autoimmune disease may follow, although the response may very well be downregulated in which case, the fleeting rheumatic ache will soon be forgotten. Propagation appears to depend on determinant spreading which occurs through the raising of an inflammatory environment by the interaction of certain determinants with particular T cells which secrete proinflammatory mediators to recruit additional cells into reactivity. Antigenic determinants must bind to the binding groove of major histocompatibility (MHC) molecules on antigen-presenting cells (APC) designed by evolution to present portions of antigens to ambient T cells. If the interaction affinity between antigenic determinant and MHC groove is high, then the resultant T cell emerging from the APC-T cell interaction will generally display a proinflammatory (Th1) set of cytokines. As part of the scenario, it must be realized that the only T cells remaining in the organism following the induction of self-tolerance will be those with (a) very low affinity for the determinant and/or (b) varying levels of affinity for determinants which are only poorly expressed on APC, so-called "cryptic determinants" [10]. Thus, if a Th1 T cell arises after the aforementioned viral infection (rather than a Th2, anti-inflammatory or regulatory T cell), and the cytokine density achieved is adequate to recruit new T cells [4], what started as a response of limited breadth may turn out to be a conflagration, viz. a broad response with a diversity of involved T cells. However, if all of the IS safeguards are in place, and regulatory cells are able to contain the initial spreading, the "autoimmune incident" can be transient and invisible to the subject.

Accordingly, if we examine a behavioral feature of the immune response such as determinant spreading, it requires the interaction of several superassemblies and subassemblies, as shown in figure 3. Only some of the necessary components are indicated for simplicity. In this IS model, if you turn to the Manual of the Immune System, a small part of which can be seen in table 1, there are several kits that need be assembled to constitute each subassembly. The individual kit and its operator can be called a "Golem" in the sense that it is semiautonomous, comprised of several components necessary to complete a necessary and circumscribed task.

3.1.2 The Golems of the Antigen Processing Subassembly. If we consider the "Protection of the MHC" Golem, we must include the invariant chain Ii, which interacts with the MHC $\alpha\beta$ heterodimer as a chaperone to protect its active site from binding to unwanted peptides until the $(\mathrm{Ii}\alpha\beta)_3$ nonamer arrives at the vesicular, endosomal-lysosomal compartments. Ii also directs intracellular trafficking of the $(\alpha\beta\mathrm{Ii})_3$ nonamer, from the endoplasmic reticulum to its vesicular compartments: once in the acidic environment of the vesicles, Ii is cleaved by a set of special cathepsins such as Cathepsin S, found throughout the system except in the thymus where the job is left to Cathepsin L. Another pair of chaperones, called DM and DO in man and H2M and H2O in mouse, help to remove the residual product of enzymatic degradation of Ii, termed CLIP, from the MHC groove. Without considering a few other intricacies about which this Golem need be concerned, let me mention several areas of ambiguity already mentioned about its activities. It must be realized that the Ii and its CLIP region, which are truly monomorphic, bind with varying affinity to different MHC molecules [11]. Likewise, adjoining the CLIP region is a 10-mer region of Ii which is important in CLIP self-release and this inter-

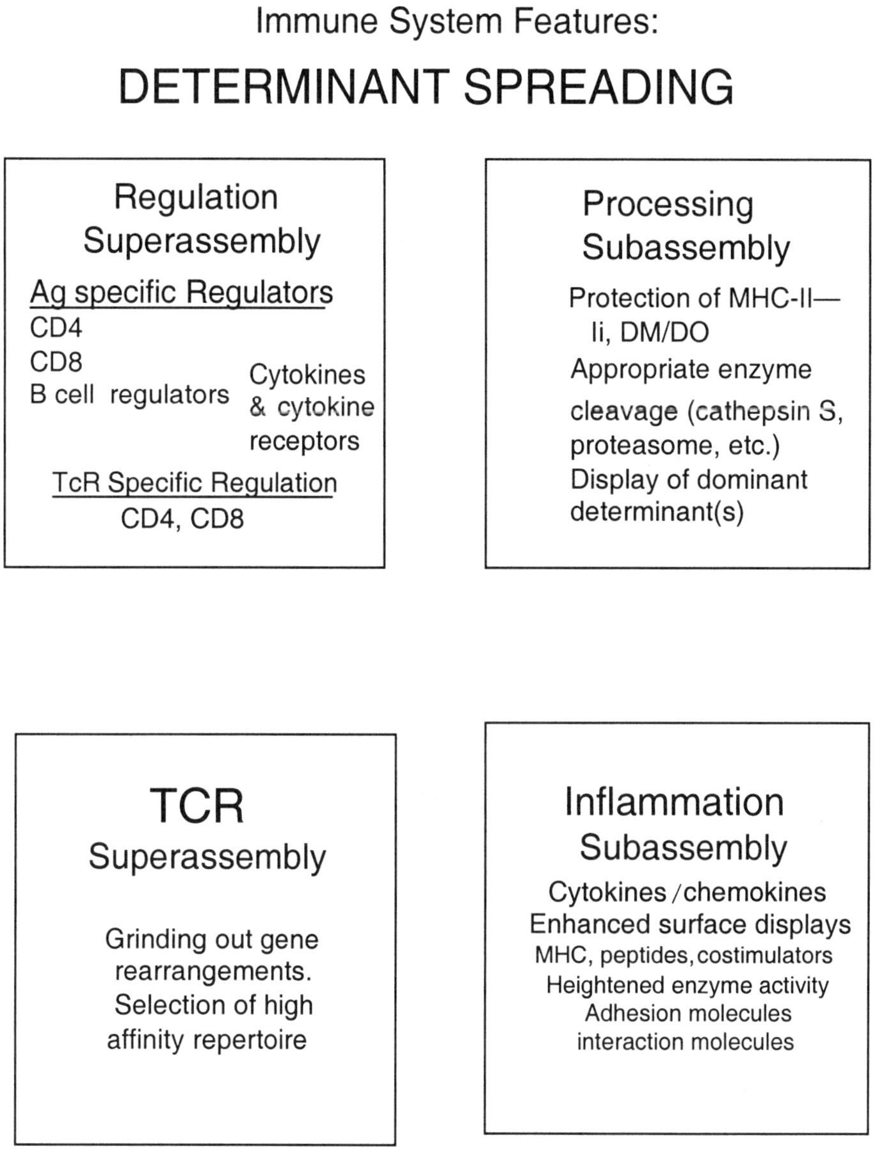

Immune System Features:
DETERMINANT SPREADING
Regulation Superassembly
Ag specific Regulators
CD4
CD8
B cell regulators
Cytokines & cytokine receptors
TcR Specific Regulation
CD4, CD8
Processing Subassembly
Protection of MHC-II—Ii, DM/DO
Appropriate enzyme cleavage (cathepsin S, proteasome, etc.)
Display of dominant determinant(s)
TCR Superassembly
Grinding out gene rearrangements.
Selection of high affinity repertoire
Inflammation Subassembly
Cytokines / chemokines
Enhanced surface displays
MHC, peptides, costimulators
Heightened enzyme activity
Adhesion molecules
interaction molecules

FIGURE 3 Determinant spreading: two superassemblies and two subassemblies are shown. The superassemblies involve the organization and interaction of several subassemblies.

TABLE 1 Manual of Immune System.

Superassemblies and Subassemblies	Kit Units (Golems)
Antigen Processing	Protection of the MHC molecule Enzymes: cathepsins, endopeptidases, exopeptidases Intracellular trafficking pH of endocytic vesicles
Intercellular Regulation	Cytokine polarization V Region interactions Suppression
Intracellular Regulation	Phosphorylation events Receptor aggregation Cytoskeletal events Rafts and the immune synapse
Generation of Repertoire	Positive selection Negative selection Unusual T-cell repertoires: CD1, $\gamma\delta$ Unusual B-cell repertoires: CD5

action between the self-release portion and the MHC also varies between MHC molecules. Further along, there is also considerable variance in the molecular interaction of the invariant DM molecule and the variable MHC molecules, leading to a broad range of interaction energies. DO, which is found largely in B cells, and not elsewhere, modulates the activity of DM through its interactions with the latter. One of the critical elements that holds the MHC molecule together is the peptide which it binds, and the DM/DO pair of chaperones also edits the final selection of peptides so that the best fitting ones win the competition for emergence to the cell surface. In summary, the supposedly simple act of protecting the MHC molecule and guiding it through to its meeting place with antigenic determinants requires an enormous amount of cooperation among different structures, which vary from MHC haplotype to haplotype. It is no wonder that a golemic system such as exists in the IS, is so unpredictable.

3.1.3 Immunodominance and the Vagaries of Antigen Processing. To reiterate, and to put the problem in perspective, when asking the "dominance question" —what determinants will emerge as victorious in the competition among those on a particular antigen, in a particular human subject—the answer cannot easily be predicted by any combination of arguments from first principles. The choice of favored determinant is essentially aleatory and requires empirical test, and even then, the given environmental circumstances may alter the results between laboratories or within the same laboratory.

Let us examine the antigen processing subassemblies more closely, as shown in figure 4. As discussed above, the protection of the MHC in all its aspects might be too coordinated a task for a single Golem, and in trying to analyze the system, questions will arise as to whether Golems should be restricted to tasks which are useful, but dispensable in the system as a whole. Knockout experiments have demonstrated that Ii, DM, and even Cathepsin S loss is nonlethal. One organizational possibility is shown in figure 4, creating separate kits (1) and (2) for the job of MHC chaperoning. The major point is that each of these Golems operate in an area which is a source of unpredictability. For example, Golem (1) in certain cells operates without DO, and in all cells has to deal with the variations (a) in DM binding affinity to the wide genetic disparity of MHC class II molecules, and (b) the tendency for different Ii molecules in the "self-release region" adjoining CLIP to self-dissociate as well as (c) the differences in intrinsic affinity for CLIP among class II molecules. Each of these variants within the population presumably has given certain organisms a decided advantage in counteracting particular infections, or avoiding certain autoimmune diseases. Thus, the golemic/soft-wired organization rather than a static, hard-wired immune apparatus appears to guarantee survival of the species.

3.2 DETAILED GOLEMIC EXAMPLES IN THE IMMUNE SYSTEM, INVOLVING PROCESSING AND REGULATORY SUBASSEMBLIES

We will explore several experimental systems in detail which provide illustrative examples of the types of unpredictability evident in everyday immunologic situations. Although most murine models involve homozygous mice, with a minimal variety of MHC molecules, in the wild most individuals are heterozygous at many genetic loci, especially humans.

3.2.1 F1 vs. Parental Responses. Thus, we undertook to examine the detailed heterogeneity of HEL (hen eggwhite lysozyme) response patterns in F1 mouse strains compared to the parental strains. Only responses to dominant and subdominant determinants were studied. Dominance is a complex and competitive function of the relative affinity and availability of different determinants on the antigen, as well as including a T-cell repertoire component. The pattern of dominance of HEL determinants for CD4 T cells is different in each mouse H-2 haplotype, and each determinant on the molecule can be dominant or cryptic in one strain or other, dependent on the nature of the class II molecules. In the F1 experiments, the pattern of recall responses among a set of seven HEL determinants was examined and classified nine days after priming mice in the hind footpad with HEL [6]. Thus, in table 2, BALB/c mice recall a response uniquely to p106–116, while B6 mice have three different patterns of response

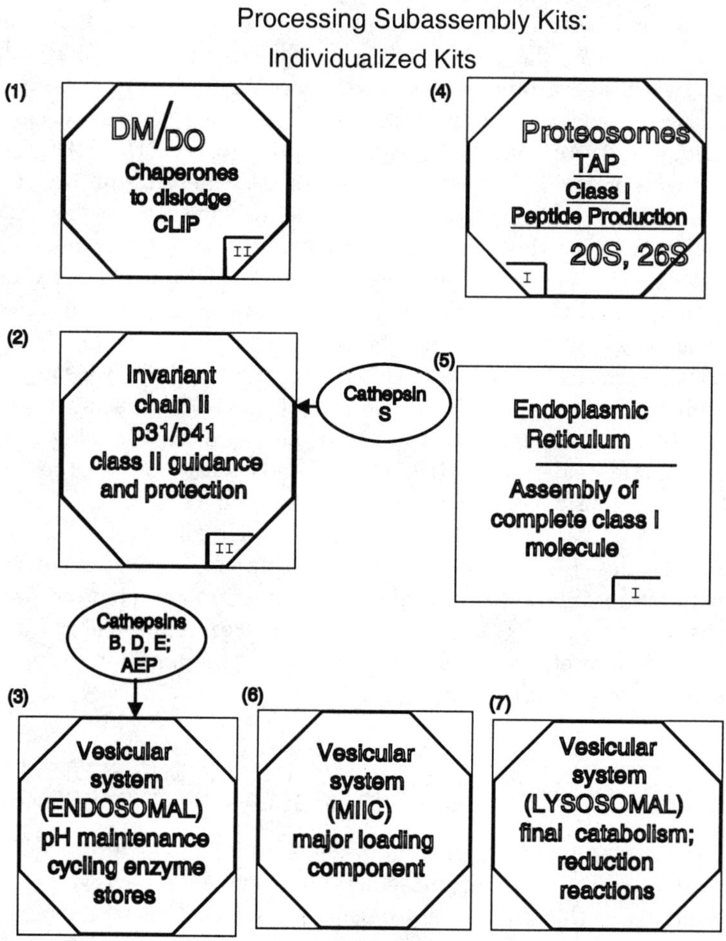

FIGURE 4 Golems responsible for certain kits would tend to be involved with other Golems (for example, kits 1 and 2).

among the 16 mice tested. However, the F1 mice display 6 different patterns among the 24 mice tested, with 14 responding like the BALB/c parent and none exactly like the B6 parent. This increased heterogeneity was seen in each of the three F1 mouse strains that were examined.

Another interesting feature was the loss or gain of response to HEL determinants in F1 mice. In considering HEL peptide 30–53 (table 3), an overwhelming 18 of 21 F1 mice tested lacked a response to this peptide, while the parental strain mice were almost all good responders! Furthermore, in a small

TABLE 2 Pattern of Responses to HEL of Parental and F1 Mice.

Strain	# Patterns	# in each pattern
B6	3	12, 2, 1
BALB/c	1	20
(B6 x BALB/c)F1	6	14, 3, 2, 2, 2, 1
B6	3	13, 2, 1
CBA/J	3	12, 5, 3
(B6 x CBA/J)F1	7	9, 3, 3, 2, 2, 1, 1

TABLE 3 Loss/Gain of Response to Determinants of HEL in F1 Mice.

Loss HEL Peptide 30–53		Gain HEL peptide 11–25	
Strain	+ve/total(%)	Strain	+ve/total(%)t
B6	14/16 (87.5%)	B6	0/16 (0%)
CBA/J	17/20 (85%)	BALB.c	0/16 (%)
F1	3/21 (14.3%)	F1	3/24 (12.5%)

group of 3 mice (out of 24), a response to p11–25 was gained where neither parent had been able to respond to this peptide.

Some of the factors responsible for these effects in F1 mice illustrate the unexpected richness of the IS (see table 4). For example, an $\alpha\beta$ heterodimer from a class II allele such as I-A will be comprised of a chain from each parent, I-$A^b_\alpha A^k_b$ or I-$A^k_\alpha A^b_\beta$, which might have a unique binding specificity. Occasionally, a mixed hybrid molecule from the I-A and I-E loci, such as $Ia^b_\alpha E^k_\beta$ arises; in one case, this species represented only 1% of the class II MHC molecules, yet was immunodominant [9]. Another cause of change in response reflects a heightened expression of one set of class II MHC molecules over that of the other haplotype.

3.2.2 Determinant Capture. Determinant capture is a very important byproduct of the way in which class II molecules bind to the unfolding antigen in the

TABLE 4 Five Factors Contributing to Heterogeneity of Response in F1 Mice.

- Determinant capture
- Alterations in expression of particular MHC molecules
- Expression of "unique haplotype hybrids" in F1
- Variations in protease levels
- Alterations in T-cell repertoire

acidic vesicular endosomal compartment of the cell. The binding occurs with the earliest available determinants that possess reasonable affinity, with the antigen still essentially unfragmented [1, 3]. Thus, there will be competition among the ambient class II binding grooves for sites along the antigen and the determinants which become available earliest will be bound. The dangling ends protruding from the groove, dotted with less competitive, subdominant and cryptic determinants will be processed by available endopeptidase and exopeptidase enzymes. As more class II molecules exist in man than in mouse, there should be even greater competition among MHC molecules leading to greater unpredictability. Although determinant capture leads to the display of new dominant determinants, it also results in the abrogation of response to other determinants that are losers in the competition for binding to MHC grooves, the cryptic determinants. Actually, this can be of benefit in the case of autoimmune diseases, where an MHC molecule of a susceptible haplotype, ready to raise a response to an autoimmunogenic peptide might have that region of the molecule stolen away by a resistant individual's MHC molecule binding at high affinity to a neighboring determinant.

With an increase in enzyme activity during inflammation, differences that occur in the pattern of the first endopeptidic cleavage will lead to the revelation of new available sites where competition can occur. The comparative level of proteases within F1 APC processing compartments may also play a role in creating heterogeneity among individuals since certain proteases may more readily render particular determinants on the antigen more visible than others.

The development of the T-cell repertoire itself is an aleatory process, occurring in each individual with the genetic elements underlying variability—named N, D, and J, rearranging alongside the variable region, V, to yield a potentially enormous universe of distinct T-cell receptors. Again, the same processing forces working to create heterogeneity in the periphery influence thymic development of the repertoire, since in the thymus, the levels of negative and positive selection to each determinant are decided by the same spectrum of antigen processing and presentation variables mentioned in earlier paragraphs.

Given the large number of aleatory components involved in determinant choice—the type and level of MHC molecules, and varying constellations of TcR specificities and processing enzymes in different individuals, it would clearly be hazardous to predict the nature of the specificities emerging from a particular antigenic challenge in an F1 individual! In figure 5 is an artist's rendition of the state of affairs—each Golem has a limited repertoire of capabilities, and a unique perspective of the endproduct, but little conception of its contribution to the whole enterprise.

FIGURE 5 Golems at work. Each Golem, working on its own kit within a subassembly has no clue as to the finished product. This is represented as a different vision of the final automobile product within the "minds" of the different Golems.

TABLE 5 An encounter between the Processing Golem and the Dominance Golem.

Starting Immunogen	Response to HEL Peptide 74–88	81–93
HEL (1–129)	3+	0
HEL (13–105)	0	3+

4 ENCOUNTERS BETWEEN GOLEMS

4.1 THE PROCESSING GOLEM AND THE DOMINANCE GOLEM

The Processing Golem, as we have seen, is a major contributor of unpredictability. This was first seen many years ago in the study of the resultant dominant determinant that emerges when different starting materials are used as immunogens. In this example of an encounter between the Processing Golem and the Dominance Golem, shown in table 5, HEL itself was compared to the large fragment HEL (13–105) in their abilities to favor determinants within the region aa. 74–96. The two overlapping peptides studied were 74–88 and 81–93, each restricted by I-A^b. In this case, the starting molecule absolutely dictated the final outcome, as is visible from the table. Presumably, once a molecule starts to be catabolized in one direction, certain routes of antigen processing are prevented, perhaps by steric hindrance of enzymatic sites by the rest of the molecule, and the final end products that appear on the cell surface in an MHC context, are distinct.

4.2 THE PROCESSING GOLEM AND A REGULATORY GOLEM

The following interesting scenario could be said to represent an encounter between the Processing Golem and a Regulatory Golem. In the disease adjuvant arthritis, the injection of complete Freund's adjuvant into the RT-1^1 Lewis rat leads to a very broad response to self heat shock protein 65 (Rhsp65). Despite the very extensive responsiveness to this self protein, there was little response to the C-terminal part of Rhsp65 in the first several months following injection but toward the end of this period, a response did appear, concomitant with the cessation of the disease process. The closely related WKY RT-1^1 strain (of the same MHC haplotype) is resistant to this form of arthritis, and remarkably, it makes an early response to the C-terminal determinants (CTD). We postulated that the CTD were regulatory determinants in Rhsp65 and injection of these peptides into the Lewis rat protected it from arthritis! It could be argued that processing of the CTD was considerably delayed in the Lewis and that this was the root cause of the susceptibility/resistance decision [5]. A third, related strain—the Fischer F344 rat with a slightly different MHC molecule, had a third phenotype. This rat arrived at our facility susceptible to disease, but acquired resistance during residence for several weeks in our animal house. If these rats were kept on antibiotic-containing, acidified water, they did not acquire resistance. To summarize the Fischer story [7], apparently a response is induced by the microbial colonizer which cross-reacts with the regulatory determinant. Another possibility is that a microbial enzyme may have substituted for the processing enzyme able to free up the CTD. Again, it appears that in a rather unpredictable way, the functioning of separate systems leads to the happenstance that resistance to arthritis is achieved. It is worth pointing out that the initial response being at the amino and central portions of the hsp65 molecule, and spreading to include the C-terminus is the first example of determinant spreading leading to protection from disease rather than its exacerbation.

5 CONCLUDING COMMENTS

5.1 A DEGREE OF ANARCHY

These few examples illustrate (perhaps inexactly and in a characteristically golemic fashion) the major point about the unpredictability of many features of immune responsiveness. It is the thesis of this chapter that this aleatory essence is an evolutionary outcome of the distributed organization of the immune system and the quasi-autonomous nature of its modular subassemblies. Within each subassembly, clever gadgetry and beautifully evolved mechanisms can accomplish many goals—no one has suggested that Golems are inadequate at what they do.

5.2 ALEATORY AUTOIMMUNE DISEASE

As we learn more of the immune system, it might become possible with a little information, to avoid such outcomes as autoimmune disease. Environmental factors and careful human control can surely diminish heart disease, stroke, and many cancers. But contracting certain diseases such as diabetes or lupus, within genetic constraints, appears to be owing to a succession of "bad breaks." First there is an encounter with an initiating agent (e.g., a virus or excessive UV light from the sun), followed by the appearance of a dominant determinant on or aroused by that agent, the finding of a suitable self-reactive T/B cell in the body which has not been rendered tolerant, the propagation of the inflammatory focus, and overlying it all, the lack of sufficient regulation to contain the broadening response. This succession of untoward events merely is the price one pays for a system which owing to its golemic nature, is supremely flexible, permitting survival of many species over extended periods of time.

5.3 ORGANIZATION OF IMMUNE TRAFFIC

Finally, we should mention the invisible regulatory elements that tie the system together. In the IS, the question of "place" has recently excited much interest—by what mixture of homing receptors, chemokines and their receptors, activation markers, do cells address themselves to wandering through the tissues, and related questions. An intricate and distributed system such as the IS can appear very disordered: e.g., consider Grand Central Station in New York at 5:00 P.M.on a Friday, where a visitor from another planet would be hard-pressed to sense any organizing principle or to find someone. A little bit of information, two coordinates such as place and time, (e.g., Track 29 at 5:30 P.M.) establishes the basis for an unambiguous contact. Once locale has been established, then identification through receptor recognition serves to provide a means to organize the various distributed threads of the IS. Clusters of APC with attached effectors and targets describe one such tactic; another is bystander suppression, where the effector molecules produced by regulatory T cells also affect bystanders in the immediate environment.

5.4 REDUNDANCY OF REGULATORY DEVICES AND THE GODS

Although these types of modulation may seem somewhat hard-wired, they actually display the chance elements of a typical golemic "tinkering" solution. Because of the often unpredictable nature of determinant choice and T cell selection, and owing to the distributed nature of the system, it may be difficult for the IS to insure reproducible regulation. This probably has led to considerable redundancy in the evolution and establishment of regulatory devices. This will no doubt be one major focus of a new generation of immunologists: for now, we can leave the regulatory details to the GODS—the Golem(s) of Distributed Systems (see fig. 6).

FIGURE 6 The Golem of Distributed Systems is represented here as making a decision about the interactive requirements among several golems.

ACKNOWLEDGMENTS

I would like to thank Rabyn Blake, M. F. A. and Stephen S. Wilson, Ph.D. for again bringing these golems to life.

REFERENCES

[1] Castellino F., F. Zappacosta, J. E. Coligan, and R. N. Germain. "Large Protein Fragments as Substrates for Endocytic Antigen Capture by MHC Class II Molecules." *J. Immunol.* **161** (1998): 4048–4057.

[2] Cohen, I. R. *Tending Adam's Garden: Evolving the Cognitive Immune Self*, 288. San Diego, CA: Academic Press, 2000.

[3] Deng, H-K., R. Apple, M. Clare-Salzler, S. Trembleau, D. Mathis, L. Adorini, and E. E. Sercarz. "Determinant Capture as a Mechanism of Protection Afforded by MHC Class II Molecules in Insulin-dependent Diabetes Mellitus." *J. Exp. Med.* **178** (1993): 1675–1680.

[4] Lehmann, P. V., E. E. Sercarz, T. Forsthuber, C. M. Dayan, and G. Gammon. "Determinant Spreading and the Dynamics of the Autoimmune T-Cell Repertoire." *Immunol. Today* **14** (1993): 203–208.

[5] Moudgil, K. D., T. T. Change, H. Eradat, A. M. Chen, R. S. Gupta, E. Brahn, and E. Sercarz. "Diversification of T-Cell Responses to Carboxy-Terminal Determinant within the 65-kD Heat-Shock Protein is Involved in Regulation of Autoimmune Arthritis." *J. Exp. Med.* **185** (1997): 1307–1316.

[6] Moudgil, K. D., J. Wang, V. P. Yeung, and E. E. Sercarz. "Heterogeneity of T-Cell Response to the Immunodominant Determinants within

Hen Eggwhite Lysozyme of Individual Syngeneic Hybrid F1 Mice: Implications for Autoimmunity and Infection." *J. Immunol.* **161** (1998): 6046–6053.

[7] Moudgil, K. D., E. Kim, O. J. Yun, H. H. Chi, E. Brahn, and E. E. Sercarz. "Environmental Modulation of Autoimmune Arthritis Involves the Spontaneous, Microbial Induction of T-Cell Responses to Regulatory Determinants within hsp65." *J. Immunol.* (2000): submitted.

[8] Partham, P., ed. "Determinant Spreading and Diversification of Immune Response." In *Immunological Reviews*, vol. 164. Copenhagen, Denmark: Munksgaard International Publishers, Ltd., 1998.

[9] Ruberti, G., R. S. Sellins, C. M. Hill, R. N. Germain, C. G. Fathman, and A. Livingstone. "Presentation of Antigen by Mixed Isotype Class II Molecules in Normal H-2^d Mice." *J. Exp. Med.* **175(1)** (1992): 157–162.

[10] Sercarz, E. E., P. V. Lehmann, A. Ametani, G. Benichou, A. Miller, and K. Moudgil. "Dominance and Crypticity of T-Cell Antigenic Determinants." *Ann. Rev. Immunol.* **11** (1993): 729–766.

[11] Villadangos, J. A., R. A. R. Bryant, J. Deussing, C. Driessen, A.-M. Lennon-Deménil, R. J. Riese, W. Roth, P. Saftig, G.-P. Shi, H. A. Chapman, C. Peters, and H. L. Ploegh. "Proteases Involved in MHC Class II Antigen Presentation." In *Pathways of Antigen Processing and Presentation. Immunological Reviews*, edited by P. Parham, vol. 172. Copenhagen, Denmark: Munksgaard International Publishers, Ltd., 1999.

[12] Winkler, G., and Y. Jones. *The Golem of Prague*, 356. New York: The Judaica Press, Inc., 1980.

Part IV: Biochemical Systems

New Approaches to Complex Chemical Reaction Mechanisms

John Ross
Marcel O. Vlad

The original paper appeared in *Ann. Rev. Phys. Chem.* **50** (1999): 51–61. ⟨http://www.AnnualReviews.org⟩. Reprinted by permission.

1 INTRODUCTION

Complex chemical reaction mechanisms, for example the citric acid cycle, consist of many elementary reaction steps; each step can be described by listing the reactants and products in that step, taken to occur in a single collision event. Further, each step may have catalysts as well as positive and negative effectors on these catalysts. The traditional approach to investigating complex reaction mechanisms consists of identifying as many participating species as possible (reactants, products, intermediates), isolating each elementary reaction step, and determining the stoichiometry and kinetics of each step. From these measurements and any others, such as isotope studies, reaction mechanisms are guessed by intuition, prior experience, analogies, etc., and the predictions of the hypothesized mechanism are checked with experiments. There exists no agreed-upon prescription for this approach, and hence it is not surprising that for many mechanisms discussions persist for years.

In this chapter we review several new approaches to the establishment of complex reaction mechanisms presented in the last few years. The goal of these approaches is the formation of methods for *deducing* reaction mechanisms, or perhaps only reaction pathways that show the connections among species due to reactions, from prescribed measurements. The question naturally arises whether such a deduced reaction mechanism is unique: the answer is "no." Scientific theories, models, and hypotheses, when confirmed by experiments, are thereby shown to be sufficient, never necessary, never unique.

2 OSCILLATORY REACTIONS

In oscillatory reactions, concentrations of several species (at least 2) vary periodically in time, not necessarily sinusoidally. In a closed system (no mass flows across the boundary of the system) such oscillations may occur when the system is far from equilibrium initially, but will disappear as the system approaches equilibrium. In an open system, with persistent mass flows in and out of the system, such oscillations can persist forever. There are about 150 known inorganic and organic reactions [16, 21], and about the same number of biochemical reactions [20, 35] in which oscillations of concentrations of chemical species take place for certain, but not necessarily all, conditions of external constraints, such as inflow rates of reactants into the system. Some examples of biochemical oscillatory reactions are glycolysis, calcium oscillations in signal transduction in cells, and cyclic AMP oscillations in slime mold cells [20].

Prior to presenting an approach to the deduction of reaction mechanisms from prescribed experiments, we need to introduce the concepts of stoichiometric network analysis, which is used as a mathematical tool.

3 STOICHIOMETRIC NETWORK ANALYSIS

A chemical reaction mechanism, sometimes also called a reaction network, consists of a sequence of elementary reaction steps, each one described by a stoichiometric equation listing the reactants and products of a single collision event. Stoichiometric network analysis (SNA), developed by B. L. Clarke [9], is an approach to the analysis of the nonlinear dynamics of reaction mechanisms: one objective of SNA is the determination of different possible dynamic behaviors that are possible with a given mechanism, such as chemical oscillations; another is the prediction of potential stability or instability of a reaction network, but the range of useful applications of SNA goes beyond that of stability studies.

We begin with a brief description of SNA. Consider a reaction mechanism of n species, X_i, and r reactions, R_j. The symbol X_i denotes both a given species and its concentration. A reversible reaction is written as one irreversible reaction in one direction, plus that in the other direction. The net production of species X_i in reaction R_j is given by the stoichiometric coefficient ν_{ij}.The set of stoichiometric coefficients ν_{ij} constitute the $n \times r$ stoichiometric matrix $\boldsymbol{\nu}$. The reaction order of species X_i in reaction R_j is denoted by κ_{ij}, and the set of all the elements κ_{ij} form the $n \times r$ kinetic matrix $\boldsymbol{\kappa}$. The rate of the jth reaction is

$$\nu_j(\boldsymbol{X}, k_j) = k_j \prod_{i=1}^{n} (X_i)^{\kappa_{ij}}, j = 1, \ldots, r, \tag{1}$$

with k_j being the rate coefficient of that reaction. The rates ν_j form the elements of a column vector $\boldsymbol{v}(\boldsymbol{X}, \boldsymbol{k})$; both $\boldsymbol{k}$ and $\boldsymbol{X}$ are always nonnegative, and hence the vector $\boldsymbol{v}(\boldsymbol{X}, \boldsymbol{k})$ is a point in the nonnegative orthant of the reaction velocity space. The deterministic evolution of the homogeneous reaction mechanism is given by the differential equations

$$\frac{d}{dt} X_i = \sum_{j=1}^{r} \left\{ \nu_{ij} k_j \prod_{i-1}^{n} (X_i)^{\kappa_{ij}} \right\} c, i = 1, \ldots, n, \tag{2}$$

$$\frac{d}{dt} X_i = \sum_{j=1}^{r} \{\nu_{ij} \nu_j\}, i = 1, \ldots, n, \tag{3}$$

$$\frac{d\boldsymbol{X}}{dt} = \boldsymbol{\nu v}. \tag{4}$$

At a nonequilibrium stationary state, $\boldsymbol{X}_0 = (X_{10}, \ldots, X_{n0})$ the time derivatives of all intermediates X_j vanish and we have

$$\boldsymbol{0} = \boldsymbol{\nu v}_o. \tag{5}$$

Since the kinetic equation for each intermediate equals zero at a stationary state, there are typically many solutions $\boldsymbol{v}_0$ to eq. (5). Each of these solutions is called a current; each current $\boldsymbol{v}_0$ is described by a (sub)network of reaction rates. All possible solution vectors form an open subspace, a convex current cone. The linearization of the rate equations close to stationary state yields the Jacobian matrix elements; the Jacobian matrix at the stationary state is

$$\boldsymbol{J} = \boldsymbol{\nu} \ (\text{diag}\boldsymbol{\nu}_o) \ \boldsymbol{\kappa}^{\tau} \ (\text{diag}\boldsymbol{h}) \text{ with } h_i = \frac{1}{X_{i_o}}; \tau = \text{transpose}. \tag{6}$$

A further useful concept is that of "extreme currents": any solution $\boldsymbol{v}_0$ of eq. (5) can be decomposed into a linear combination of extreme currents with nonnegative coefficients (weighting factors). The extreme currents are systematically determined: any current vector $\boldsymbol{v}_0$ is a linear combination of rates lying on the boundary of the current cone; these boundary rates are part of one or more coordinate hyperplanes. Hence, one or more components of the boundary rates are zero and, thus, the boundary rates represent a simpler subnetwork. This process is repeated until no further simplification is obtained and the remaining rates are extreme currents, which are the smallest subsets of the original reaction mechanism for which eq. (5) is fulfilled. Thus extreme currents form the edge of the current cone. For further details see Eiswirth et al. [13] and Strasser et al. [50].

Clarke showed that the stability of a reaction network can be investigated by determining the stability of the extreme current subnetworks as related to properties of subdeterminants of the Jacobian of the linearized system. Thus he provided a procedure for establishing whether an extreme network has an unstable feedback cycle and, if so, where that cycle is located. SNA provides a method of assessing a given reaction mechanism for a potential instability, say, leading to chemical oscillation, somewhere in the parameter space of the rate coefficients, that is without explicit knowledge of those coefficients.

An example of the application of SNA [50] is an analysis of the mechanism of the chlorite-iodide reaction proposed by Citri and Epstein [8]. SNA was used to determine the complete set of the major reaction pathways, the extreme currents. Two qualitatively different types of unstable extreme currents were found within this mechanism. Further the dominant extreme currents were determined for each of three stationary states and the different phases of the cycle of oscillations beyond the Hopf bifurcation. The study provided an illustration of considerable insight offered by SNA into the functioning of a reaction mechanism. The prediction of bifurcation structures requires the input of chosen values of rate coefficients and this was done in Strasser et al. [50]. Clarke and Jiang [10] have presented a method for obtaining approximate equations for Hopf and saddle-node bifurcations based on SNA, and showed the effects of adding or deleting certain elementary reactions from the mechanism on the bifurcation structures.

4 CLASSIFICATION OF CHEMICAL OSCILLATORS AND FORMULATION OF THEIR MECHANISMS BY DEDUCTION FROM EXPERIMENTS

Eiswirth, Freund, and Ross [13] have proposed a classification of oscillatory reactions based in part on SNA, but in addition the approach is closely tied to experiments. Chemical species are first identified as either essential or nonessential: if the concentration of an essential species is held constant, then oscillations in concentrations of all other species stop. If, however, the concentration of an inessential species is held constant, then oscillations of other species do not stop. In examining more than 30 oscillatory reaction mechanisms it was found that all could be classified into just two categories: the first has three subcategories, and the second two. If we specify species X as an autocatalytic or cycle species, species Y as an exit species, species Z as a negative feedback species, and species W as a recovery species, then all oscillatory reactions investigated so far can be classified as shown in figure 1. For each (sub) category there is given a skeleton of elementary reaction steps and the corresponding reaction network. The networks show the connectivity of the essential species; the number of barbs (total number of feathers) of an arrow at a product (reactant) equals the stoichiometric coefficient of this product (reactant) in the respective reaction, and the number of left feathers equals the kinetic exponent of the reactant. For example, an analysis of the Field et al. [17] mechanism of the Belusov-Zhabotinsky reaction identifies the nonessential and essential species, their connectivities, and the classification 1B (fig. 2). The categorization of enzymatic oscillatory reactions is presented in Schreiber et al. [45].

A series of experiments have been designed to achieve the identification of the species, the connectivities of the essential species and the assignment to a given classification. These experiments, with a few representative references, include.

1. Characterization of time series of oscillatory species [8, 15].
2. Relations of amplitudes of oscillatory species [22, 29].
3. Phase relations [7, 22].
4. Jacobian matrix elements [7, 39, 52].
5. Concentration shift regulation [7, 51].
6. Concentration shift destabilization [7, 13].
7. Pulsed species response [4, 7, 39, 47, 51, 52].
8. Time delay experiments [6, 14].
9. Quenching experiments [31, 32, 46, 55].
10. Periodic phase response experiments [12, 18, 44].
11. External periodic perturbations [12, 34, 41, 53].
12. Nonlinear bifurcation analysis [11, 42, 47].

For a discussion of all of these experiments see Stemwedel et al. [48]. In general each experiment yields different information; however, near a Hopf bifurcation, experiments [8, 9, 13, 50] yield the same information [54]. We describe next the application of only a few experiments.

Nonessential species are recognized by smaller relative amplitudes of oscillations compared to those of essential species, and for temporary quenching of oscillation, they require larger additions of the species (larger quench vectors). On periodic external perturbation of a species there occur entrainment bands for certain frequency ranges of the perturbation and the response of the species in the reaction system is phase shifted with respect to the periodic perturbation. The total change in the phase shift with respect to a sinusoidal perturbation in a given species from one edge of the fundamental entrainment band to the other is π for essential species, but smaller than that for all nonessential species. There are other tests for distinguishing essential from nonessential species, but the examples suffice the operational distinction, that is, based on experiments [13].

The identification and connectivity of essential species can be deduced from several experiments (numbers 3–9 listed above) and we illustrate with two examples. The first consists of a listing of the phase shifts of the oscillation of species j with respect to species i (see table 1), from Schreiber et al. [45],

Category 1B:

$(\ ) + X \rightarrow 2X + 2Z$

$Z \rightarrow Y$

$Y \rightarrow X$

$X + Y \rightarrow (\ )$

$2X \rightarrow (\ )$

$\rightarrow Y$

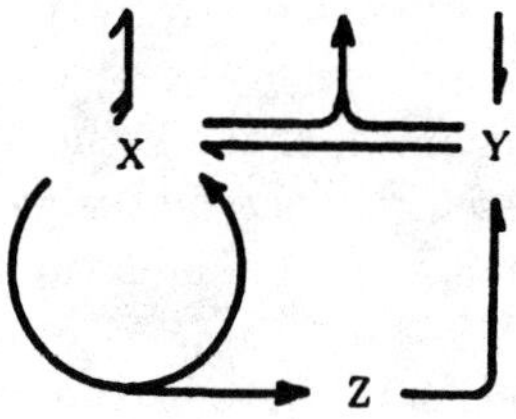

Category 1CX:

$Z + X \rightarrow 2X$

$X + Y \rightarrow (\ )$

$\rightarrow X$

$2X \rightarrow (\ )$

$\rightleftharpoons Y$

$\rightleftharpoons Z$

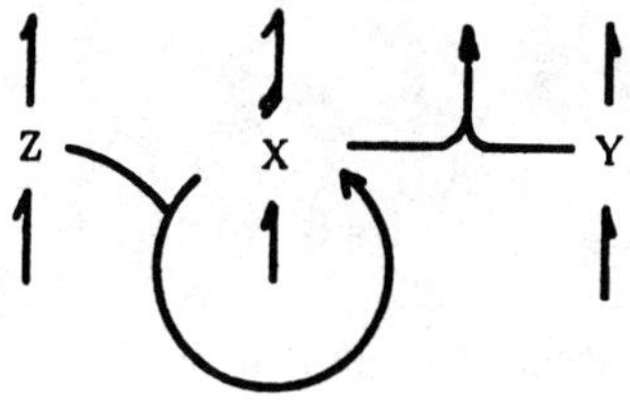

Category 1CW:

$Z + X \rightarrow 2X$

$X + Y \rightarrow 2W$

$W + Y \rightarrow (\ )$

$Y \rightarrow X$

$2X \rightarrow (\ )$

$\rightarrow Y$

$\rightleftharpoons Z$

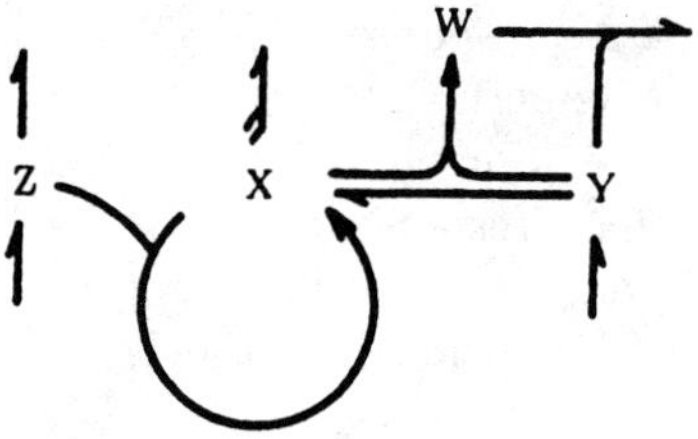

Category 2B:

$(\ ) + 2X \rightarrow 3X + Z$

$Z + X \rightarrow (\ )$

$\rightleftharpoons X$

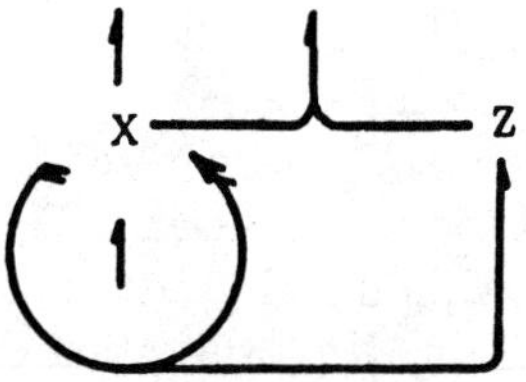

Category 2C:

$Z + 2X \rightarrow 3X$

$\rightleftharpoons X$

$\rightleftharpoons Z$

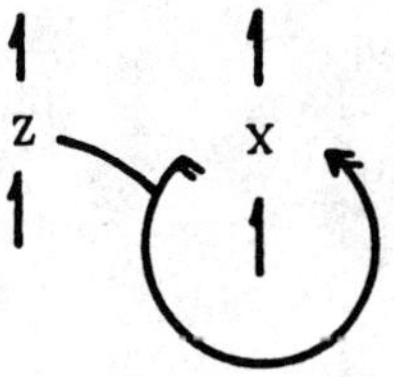

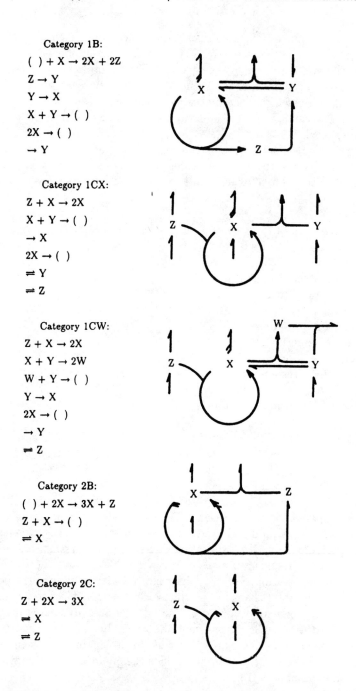

FIGURE 1 List of reactions and corresponding network diagrams of prototypes of basic categories and subcategories. From Eiswirth et al. [13] and Schreiber et al. [45]. Printed by permission of ACS Publications.

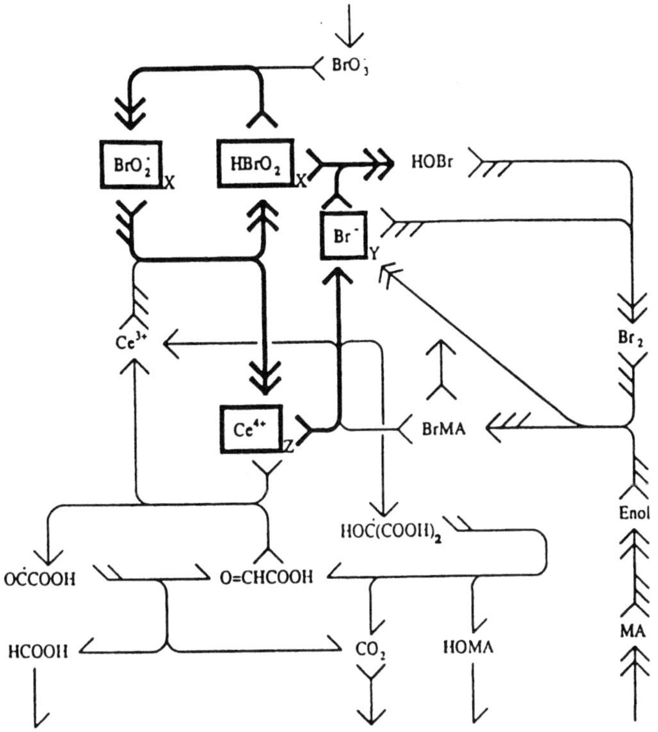

FIGURE 2 (*Boxes*) The essential species; (*bold lines*) their reactions drawn in bold lines. The indices of the boxes denote the roles of the species. (*Vertical arrows*) Flows are not distinguished from reactions to (formation by) otherwise inert products (reactants). From Eiswirth et al. [13].

the top entries in each category. The symbols I, A, $+$, $-$, denote in-phase, antiphase, advanced, and delayed oscillations, respectively. Species i is listed in the first column and species j in the top row. The entries are different for the various categories and, thus, measurements of phase shifts lead to identification of essential species and the classification of the oscillatory reaction. The second example consists of a listing of the sign ($+$, increase; $-$, decrease) of the change in the stationary concentration of species i upon an increase in the inflow concentration of species j; see the second set of entries in each category in table 1. Again the entries differ and, thus, measurements of concentration shift regulation (item 5 above) lead to identification and connectivity of essential species, and hence to the appropriate classification.

The operational approach to the formulation of oscillatory reactions outlined here has been tested on two chemical systems. In the chlorite-iodide reaction Stemwedel and Ross [47, 49] made measurements on two species, and used

TABLE 1 Symbolic phase shifts $\Delta\Phi_{ij}^{\text{symb}}$ and sign symbolic concentration shifts, $\text{Sign}(\partial x_{i,j}/\partial x_{0,j})$ of prototypes of the categories 1B, 1CX, 2B, 2C. The symbols I, A, $+$, and $-$ represent the phase shifts which correspond to in-phase, antiphase, advanced, and delayed oscillations, respectively, of species j with respect to species i. The symbols $+$ and $-$ represent concentration shifts which correspond to an increase and decrease, respectively, in steady-state concentration of species i upon an increase in inflow of species j. Notice that the shifts for corresponding species in 1B and 2B (1CX and 2C, respectively) are the same. The corresponding species in category 1CW have the same shift behavior as in 1CX, and the additional species W has the same shift behavior as species X.

Category 1B

i	j	X	Y	Z
X		I	$-$	$-$
Y	$\Delta\Phi_{ij}^{\text{symb.}}$	$+$	I	$+$
Z		$+$	$-$	I
X		$+$	$-$	$-$
Y	$\text{Sign}(\partial x_{s,i}/\partial x_{0,j})$	$+$	$+$	$+$
Z		$+$	$-$	$-$

Category 1CX

i	j	X	Y	Z
X		I	A	$+$
Y	$\Delta\Phi_{ij}^{\text{symb.}}$	A	I	$-$
Z		$-$	$+$	I
X		$+$	$-$	$+$
Y	$\text{Sign}(\partial x_{s,i}/\partial x_{0,j})$	$-$	$+$	$-$
Z		$-$	$+$	$-$

Category 2B

i	j	Y	Z
X		I	$-$
Z	$\Delta\Phi_{ij}^{\text{symb.}}$	$+$	I
X		$+$	$-$
Z	$\text{Sign}(\partial x_{s,i}/\partial x_{0,j})$	$+$	$-$

Category 2C

i	j	X	Z
X		I	$+$
Z	$\Delta\Phi_{ij}^{\text{symb.}}$	$-$	I
X		$+$	$+$
Z	$\text{Sign}(\partial x_{s,i}/\partial x_{0,j})$	$-$	$-$

measurements from Citri and Epstein [8], to obtain phase relations, concentration shift destabilization, and parts of a bifurcation diagram to determine the essential and nonessential species, and the reaction category; they thus established by deduction the basic reaction network hypothesized in Epstein [8]. In the peroxidase-oxidase reaction Hung and Ross made measurements, on four species, of relative amplitudes, relative phase shifts, concentration shift regulation and destabilization, qualitative pulsed species response, and quench vectors. Again essential and nonessential species were identified and parts of this complex reaction mechanism were deduced from these measurements [30].

Measurements near a Hopf bifurcation of the amplitude of an applied concentration change, and the phase of oscillation at the instant of application to achieve a temporary quenching of the oscillation yield information about matrix elements of the transpose of the Jacobian. Hynne and Sorenson have developed the theory and reported measurements [31, 46, 55]. Further they have outlined a method of combining quenching measurements and representations of stationary states by extreme current diagrams to optimize sets of rate coefficients for a given reaction, and they applied this approach to the BZ reaction [32].

5 CORRELATION METRIC CONSTRUCTION OF REACTION PATHWAYS FROM MEASUREMENTS

The origins of this work may be of interest. The implementation of logic gates, sequential (universal Turing) computing machines, and parallel machines by means of macroscopic kinetics has been developed and discussed in a number of articles [5, 23, 24, 25, 26, 27, 28, 33, 36, 37, 56]. From these studies came the question whether computational functions can be recognized in established reaction networks, such as glycolysis. Arkin and Ross [1] found that a part of glycolysis, that involving fructose-6-phosphate and the fructose 1,6 and 2,6 biphosphates, functions as a fuzzy logic AND gate which controls the switch from glycolysis to gluconeogenesis.

If, then, computational functions are built into macroscopic reaction pathways, there may be available new approaches to the determination of such pathways, approaches used in circuit theory, system analysis, multivariate statistics, etc. We outline briefly first the theory [2] of such a new approach, based on the statistical construction of reaction pathways from measured time series of concentrations of chemical species, and derived concentration correlations.

Consider a chemical reaction in an open system at a nonequilibrium stationary state and the concentrations of j species can be measured as a function of time. Choose a subset of species, say two, and perturb the system away from the stationary state by additions of random amounts of the two chosen species. The temporal response of all species is measured as the system relaxes back to the stationary state. Prior to return to the stationary state another perturbation is applied and this process is repeated many times. The measured time series are then used to form correlation functions, for example, the correlation of the concentration of species I, $x(t)$ at time t, minus its average over all measurements, with the concentration of species j at time $t+\tau$

$$S_{ij}(\tau) = \langle (x_i(t) - \langle x_i \rangle)(x_j(t+\tau) - \langle x_j \rangle) \rangle \,, \tag{7}$$

where the interval τ can be positive, negative, or zero. We introduce the reduced correlation $r_{ij}(\tau)$

$$r_{ij}(\tau) = \frac{S_{ij}(\tau)}{\sqrt{S_{ii}(\tau)S_{jj}(\tau)}} \,, \tag{8}$$

and its maximum value

$$c_{ij} = \max |r_{ij}(\tau)|_\tau \,; \tag{9}$$

we define a distance

$$d_{ij} = (c_{ii} + c_{jj} - 2c_{ij})^{1/2} = \sqrt{2}(1 - c_{ij})^{1/2} \,, \tag{10}$$

and thereby convert the matrix of correlations, with elements S_{ij}, into a Euclidian distance matrix. We thus have $j(j-1)/2$ distances (sticks), each with the numbers of two species marked on it, say, 7 on one end of a stick and 3 on the other. If the correlation between two species is small, then the stick is large, ($\sqrt{2}$ for zero correlation), and if the correlation is large then the stick is small (zero length for two fully correlated species). With these sticks, and we may have to deal only with the most importantly correlated species, we build an object: all sticks with the number one on one end of a stick must be at one point, and similarly for all the other numbers. To build the object we require a multidimensional space; mathematically this is accomplished with multidimensional scaling analysis with which we find the eigenvalues and eigenvectors of the centered inner product matrix. The method also gives the reliability of projections of this object unto lower dimensions, say, a plane for easy visualization. For several model systems studied, this optimized projection printed out the reaction pathway! For reaction mechanisms which have subsets of elementary reactions with different time scales the correlation metric construction as described shows a clear separation of the chemical species into those subsets. As a corollary, this approach also readily indicates whether there is a rate-determining step in the reaction mechanism, or whether the timescales of the elementary steps are approximately all the same, so that perhaps a stationary state hypothesis may be applied. From the correlation metric analysis there can be constructed a hierarchically clustered dendogram, which gives a good representation of the flow of control from the randomly perturbed input concentrations to the remaining species.

The theory of correlation metric construction has been tested against experiments, by Arkin, Shen, and Ross [3], on a part of glycolysis shown in figure 3. The experiment was run in a continuous-flow stirred reactor; the

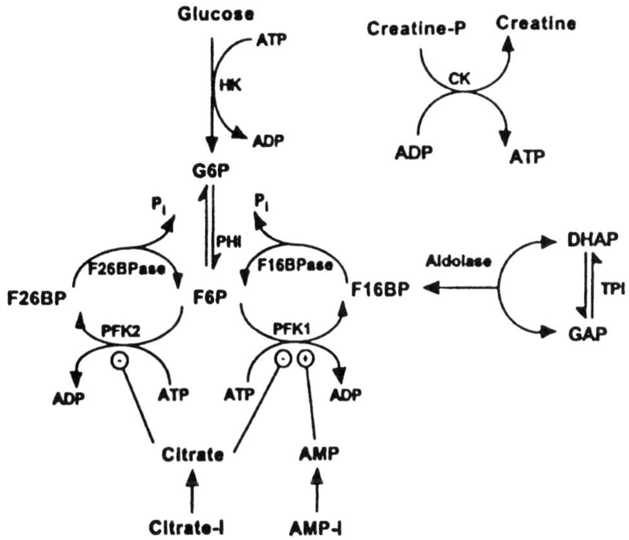

FIGURE 3 The first few reaction steps of glycolysis. Regulatory interaction: (*minus signs*), a negative effector; (*plus sign*), a positive effector. Creatine-P and CK keep the concentrations of ATP and ADP constant. Pi, Inorganic phosphate; HK, hexokinase; G6P, glucose-6-phosphate; PHI, phosphoglucose isomerase; F6P, fructose-6-phosphate; F26BPase, fructose-2,6-biphosphatase; F26BP, fructose-2,6-biphosphate; PFK, phosphofructokinase; TPI, triose-phosphate isomerase; GAP, glyceraldehyde phosphate.

enzymes were confined to the reactor by a membrane. The outflow from the reactor was sampled periodically and analyzed quantitatively with capillary electrophoresis. The input concentrations of citrate and AMP were varied randomly over time to perturb the system away from its nonequilibium stationary state. The concentration of citrate and AMP, as well as the responses in the concentrations of six other metabolites were measured in time. From these measurements, two-species time-lagged correlation functions were calculated. Multidimensional scaling analysis and heuristic arguments derived from the time dependence of the correlation functions yielded the two-dimensional projection of the MDS diagram shown in figure 4(a), and from that diagram was deduced the reaction pathway, in figure 4(b). Both the MDS diagram and the predicted new reaction pathway resemble closely the classically determined reaction pathway, in figure 3. The connectivities of the species are predicted correctly, and so are the locations of the positive and negative effectors on enzymes.

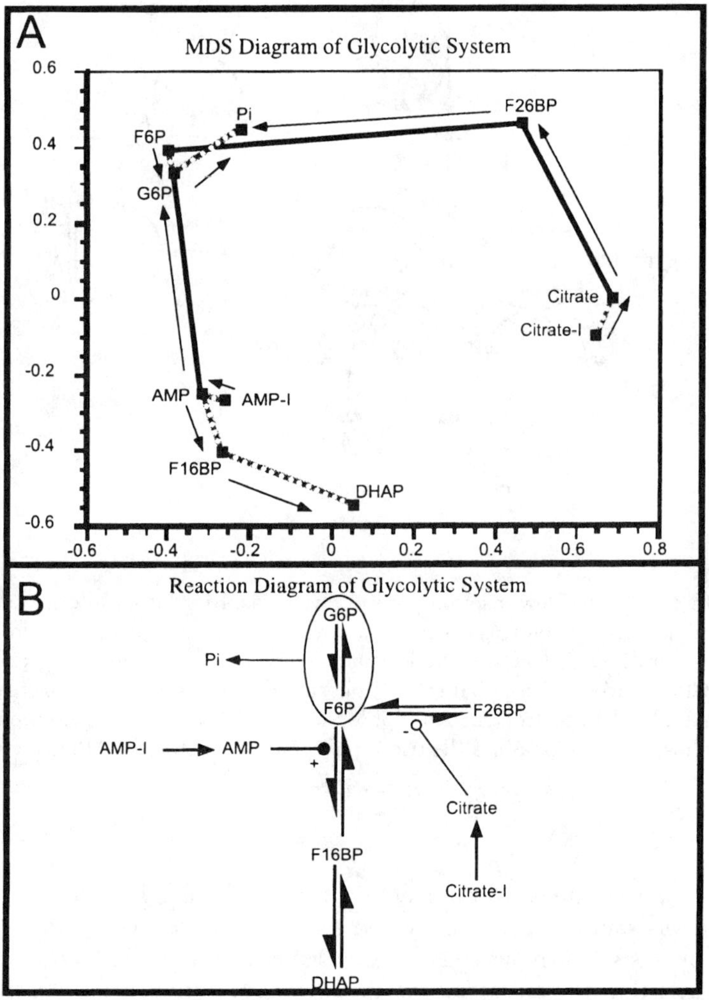

FIGURE 4 (A) The two-dimensional projection of the multidimensional scaling diagram for the measured time series. Each point represents the time series of a single species. The closer the two points are, the higher the correlation between the respective time series. Negative (*black lines*) and positive (*dotted lines*) correlation between the respective species. (*Arrows*) temporal ordering among species based on the lagged correlations between their time series. (B) The predicted reaction pathway derived from the correlation metric construction diagram. Its correspondence to the known mechanism (fig. 3) is high. From Arkin, A., P. Shen, and J. Ross. "A Test Case of Correlation Metric Construction of a Reaction Pathway from Measurements." *Science* **77** (1997): 1275–1279. Reprinted with permission. Copyright 1997, American Association for the Advancement of Science.

6 USE OF GENETIC ALGORITHMS TO OPTIMIZE A REACTION MECHANISM

Suppose we have some information available on a given reaction mechanism: we know some but perhaps not all species participating; we know some but not all elementary steps and perhaps some rate coefficients. We seek a parallel computational procedure for the construction of a reaction mechanism with rate coefficients and enzymatic activities, that fulfills some imposed requirement; for example, the mechanism should predict chemical oscillations of some species. Genetic algorithms are methods of achieving stated goals in an optimal manner. The method is mathematical and its structure has some similarities to biological evolution, hence its name. In the first study to chemical reaction theory based on this approach, Gilman and Ross [19] started with a given reaction mechanism and sought by GA techniques to maximize the achievement of a task assigned to this mechanism by systematic variation of the activity of four enzymes. They used a genetic algorithm to optimize kinetic parameters, rate coefficients and efficiencies of enzymes, in a model of a futile metabolic cycle in which two metabolites F and T are interconverted by two irreversible enzymatic reactions. Each metabolite is connected to its own reservoir by a simple chemical reaction (fig. 5). The optimization criterion chosen for this system is an integrated measure of system performance. Energy cost is included, but the most important feature of this measure is the desirability of the appropriate direction of net flux, toward F and T, respectively, given an imposed variation of the concentrations of these two reservoirs around their assigned optimal concentrations. The reaction mechanism for this task is an idealization of an animal cell that metabolizes glucose for energy as well as maintains a sufficiently high glucose level F in the blood, but synthesizes glucose for export into the blood stream if the concentration of glucose there drops too low. The best performing reaction mechanisms selected by the genetic algorithm have sets of kinetic parameters for which the reaction mechanism switches the direction of the net flux from F to T, or the reverse, according to need. The mechanisms thus selected for the desired control all have the intuitively expected features of negative feedback and reciprocal regulation, purely as a consequence of the specification of the task of flux direction. These features were not put in; they evolved as a result of the genetic algorithm.

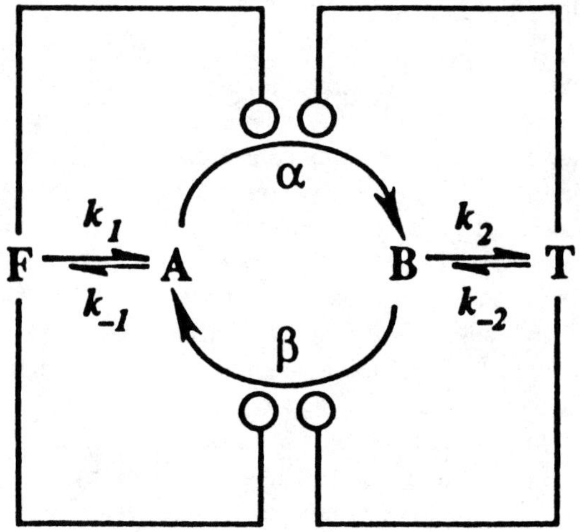

FIGURE 5 Diagram of the model. F and T are reservoir species. A and B are reaction intermediates, interconverted by enzymes α and β. (*Arrows*) reactions, (*knobs*) regulation. From Gilman and Ross [19]. Reprinted by permission of Biophysical Society.

REFERENCES

[1] Arkin, A, and J. Ross. "Computational Functions in Biochemical Reaction Networks." *Biophys. J.* **67** (1994): 560–578.

[2] Arkin, A, and J. Ross. "Statistical Construction of Chemical Reaction Mechanisms from Measured Time Series." *J. Phys. Chem.* **99** (1995): 970–979.

[3] Arkin, A., P. Shen, and J. Ross. "A Test Case of Correlation Metric Construction of a Reaction Pathway from Measurements." *Science* **277** (1997): 1275–1279.

[4] Bar-Eli, K. "Coupling of Identical Chemical Oscillators." *J. Phys. Chem.* **94** (1983): 2368–2374.

[5] Blittersdorf, R., J. Müller, and F. W. Schneider. "Chemical Visualization of Boolean Functions: A Simple Chemical Computer 1995." *J. Chem. Educ.* **72** (1995): 760–763.

[6] Chevalier, T., A. Freund, and J. Ross. "The Effects of a Nonlinear Delayed Feedback on a Chemical Reaction." *J. Chem. Phys.* **95** (1991): 308–316.

[7] Chevalier, T., I. Schreiber, and J. Ross. "Towards a Systematic Determination of Complex Reaction Mechanisms." *J. Phys. Chem.* **97** (1993): 6776–6787.

[8] Citri, O., and I. R. Epstein. "Dynamical Behavior of the Chlorite-Iodide Reaction. A Simple Mechanism." *J. Phys. Chem.* **91** (1987): 6034–6042.

[9] Clarke, B. L. "Stability of Complex Reaction Networks." *Adv. Chem. Phys.* **43** (1983): 1–216. Also in *Chemical Applications of Topology and Graph Theory*, edited by R. B. Kind, 322–357. Amsterdam: Elsevier, 1980.

[10] Clarke, B. L, and W. Jiang. "Method for Deriving Hopf and Saddle Bifurcation Hypersurfaces and Application to a Model of the B-Z System." *J. Chem. Phys.* **99** (1993): 4464–4478.

[11] DeKepper, P., J. Boissonade, and I. R. Epstein. "Chlorite-Iodide Reaction, a Versatile System for the Study of Nonlinear Dynamic Behavior." **94** (1990): 6525–6536.

[12] Dolnik, M., and M. Marek. "Phase Excitation Curves of Forced Excitable Reaction Systems." *J. Phys. Chem.* **95** (1991): 7267–7272.

[13] Eiswirth, M., A. Freund, and J. Ross. "Mechanistic Classification of Chemical Oscillators and the Role of Species." *Adv. Chem. Phys.* **53** (1991): 127–199.

[14] Epstein, I. R. "Difference Delay Equations in Chemical Kinetics: Some Simple Linear Model Systems." *J. Chem. Phys.* **92** (1990): 1702–1712.

[15] Epstein, I. R., and K. Kustin. "A Mechanism of Dynamical Behavior in the Oscillatory Chlorite-Iodide Rreaction." *J. Phys. Chem.* **89** (1985): 2275–2282.

[16] Epstein, I. R., and K. Showalter. "Nonlinear Chemical Dynamics: Oscillations, Patterns and Chaos." *J. Phys. Chem.* **100** (1996): 13132–13147.

[17] Field, J. R., E. Körös, and R. M. Noyes. "Oscillations in Chemical Systems II. Thorough Analysis of Temporal Oscillation of the Bromate-Cerium-Malonic Acid System." *J. Amer. Chem. Soc.* **94** (1972): 8649–8664.

[18] Finkeova, J., M. Dolnik, B. Hrudka, and M. Marek. "Excitable Chemical Reaction Systems in a CSTR." *J. Phys. Chem.* **94** (1990): 4110–4115.

[19] Gilman, A., and J. Ross. "Genetic-Algorithm Selection of a Regulatory Structure that Directs Flux in a Simple Metabolic Model." *Biophys. J.* **69** (1995): 1321–1333.

[20] Goldbeter, A. *Biochemical Oscillations and Cellular Rhythms: The Molecular Bases of Periodic and Chaotic Behaviour.* Cambridge, MA: Cambridge University Press, 1997.

[21] Gray, P., and S. K. Scotts. *Chemical Oscillations and Instabilities.* Oxford: Oxford University Press, 1990.

[22] Hess, B., and A. Boiteux. "Oscillatory Phenomena in Biochemistry." *Ann. Rev. Biochem.* **40** (1971): 237–258.

[23] Hjelmfelt, A., E. D. Weinberger, and J. Ross. "Chemical Implementation of Neural Networks and Turing Machines." *PNAS* **88** (1991): 10983–10987.

[24] Hjelmfelt, A., E. D. Weinberger, and J. Ross. "Chemical Implementation of Finite-State Machines." *PNAS* **89** (1992): 383–387.

[25] Hjelmfelt, A, and J. Ross. "Chemical Implementation and Thermodynamics of Collective Neural Networks." *PNAS* **89** (1992): 388–391.

[26] Hjelmfelt, A., F. W. Schneider, and J. Ross. "Pattern Recognition in Coupled Chemical Kinetic Systems." *Science* **260** (1993): 335–337.

[27] Hjelmfelt, A., and J. Ross. "Mass Coupled Chemical Systems with Computational Properties." *J. Phys. Chem.* **97** (1993): 7988–7992.

[28] Hjelmfelt, A., and J. Ross. "Pattern Recognition, Chaos, and Multiplicity in Neural Networks of Excitable Systems." *PNAS* **91** (1994): 63–67.

[29] Hung, Y. F., and J. Ross. "Further Experimental Studies on the Horseradish Peroxidase-Oxidase Reaction." *J. Phys. Chem.* **96** (1992): 7338–7342.

[30] Hung, Y. F., and J. Ross. "New Experimental Methods Toward the Deduction of the Mechanism of the Oscillatory Peroxidase-Oxidase Reaction." *J. Phys. Chem.* **99** (1995): 1974–1979.

[31] Hynne, F., and P. G. Srensen. "Quenching of Chemical Oscillations." *J. Phys. Chem.* **91** (1987): 6573–6575.

[32] Hynne, F., P. G. Sørensen, and T. Møller. "Current and Eigenvector Analyses of Chemical Reaction Networks at Hopf Bifurcation." *J. Chem. Phys.* **98** (1992): 211–218.

[33] Laplante, J. P., M., Pemberton, A. Hjelmfelt, and J. Ross. "Experiments on Pattern Recognition by Chemical Kinetics." *J. Phys. Chem.* **99** (1995): 10063–10065.

[34] Lazar, J. G., and J. Ross. "Changes in Mean Concentrations, Phase Shifts, and Dissipation in a Forced Oscillatory System." *Science* **247** (1990): 189–192.

[35] Larter, R. "Oscillations and Spatial Nonuniformities in Membranes." *Chem. Rev.* **90** (1990): 355–381.

[36] Lebender, D., J. Müller, and F. W. Schneider. "Control of Chemical Chaos and Noise: A Nonlinear Neura-Net-Based Algorithm." *J. Phys. Chem.* **99** (1995): 4992–5000.

[37] Lebender, D., and F. W. Schneider. "Logical Gates using a Nonlinear Chemical Reaction." *J. Phys. Chem.* **98** (1994): 7533–7537.

[38] Lengyel, I., and I. R. Epstein. "Turing Structures in Simple Chemical Reactions." *Acc. Chem. Res.* **26** (1993): 235–240.

[39] Luo, Y., and I. R. Epstein. "Feedback Analysis of Mechanisms for Chemical Oscillators." *Adv. Chem. Phys.* **79** (1990): 269–299.

[40] Ortoleva, P. *Nonlinear Chemical Waves*. New York: John Wiley and Sons, 1992.

[41] Pugh, S. A., M. Schell, and J. Ross. "Effects of Periodic Perturbations on the Oscillatory Combustion of Acetaldehyde." *J. Chem. Phys.* **85** (1986): 868–878.
[42] Ringland, J. "Rapid Reconnaissance of a Model of a Chemical Oscillator by Numerical Continuation of a Bifurcation Feature of Codimension Two." *J. Chem. Phys.* **95** (1991): 555–562.
[43] Olson, R. J., and I. R. Epstein. "Bifurcation Analysis of Chemical Reaction Mechanisms. Steady State Bifurcation Structure." *J. Chem. Phys.* **94** (1991): 3083–3095.
[44] Ruoff, P., H. D. Forsterling, L. Gyorgi, and R. M. Noyes. "Bromous Acid Perturbations in the B-Z Reaction." *J. Phys. Chem.* **95** (1991): 9314–9320.
[45] Schreiber, I., Y. F. Hung, and J. Ross. "Categorization of Some Oscillatory Enzymatic Reactions." *J. Phys. Chem.* **100** (1996): 8556–8566.
[46] Sørensen, P. G., and F. Hynne. "Amplitudes and Phases of Small Amplitude B-Z Reaction, Derived from Quenching Experiments." *J. Phys. Chem.* **93** (1989): 5467–5474.
[47] Stemwedel J. D., and J. Ross. "Experimental Determination of Bifurcation Features of the Chlorite-Iodide Reaction." *J. Phys. Chem.* **97** (1993): 2863–2867.
[48] Stemwedel, J. D., I. Schreiber, and J. Ross. "Formulation of Oscillatory Reaction Mechanisms by Deduction from Experiments." *Adv. Chem. Phys.* **89** (1995): 327–388.
[49] Stemwedel, J., and J. Ross. "New Measurements on the Chlorite-Iodide Reaction and Deduction of Roles of Species and Categorization." *J. Phys. Chem.* **99** (1995): 1988–1994.
[50] Strasser, P., J. D. Stemwedel, and J. Ross. "Analysis of a Mechanism of the Chlorite-Iodide Reaction." *J. Phys. Chem.* **97** (1993): 2851–2862.
[51] Turány, T. "Sensitivity Analysis of Complex Kinetic Systems. Tools and Applications." *J. Math. Chem.* **5** (1990): 203–248.
[52] Tyson, J. J. "Classification of Instabilities in Chemical Reaction Systems." *J. Chem. Phys.* **62** (1975): 1010.
[53] Vance, W., and J. Ross. "Experiments on Bifurcation of Periodic States into Tori for a Periodically Forced Chemical Oscillator." *J. Chem. Phys.* **88** (1988): 5536–5546.
[54] Vance, W., and J. Ross. "Entrainment, Phase Resetting, and Quenching of Chemical Oscillations." *J. Chem. Phys.* **103** (1995): 2472–2481.
[55] Vukojevic, V., P. G. Sørensen, and F. Hynne. "Predictive Value of a Model of the Briggs-Rauscher Reaction Fitted to Quenching." *J. Phys. Chem.* **100** (1996): 17175–17196.
[56] Zeyer, K. P., G. Dechert, W. Hohmann, R. Blittersdorf, and F. W. Schneider. "Coupled Bistable Chemical Systems—Experimental Realization of Boolean Functions using a Simple Feedforward Net." *Naturforsch.* **49(a)** (1994): 953–963.

Part V: Social Insects

Control Mechanisms for Distributed Autonomous Systems: Insights from the Social Insects

Eric Bonabeau

1 INTRODUCTION

Social insect societies are distributed autonomous systems that provide us with fascinating examples of functional collective behavior. In this chapter I describe three collective control (or coordination) mechanisms—self-organization, response thresholds, and templates—which I believe are fundamental not only to social insects but also to many, if not all, other distributed autonomous systems.

A social insect colony functions as an integrated unit that possesses the abilities to process a large amount of information in a distributed manner, to make decisions about how to allocate individuals to various tasks, to coordinate the activities of tens to thousands of workers, and to undertake enormous construction projects. A colony also exhibits flexibility and robustness in response to external challenges and internal perturbations [33]. Every insect in a social insect colony seems to have its own agenda—each insect is an autonomous agent—and yet an insect colony is remarkably efficient as a whole. The seamless integration of all individual activities does not seem to require any supervisor. This lack of "central" control or supervision may even be the secret behind the social insects' ecological success.

Design Principles for Immune System & Other Distributed Autonomous Systems. L. A. Segel and I. R. Cohen, eds. Oxford Univ. Press, 2001

Here are a few selected examples of tasks that social insects perform collectively:

- Leafcutter ants (*Atta*) cut leaves from plants and trees to grow fungi. Workers forage for leaves hundreds of meters away from their nest, literally organizing highways to and from their foraging sites [18].
- Weaver ant (*Oecophylla*) workers form chains of their own bodies, allowing them to cross wide gaps and pull stiff leaf edges together to form a nest. Several chains can join to form a bigger one over which workers run back and forth. Such chains create enough force to pull leaf edges together. When the leaves are in place, the ants connect both edges with a continuous thread of silk emitted by a mature larva held by a worker [17, 18].
- In their moving phase, army ants (such as *Eciton*) organize impressive hunting raids, involving up to 200,000 workers, during which they collect thousands of prey [7, 27, 30].
- In a social insect colony, a worker usually does not perform all tasks, but rather specializes in a set of tasks, according to its morphology, age, or chance. This division of labor among nestmates, whereby different activities are performed simultaneously by groups of specialized individuals, is believed to be more efficient than if tasks were performed sequentially by unspecialized individuals [20, 29]. In polymorphic species of ants, two (or more) physically different types of workers coexist. For example, in *Pheidole* species, minor workers are smaller and morphologically distinct from major workers. Minors and majors tend to perform different tasks: whereas majors cut large prey with their large mandibles or defend the nest, minors feed the brood or clean the nest. Removal of minor workers stimulates major workers into performing tasks usually carried out by minors [32]. This replacement takes place within two hours of minor removal. More generally, it has been observed in many species of insects that removal of a class of other workers quickly compensate after the removal of a class of workers: division of labor exhibits a high degree of plasticity.
- Honey bees (*Apis mellifica*) build series of parallel combs by forming chains that induce a local increase in temperature. The wax combs can be more easily shaped thanks to this temperature increase [11]. With the combined forces of individuals in the chains, wax combs can be untwisted and be made parallel to one another. Each comb is organized in concentric rings of brood, pollen, and honey. Food sources are exploited according to their quality and distance from the hive. At certain times, a honey bee colony divides: the queen and approximately half of the workers leave the hive in a swarm, at first forming a cluster on the branch of a nearby tree. Potential nesting sites are carefully explored by scouts. The selection of the nesting site can take up to several days, during which the swarm precisely regulates its temperature [16].
- Nest construction in the wasp *Polybia occidentalis* involves three groups of workers, pulp foragers, water foragers, and builders. The size of each group

is regulated according to colony needs through some flow of information among them [21].

- Tropical wasps (for example, *Parachartergus, Epipona*) build complex nests, comprised of a series of horizontal combs protected by an external envelope and connected to each other by a peripheral or central entrance hole [19].
- Termites (*Macrotermes*) build even more complex nests, comprised of roughly cone-shaped outer walls that often have conspicuous ribs containing ventilation ducts which run from the base of the mound toward its summit, brood chambers within the central "hive" area, which consists of thin horizontal lamellae supported by pillars, a base plate with spiral cooling vents, a royal chamber, which is a thick-walled protective bunker with a few minute holes in its walls through which workers can pass, fungus gardens, draped around the hive and consisting of special galleries or combs that lie between the inner hive and the outer walls, and, finally, peripheral galleries constructed both above and below ground which connect the mound to its foraging sites [22, 23].

If no one is in charge in social insect colonies, how can one explain the complexity and sophistication of their collective behavior? An insect is a complex creature: it can process a lot of sensory inputs, modulate its behavior according to many stimuli, including interactions with nestmates, and make decisions on the basis of a large amount of information. However, the complexity of an individual insect is still not sufficient to explain the complexity of what social insect colonies can do. Perhaps the most difficult question is: how does individual behavior connect to collective performance? In other words, how does cooperation arise, how do many insects coordinate their activities *in a meaningful way*? Coordination *per se* is not worth much in a biological system: it is important that coordination be "meaningful," as what social insects do has to be connected to their environments and to the colony's needs, and finally, in one way or another, to the colony's and its members' reproductive success. Social insect colonies are complex adaptive, distributed systems whose constituent units are autonomous agents [3]. In this chapter I will indicate a few coordination or "control" mechanisms for such systems. I believe these mechanisms are rather general in autonomous (and mobile) distributed biological systems, such as the immune system.

Many aspects of the collective activities of social insects are *self-organized.* Theories of self-organization [15, 24], originally developed in the context of physics and chemistry to describe the emergence of macroscopic patterns out of processes and interactions defined at the microscopic level, have been extended to social insects. Self-organization shows that complex collective behavior may emerge from interactions among individuals that exhibit simple behavior: in these cases, there is no need to invoke individual complexity to explain complex collective behavior. Recent research suggests that self-organization is indeed a major component of a wide range of collective phenomena in social insects [4, 13].

In the context of social insects the self-organization approach consists of viewing complex colony-level behavior as resulting from the interplay of interactions among individual insects and interactions between insects and their environment. An important insight of this approach has been the suggestion that it is not always necessary to invoke individual complexity (the ability to take into account numerous parameters to modulate one's behavior) in order to explain complex colony-level phenomena or to explain complex spatiotemporal patterns, the time and length scales of which go far beyond the characteristic time and length scales of individual insects.

Other mechanisms than self-organization obviously play a role in shaping collective behavior in social insects. In particular, social insect societies self-organize within a set of constraints. It is important to acknowledge that global order in social insects can arise as a result of internal interactions among insects. But it is equally important to keep external factors and constraints into the picture, all the more as the colony and its environment influence one another through interactions among internal and external factors.

The "control" mechanisms that I will describe in the rest of the chapter are based on self-organization and on two "alternative mechanisms," response thresholds, and templates. These mechanisms all provide feedback and constraints from the environment, but it is important to remember that part of the environment may have been shaped by the colony's past activities. I will use the word "control" in a rather loose sense; perhaps a more appropriate word in the present context would be coordination.

2 SELF-ORGANIZATION

Self-organization is a set of dynamical mechanisms whereby structures appear at the global level of a system from interactions among its lower-level components. The rules specifying the interactions among the system's constituent units are executed on the basis of purely local information, without reference to the global pattern, which is an emergent property of the system rather than a property imposed upon the system by an external ordering influence. Self-organization relies on three basic ingredients:

1. Positive feedback, or amplification, that promotes the creation of structures. Examples of positive feedback include recruitment and reinforcement. Self-organization relies on the amplification of fluctuations due to random walks, errors, random task switching, and so forth: fluctuations, such as preexisting or behavior-induced heterogeneities in the environment, can act as seeds from which structures nucleate and grow.
2. Negative feedback that counterbalances positive feedback and helps stabilize the collective pattern: it may take the form of saturation, exhaustion, or competition.

3. In social insects, self-organization relies on multiple interactions, either directly among individuals, or among elements that can be manipulated by them, such as soil pellets, seeds, corpses, eggs, larvae, etc.

Self-organization usually results in three important properties or signatures:

1. The emergence of spatiotemporal structures in an initially homogeneous medium.
2. The possible coexistence of several stable states, or multistability: structures emerge by amplification of random deviations, and any such deviation can be amplified, so that the system converges to one among several possible stable states, depending on initial conditions (path dependency).
3. The existence of (parameter driven) bifurcations, where the behavior of a self-organized system changes dramatically.

The double-bridge experiment [13] is one of the simplest examples of self-organization in social insects. In experiments with the ant *Linepithema humile*, a food source is separated from the nest by a bridge with two equally long branches A and B. Initially, both branches have the same probability of being selected: choices are made at random. But a few more ants randomly select, say, branch A, where they deposit pheromone, a chemical that attracts nestmates. The greater amount of pheromone on A stimulates more ants to select A, and so forth.

When the bridge's branches are not the same length, the shorter branch is selected more frequently by the same mechanism, that is, the amplification of initial fluctuations: the first ants returning to the nest take the shorter path twice, from the nest to the source and back, and therefore influence outgoing ants toward the short branch. This example illustrates the notions of positive feedback, amplification of fluctutations, and multiple interactions. If the experiment lasts for several hours, negative feedback also comes into play in the form of food source exhaustion or satiation, preventing foraging from going on. Note that negative feedback from food source exhaustion comes from the environment.

Two of the three signatures of self-organization can be observed in this example:

1. *Emergence of structure.* In the case of equally long branches, the environment is initially homogeneous in that both branches are equally likely to be selected. The environment acquires structure, or loses its homogeneity, when one of the branches "wins."
2. *Multistability.* Depending on which branch is favored by initial fluctuations, either branch may eventually win. The system, therefore, has two stable states. This is true when both branches are the same length and also when they are not: in this latter case, initial fluctuations favor the shorter branch.

A similar example can be found in honey bees. When a bee finds a nectar source, she goes back to the hive and relinquishes her nectar to a hive bee. Then she can either start to dance to indicate to other bees the direction and the distance to the food source, or continue to forage at the food source without recruiting nestmates, or she can abandon her food source and become an uncommitted follower herself. If the colony is offered two identical food sources at the same distance from the nest, the bees exploit the two sources symmetrically. However, if one source is better than the other, the bees are able to exploit the better source, or to switch to this better source even if it is discovered later.

Let us consider the following experiment. Two food sources are presented to the colony at 8:00 A.M. at the same distance from the hive: source A is characterized by a sugar concentration of 1.00 mol/l and source B by a concentration of 2.5 mol/l. Between 8:00 and noon, source A has been visited 12 times and source B 91 times. At noon, the sources are modified: source A is now characterized by a sugar concentration of 2.5 mol/l and source B 0.75 mol/l. Between noon and 4:00 P.M., source A has been visited 121 times and source B only 10 times. It has been shown experimentally that a bee has a relatively high probability of dancing for a good food source and abandoning a poor food source. These simple behavioral rules allow the colony to select the better quality source. With the aid of a simple mathematical model based on these observations, Seeley et al. [31] and Camazine et al. [9] have confirmed that foragers can concentrate on the best food source through a positive feedback created by differential rates of dancing and abandonment based upon nectar source quality.

Another interesting example of self-organization is the construction of pillars in termites [12]. The termite *Macrotermes* uses soil pellets impregnated with pheromone to build pillars. Two successive phases take place [14]. A first, noncoordinated phase is characterized by a random deposition of pellets. This phase lasts until one of the deposits reaches a critical size. Then, the coordination phase starts if the group of builders is sufficiently large: pillars emerge. The existence of an initial deposit of soil pellets stimulates workers to accumulate more material through a positive feedback mechanism, since the accumulation of material reinforces the attractivity of deposits through the diffusing pheromone emitted by the pellets [6]. This autocatalytic effect leads to the coordinated phase. If the spatial density of builders is too small, the pheromone disappears between two successive passages by the workers and the amplification mechanism cannot work. The system undergoes a bifurcation at this critical density: no pillar emerges below it, but pillars can emerge above it. This example illustrates several ingredients and signatures of self-organization:

1. *Positive feedback.* The accumulation of pheromone-impregnated material creates a snowball effect. Initial fluctuations are amplified: pillars tend to emerge where the first pellets have been dropped, because the first deposits attract more deposits.

2. *Negative feedback.* The decay of the pheromone limits the snowball effect.
3. *Multiple interactions.* Pillars emerge thanks to multiple indirect interactions among termites: one termite deposits a pellet that stimulates another termite to deposit another pellet.
4. *Emergence of structure.* The initial spatial distribution of soil pellets is random. The activity of the termites transforms this random distribution into pillars.
5. *Bifurcation.* The spatial density of termites acts as a bifurcation parameter.

3 ALTERNATIVE COORDINATION MECHANISMS

The recognition that complex colony-level behavior need not be rooted in complex individual behavior is one of the great advances that self-organization has permitted. Alternative approaches tend to assume that individuals have the ability to process huge amounts of information and make complex decisions, and that colony-level complexity is the phenotypic result of fine-tuned genotypic characteristics. For example, it was not rare for students of social insects, until recently, to assume that the queen, one way or another, gives orders and centralizes information, or that rigid caste ratios (that is, the number of workers in each caste) have been optimized by evolution [25], with caste determination being largely genetic. It has now become clear that the queen, although she certainly plays a role in regulating some of the colony's activities, rarely gives direct orders and is unlikely to centralize information (for an exception, see Reeve and Gamboa [26]), and that caste ratios in social insects are flexible rather than rigid [8].

However self-organization alone usually does not provide sufficient explanation. I would like to give two examples of crucial alternative control mechanisms in social insects: response thresholds and templates.

3.1 RESPONSE THRESHOLDS

Response thresholds have been invoked as a control mechanism in the context of division of labor in social insects [28]. The underlying idea is very simple: when some stimulus exceeds the response threshold of an individual, that individual is likely to respond to the stimulus. Responding to a task-associated stimulus by performing the task may reduce the intensity of the stimulus. Therefore, individuals with high thresholds are unlikely to perform the task when other individuals, with lower thresholds, maintain the stimulus intensity below their thresholds. When, however, individuals with low thresholds fail to perform the task, because of predation, swarming or other perturbations, those individuals that have high thresholds may engage in task performance because stimulus intensity exceeds their thresholds. This shows, rather informally, that response thresholds could explain the flexibility and robustness observed in many species of social insects.

Bonabeau et al. [1] have shown that the assumption of response thresholds can reproduce experimental results of Wilson [32]. Wilson artificially varied the ratio of majors to minors in several polymorphic ant species (*Pheidole*) and observed a dramatic increase in task performance by previously inactive majors as the ratio exceeded some value; the involvement of majors occurred within an hour of minors' removal. Assuming differential response thresholds for minors and majors, this phenomenon can be readily explained by the reasoning I just described. Not only is this model in qualitative agreement with Wilson's observations, it can also reproduce his results quantitatively [1].

The complexity of threshold models directly results from the complexity of assumed information sampling techniques. A threshold model can be quite complex if one takes into account spatial, causal, or topological relationships among tasks. Tasks are not uniformly distributed in space (for example, brood care occurs within the nest, whereas foraging occurs outside the nest), so that performing a given task may enhance or prevent contacts with other task-associated stimuli (including nestmates). Some tasks are causally related (for example, foraging and storing), so that individuals performing a task may switch to the next, etc. In other words, sampling is unlikely to be homogeneous, and some workers may end up being specialized simply because they always encounter the same stimuli, and do not encounter the stimuli that would induce them to perform other tasks. Additionally, there may be learning processes so that, for example, workers tend to become more and more sensitive to stimuli associated with the tasks they are currently performing.

3.2 TEMPLATES

Environmental factors, acting as constraints or templates, very often play an essential role in determining what kind of colony-level organization is to be expected. For example, many ant species (including *Acantholepsis custodiens* [5], *Formica polyctena*, and *Myrmica rubra* [10]) make use of temperature and humidity gradients to build their nests and spatially distribute eggs, larvae, and pupae. Another obvious template is light: darkness and daylight certainly influence patterns of activity in most species of social insects and, for that matter, most animal species. More generally, the behavior of most insects is influenced by heterogeneities present in the environment: they tend to walk, build, store, or lay eggs along such heterogeneities. By heterogeneity, I mean any perceptible deviation from a uniform distribution or constant quantity. This includes irregular soil levels, obstacles, gradients, and also predictably varying quantities such as temperature or light intensity, etc. Sometimes, an individual can directly provide a template, as illustrated by the construction of the royal chamber in termites (*Macrotermes subhyalinus*). The physogastric (filled with eggs) queen of *Macrotermes subhyalinus* emits a pheromone that diffuses and creates around her a pheromonal template in the form of a decreasing gradient. It has been shown experimentally that either a concentration window or a threshold exists that controls the workers' building activities: a worker deposits a soil pellet if the concentration of pheromone lies within this window or exceeds the threshold [2, 4, 6]. Otherwise, they do not deposit any pellet and destroy existing walls.

There may also exist more complex types of templates: those resulting from the colony's activities, which in turn influence the colony's future activities. Indeed, a single action by an insect results in a small modification of the environment that influences the actions of other insects. We have already met this mechanism, called stigmergy. A good example of a template that results from the stigmergic actions of individuals is the building of galleries along pheromone trails:

1. A trail network emerges because of the trail-laying and trail-following behavior of individual termites, which is an example of stigmergy.
2. The trail pheromone diffuses away from the center of a reinforced portion of trail, thereby creating a chemical template, very similar in function to the queen's chemical template, along which walls are built.
3. Gallery size is adapted to traffic: the more termites, the higher the pheromone concentration, and the further away from the trail center walls are built.

The important point is that this chemical template results from the termites' behavior and not merely from a preexisting heterogeneity. Self-organization based on stigmergy and combined with templates is a powerful complexity-generating mechanism. Imagine a homogeneous medium in which structure emerges through self-organization and stigmergy—for example, pillars in termites. Once it has emerged, this structure is an heterogeneity that serves as a template that directs individuals' actions. These actions create, in turn, new stimuli that trigger new building actions, either based on self-organization or templates, or both. And so forth. For example, nest building in termites is a morphogenetic process whereby complexity unfolds progressively [4]: more and more complex structures appear as stimuli become more and more complex due to past construction.

4 CONCLUSION

In this chapter I have described what I believe are fundamental distributed coordination and control mechanisms in social insect colonies: self-organization, response thresholds and templates. To summarize, self-organization is a pattern formation mechanism, response thresholds are a regulation mechanism and templates provide "operating" constraints for the other two mechanisms. These mechanisms are complementary and often observed together; they rely on cues from the environment and at the same time result in changes in the environment. Evolution has tuned these mechanisms in such a way that the interplay between them and the environment produces meaningful (= adaptive) coordination and collective behavior.

The behaviors produced by combinations and cascades of these mechanisms can be arbitrarily complex. Such powerful combinations of mechanisms seem difficult to avoid if one wants to control or coordinate a distributed system whose relatively simple constituent units (the autonomous agents) do not have access to global information. Similar mechanisms are likely to play a coordinating role in other distributed autonomous systems.

REFERENCES

[1] Bonabeau, E., G. Theraulaz, and J.-L. Deneubourg. "Quantitative Study of the Fixed Threshold Model for the Regulation of Division of Labor in Insect Societies." *Proc. Roy. Soc. Lond. B* **263** (1996): 1565–1570

[2] Bonabeau, E., G. Theraulaz, J.-L. Deneubourg, S. Aron, and S. Camazine. "Self-Organization in Social Insects." *Trends in Ecol. & Evol.* **12** (1997): 188–193.

[3] Bonabeau, E. "Social Insect Colonies as Complex Adaptive Systems." *Ecosystems* **1** (1998): 437–443.

[4] Bonabeau, E., G. Theraulaz, J.-L. Deneubourg, N. R. Franks, O. Rafelsberger, J.-L. Joly, and S. Blanco. "A Model for the Emergence of Pillars, Walls, and Royal Chambers in Termite Nests." *Phil. Trans. Roy. Soc. London* **353** (1998): 1561–1576.

[5] Brian, M. V. *Social Insects: Ecology and Behavioural Biology.* New York: Chapman & Hall, 1983.

[6] Bruinsma, O. H. "An Analysis of Building Behaviour of the Termite *Macrotermes subhyalinus* (Rambur)." Thesis, Landbouwhoge School, Wageningen, The Netherlands, 1979.

[7] Burton, J. L., and N. R. Franks. "The Foraging Ecology of the Army ant *Eciton rapax*: An Ergonomic Enigma?" *Ecol. Entomol.* **10** (1985): 131–141.

[8] Calabi, P. "Behavioral Flexibility in Hymenoptera: A Re-examination of the Concept of Caste." In *Advances in Myrmecology*, edited by J. C. Trager, 237–258. Leiden: Brill Press, 1988.

[9] Camazine, S., and J. Sneyd. "A Model of Collective Nectar Source Selection by Honey Bees: Self-Organization through Simple Rules." *J. Theor. Biol.* **149** (1991): 547–571.

[10] Ceusters, R. "Simulation du nid naturel des fourmis par des nids artificiels placés sur un gradient de température." *Actes des Colloques Insectes Sociaux* **3** (1986): 235–241.

[11] Darchen, R. "Les techniques de la construction chez *Apis mellifica.*" Ph.D. Dissertation, Université de Paris, 1959.

[12] Deneubourg, J.-L. "Application de l'ordre par fluctuations à la description de certaines étapes de la construction du nid chez les termites." *Insectes Sociaux* **24** (1977): 117–130.

[13] Deneubourg, J.-L., and S. Goss. "Collective Patterns and Decision Making." *Ethol., Ecol. & Evol.* **1** (1989): 295–311.

[14] Grassé, P.-P. "La reconstruction du nid et les coordinations interindividuelles chez *Bellicositermes natalensis et cubitermes sp.* La théorie de la stigmergie: essai d'interprétation du comportement des termites constructeurs." *Insectes Sociaux* **6** (1959): 41–84.

[15] Haken, H. *Synergetics.* Berlin: Springer-Verlag, 1983.

[16] Heinrich, B. "The Regulation of Temperature in the Honeybee Swarm." *Sci. Am.* **244** (1981): 146–160.

[17] Hölldobler, B., and E. O. Wilson. "The Multiple Recruitment Systems of the African Weaver Ant *Oecophylla longinoda* (Latreille)." *Behav. Ecol. Sociobiol.* **3** (1978): 19–60.

[18] Hölldobler, B., and E. O. Wilson. *The Ants.* Cambridge, MA: Harvard University Press, 1990.

[19] Jeanne, R. L. "The Adaptativeness of Social Wasp Nest Architecture." *Quart. Rev. Biol.* **50** (1975): 267–287.

[20] Jeanne, R. L. "The Evolution of the Organization of Work in Social Insects." *Monit. Zool. Ital.* **20** (1986): 119–133.

[21] Jeanne, R. L. "Regulation of Nest Construction Behaviour in *Polybia occidentalis.*" *Anim. Behav.* **52** (1996): 473–488.

[22] Lüscher, M. "Der lufterneuerung im nest der termite *Macrotermes natalensis* (Hav.)." *Insectes sociaux* **3** (1956): 273–276.

[23] Lüscher, M. "Air-Conditioned Termite Nests." *Sci. Am.* **205** (1961): 138–145.

[24] Nicolis, G., and I. Prigogine. *Self-Organization in Non-equilibrium Systems.* New York: Wiley & Sons, 1977.

[25] Oster, G., and E. O. Wilson. *Caste and Ecology in the Social Insects.* Princeton, NJ: Princeton University Press, 1978.

[26] Reeve, H. K., and G. J. Gamboa. "Queen Regulation of Worker Foraging in Paper Wasps: A Social Feedback Control System (*Polistes fuscatus*, Hymenoptera: Vespidae)." *Behaviour* **102** (1987): 147–167.

[27] Rettenmeyer, C. W. "Behavioral Studies of Army Ants." *Univ. Kans. Sci. Bull.* **44** (1963): 281–465.

[28] Robinson, G. E. "Modulation of Alarm Pheromone Perception in the Honey Bee: Evidence for Division of Labour Based on Hormonally Regulated Response Thresholds." *J. Comp. Physiol. A* **160** (1987): 613–619.

[29] Robinson, G. E. "Regulation of Division of Labor in Insect Societies." *Ann. Rev. Entomol.* **37** (1992): 637–665.

[30] Schneirla, T. C. "Army Ants." In *Army Ants: A Study in Social Organization*, edited by H. R. Topoff. San Francisco, CA: W. H. Freeman, 1971.

[31] Seeley, T. D., S. Camazine, and J. Sneyd. "Collective Decision-Making in Honey Bees: How Colonies Choose among Nectar Sources." *Behav. Ecol. Sociobiol.* **28** (1991): 277–290.

[32] Wilson, E. O. "The Relation between Caste Ratios and Division of Labour in the Ant Genus *Pheidole* (Hymenoptera: Formicidae)." *Behav. Ecol. Sociobiol.* **16** (1984): 89–98.

[33] Wilson, E. O., and B. Hölldobler. "Dense Heterarchies and Mass Communications as the Basis of Organization in Ant Colonies." *Trends in Ecol. & Evol.* **3** (1988): 65–68.

Task Allocation in Ant Colonies

Deborah M. Gordon

Immune systems and ant colonies are different kinds of system. The immune system is our name for those parts of a body that distinguish self from non-self and respond to threatening infractions of this boundary. An ant colony, by contrast, is a distinct, complete, reproducing individual. The two systems, though different, have a great deal in common. Both are composed of simple parts whose interactions produce global responses. They have another, more specific feature in common. In both, the units interact through physical contact, and the rate of interaction affects the function of the whole system. Ants communicate using chemical cues, most of which can be perceived only at small distances. Ants interact when they are very close, usually touching one another. Cells in the immune system interact by physical contact, formed by the bonds between receptors on cells and molecules released by other cells. In an ant colony, ants move around, meet each other, and take on certain functions according to their recent encounter history; in the immune system, cells move around the body, meet pathogens and each other, and take on certain functions in response to those encounters.

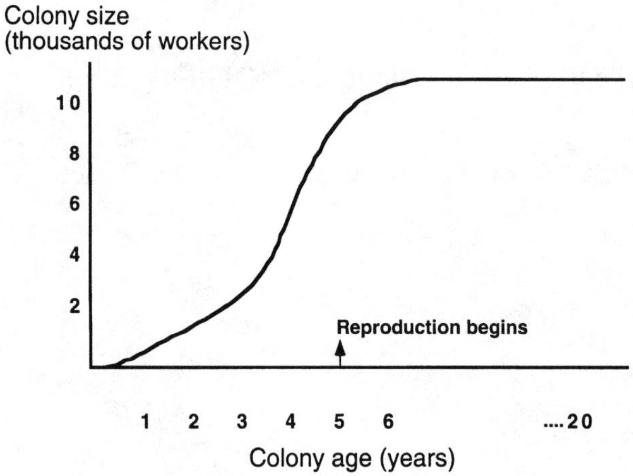

FIGURE 1 Colony age and colony size. The figure shows how colony size, in thousands of workers, changes as a function of colony age. When a colony is about five years old, it reaches a stable size and begins to produce reproductives who leave the colony to mate and find new colonies.

1 HARVESTER ANT BEHAVIOR

I study the behavior and ecology of harvester ants in the desert of southeastern Arizona [2, 3]. A harvester ant colony consists of one queen, a reproductive female who lays the eggs, and many sterile females called "workers." Reproduction occurs once a year, when the queen produces new, virgin queens. Males are haploid and grow from unfertilized eggs; these eggs can be laid by the mated queen or by unmated workers. Each year, on the same day, many colonies send out winged reproductives, virgin queens and males, to a mating aggregation. After mating, the males die and the newly mated queens fly off from the aggregation site to start new colonies. Once a newly mated queen has dug a nest, she begins to lay eggs. She may live for 15 or 20 years, continuing to produce ants using sperm from the original mating, and she will never leave the nest again. When the queen and the colony are about five years old, the queen will begin to produce new queens and males to send to the annual mating flight.

At about five years, when the colony begins to reproduce, it has grown to a size of about 10,000 workers. Workers live only a year, but by replacing all the ants each year the colony stays at a stable size of about 10,000 workers until it dies at 15 or 20 years. When the queen dies, and all of the remaining workers have died, the colony is dead. In this species, colonies do not adopt new queens when the original queen dies.

A colony performs a variety of tasks. The queen and the brood (eggs, larvae, and pupae) stay inside the nest, and some workers inside the nest care for the brood, moving it from cooler to warmer chambers at times, piling it, grooming it, and feeding the larvae. Harvester ants eat mostly seeds and, inside the nest, ants store and process the seeds brought in by the foragers. Outside the nest some ants forage, traveling 10 to 20 meters from the nest to search for seeds. Each morning before the foragers emerge another group of ants, the patrollers, chooses the direction to be foraged that day. The patrollers also respond to incursions on the nest mound by other species of ants. Nest maintenance workers build and clean out the chambers inside the nest. They plaster the walls of the chambers with moist soil, which dries to an adobe-like surface, and come out of the nest carrying dry soil to discard outside. Midden workers sort out the refuse pile, or midden, where the ants put the husks of the seeds they eat.

Foraging, nest maintenance, patrolling, and midden work are performed each day in a characteristic sequence: first, the nest maintenance workers appear, carrying out small bits of dirt. Then the patrollers move around the nest mound, inspecting the ground with their antennae and touching antennae with other ants they meet. Eventually the patrollers move off the nest mound and out into the surrounding area. When the patrollers start coming back on one of the colony's trails, the foragers leave the nest. Foragers go out on the trails in the direction chosen earlier by the patrollers. Midden workers move refuse from one pile to another. Foraging peaks in the middle of the morning, and by the end of the morning, the only ants active are returning foragers. When the last foragers run inside, moving fast in the midday heat, there is a final burst of nest maintenance work before the colony shuts down for the day.

Though this characteristic sequence of activities is preserved, the colony adjusts the numbers of workers engaged in each task, as conditions require. On some days, more ants forage, presumably when more food is available. On other days, such as the day after a storm has caused flooding, there is little foraging and a large number of ants do nest maintenance work, clearing debris and reopening the clogged nest entrance.

Task allocation is the process that adjusts the numbers of ants engaged in each task, in a way appropriate to the current situation [1]. There is no central control of task allocation. No ant directs the work of others.

I did a series of perturbation experiments to find out how the various task groups are related. The experiments consisted of changing the number of ants engaged in one task, to see how ants engaged in other tasks would react. For example, I put out piles of toothpicks early in the morning when the nest maintenance workers were first active. The nest maintenance workers moved the toothpicks to the edge of the nest mound and discarded them there. This required an increase in the number of nest maintenance workers, because additional workers were needed to move toothpicks as well as complete the usual nest maintenance tasks of that morning. Ants engaged in other tasks besides nest maintenance were not affected directly by the experiment, because by the time any other task groups were active, the nest maintenance workers had already moved the toothpicks out of the way. But there were clear indirect effects: the perturbation experiments showed that ants respond to changes in numbers engaged in another task. For example, when more ants were recruited to do extra nest maintenance work, the numbers foraging decreased.

These perturbation experiments showed that task groups are interdependent. I then repeated the experiments using marked individuals; foragers, patrollers, midden workers, and nest maintenance workers were each marked with a unique color of paint. These experiments showed that individuals switch tasks. For example, when more foragers are needed, then ants previously performing other tasks, such as patrolling, midden work, or nest maintenance, will switch tasks to foraging. However, task switching does not occur in all possible directions. For example, when more nest maintenance workers are needed, they are recruited from ants inside the nest that have not previously worked outside. Once an ant leaves nest maintenance work, she does not go back to it. Thus, nest maintenance acts as a source: this task is performed by the youngest of the exterior workers, recently recruited from inside the nest. Nest maintenance is the task performed by the younger ants who then leave it for more exterior tasks. Foraging acts as a sink for exterior workers; all workers will eventually become foragers if needed. This seems appropriate in a species which searches for sparsely distributed seeds; if ever there is a windfall of abundant food, the system quickly allocates more ants to retrieve it.

Task switching alone does not account for the interdependence of task groups. A change in the numbers performing one task leads to a change in numbers performing other tasks. This must be because of two ways that ants can adjust their behavior: first, by changing tasks, and second, by staying inactive, inside the nest, or going out to perform a task. For example, the reduced numbers foraging when I experimentally increased numbers performing nest maintenance could be due to task switching. Foragers might have switched to nest maintenance, which would simultaneously increase the numbers of nest maintenance workers and decrease the numbers of foragers. However, foragers never switch to nest maintenance. This means that when more nest maintenance workers become active, foragers simply become inactive. The foragers are a distinct group of workers whose activity is affected by the numbers currently performing nest maintenance work.

Thus, ants make two kinds of decisions: which task to perform, and whether to be active or inactive. These decisions are somehow mediated by simple, local information.

2 DEVELOPMENT OF TASK ALLOCATION

Task allocation changes as a colony grows older and larger. I repeated the perturbation experiments with young colonies, two years old with about 3,000 ants. The response of young, small colonies differed from that of old, large colonies, five years old or more with about 10,000 ants. The behavior of old, large colonies was more stable and more consistent than that of young, small colonies. When the same perturbation was made, week after week, older colonies tended to respond in the same way. Younger colonies, however, responded differently to the same perturbation from week to week. This suggests that younger, smaller colonies are more sensitive to differences in conditions from week to week.

While colonies live 15 to 20 years, individual ants live only a year. Thus, the behavior of an older colony is not due to the experience of older ants. There is no evidence that the ants of an older colony are any different from the ants of a younger one. Each year the colony consists of another cohort of sisters; the ants of an older colony are later daughters of the same queen, and thus the younger sisters of the ants, long dead, of the younger colony.

An obvious difference between older and younger colonies is colony size. Although it is possible that old colonies accumulate information passed on from one cohort to the next, this seems to me probably beyond the abilities of ants. A simpler explanation, requiring less of the ants, is that ants in younger colonies act according to the same rules as ants in older ones, but the outcome depends on colony size. This has led me to investigate rules based on interaction rate, because an ant in a small colony would experience a different interaction rate from an ant in a large one.

To test whether size accounts for age-dependent differences in colony behavior, we are using a modeling approach. We attempt to construct a mathematical description of rules which ants use in task decisions, and to see whether the model responds to a change in colony size in the way that real colonies do.

It is difficult to perform the obvious experiments. It is not possible to transform a young, small colony into a young, large one by adding ants, because ants from another colony will be recognized as non-nest mates by their odor and will be killed or excluded from the nest. It is, however, possible to transform an old, large colony into an old, small one by killing some of the ants. Unfortunately, in the field, ants respond to this experiment the same way they do to intense natural predation: by remaining inactive inside the nest for some weeks. When we transfer colonies to the laboratory from the field, many ants die, so there is a drastic reduction in colony size. In some ways, the behavior of these depleted colonies does resemble that of younger colonies. However, behavior differs greatly in the laboratory and the field, and we have not yet succeeded in maintaining large colonies of 10,000 ants in the laboratory.

3 INTERACTION RATES

To say that ants use rules based on interaction rates in task decisions is to imagine that certain threshold levels of interaction rate provide positive or negative feedback in the performance of certain tasks. The hypothesis is that the ant behaves as follows. The ant performs task x with a certain probability. If it meets ants of task x (or z) at a certain rate, its probability of task performance changes. Positive feedback means that interaction of a certain type causes it to become more likely to perform a certain task; negative feedback means that interaction causes it to be come less likely to perform a task.

This is a process in ant colonies analogous to a basic feature of the immune system. In the ant colony, ants of one task group respond to changes in the number of active ants in another task group. In the immune system, cells of type i respond to changes in the number of cells of type j. In both ant colonies and the immune system, rates of interaction between different types probably influence the responses of individual ants or individual cells.

4 MODELS OF TASK ALLOCATION

We model task allocation to help specify how simple decisions by individuals add up to the behavior we see in colonies. One way to do this is to imagine a set of rules that determine the task decisions of individuals. We try to find the simplest possible rules, because we assume that ants are not capable of complicated assessments. Two models that we have used are a neural network [5] and a deterministic one based on differential equations [6]. We are currently developing an agent-based model, in which we hope to include sufficient detail that we can match the model's predictions to empirical data. Discrepancies between the model's predictions, and observations, may indicate where we are mistaken about the rules that ants use.

Our models of task allocation assume that an ant uses the recent history of its interactions with other ants in decisions about whether to perform a task. An ant may assess the rate at which it meets ants of a particular task group. An ant's antennae are its organs of chemical perception. In the course of a brief antennal contact with another ant, an ant can determine if the other one is a nest mate. Nest mate recognition depends on a colony-specific odor which is characteristic of the hydrocarbons smeared over the ant's body in the course of grooming. We find that these cuticular hydrocarbons vary, not only from one colony to the next, but also within colonies. Ants of one task group, such as foragers, have a different chemical profile from ants of other task groups, such as nest maintenance workers [7]. In the course of brief antennal contacts between harvester ants, these task-specific chemical profiles may allow one ant to recognize not only whether the other ant is a nest mate, but what task it is performing. In current work we are examining how cuticular hydrocarbons change as an ant switches task, possibly because of the particular environmental conditions associated with each task. For example, some of the hydrocarbons are known to be sensitive to temperature and humidity. A forager who spends a long time outside the nest, in the dry air and hot sun, may come to have a characteristic odor because of the effect of the sunlight on its cuticular hydrocarbons.

Laboratory studies confirm that there is an association between an ant's interaction rate and the probability it engages in midden work [4]. We are currently investigating the effects of interaction rate on other tasks.

It is clear that environmental conditions, as well as social interactions, affect the task decisions of individual ants. If the discovery of a picnic had no effect on an ant's behavior, none of its nest mates would eventually appear to share the picnic.

Our modeling and empirical work complement each other. The simulations help us determine what kinds of rules at the individual level might explain the behavior we observe in real colonies. Suppose that each ant has some probability of leaving the nest to do the same task it last did when it left the nest. That is, let us start with a pool of workers in each task, and consider the rules that determine whether each ant will be active and whether each ant will switch tasks in the next time step.

For example, perturbation experiments showed a reciprocal relation of nest maintenance work and foraging. When nest maintenance was experimentally increased by putting out toothpicks that had to be moved out of the way, foraging decreased. When foraging was experimentally decreased by putting out barriers on the foraging trails, nest maintenance increased. What kind of simple rules might ants use, that would lead to this result?

A rule that explains why diminished foraging leads to increased nest maintenance might be as follows. This hypothesized rule is based on an additional observation: once an ant has left the nest as a forager, it rarely returns without a seed. Searching foragers stay outside the nest until they find a food item. This means that just inside the nest entrance, a forager with a seed is a returning forager. A forager without a seed is a forager on its way out of the nest. Suppose that a nest maintenance worker's probability of leaving the nest to perform nest maintenance work decreases when it meets foragers without seeds. Then if some environmental condition decreases the numbers of foragers leaving the nest, the nest maintenance workers will meet fewer foragers leaving the nest. The inhibiting effect on nest maintenance of foragers leaving the nest will be diminished, and the result is an increase in numbers of nest maintenance workers leaving the nest. This hypothesis can be tested by examining how nest maintenance workers respond to the rate of interaction with foragers without seeds.

5 THE STUDY OF IMMUNE SYSTEMS AND ANT COLONIES

To investigate "design principles" in task allocation, the approach we are taking is to postulate what rules, at the level of individual ant, might lead to the trends we see in the behavior of colonies. In the example above, the colony-level behavior is that when numbers foraging decrease, numbers performing nest maintenance increase. The individual-level rule is: don't do nest maintenance when you meet foragers on their way outside the nest. The empirical challenge is to find out whether this is the correct individual-level rule.

This approach seems to me the opposite of the one usually employed in immunology. To make the comparison between approaches in the study of ant behavior and immune systems, we have to forget for a moment that the immune system has many different types of players, in fact many different classes of types of players, whereas in the harvester ant colony I have so far divided all exterior workers into four simple types: an ant is either a forager, a nest maintenance worker, a patroller, or a midden worker. Leaving aside this important distinction between the two systems, the approaches used are opposite in the following sense. The players in the immune system are defined by what they do when they meet; for example, one type of cell is known to bind to a particular antigen. The problem is to find out what function these interactions among the components have in the behavior of the whole system. In the ant colony, the players are defined by their function; for example, some ants collect food. The problem is to find out how interactions among ants generate the functions we observe.

The SFI workshop left me with the impression that what's known about immunology is the reciprocal of what's known about ant colony behavior. Observation of ant colonies reveals patterns of behavior: for example, certain tasks are performed at certain times of day. These patterns have characteristic dynamics: for example, when numbers foraging are low, numbers doing nest maintenance work are high. We know relatively little about what produces these dynamics. By contrast, to observe an immune system the way we observe an ant colony, you would need an animal built like an aquarium, with all the parts of the immune system labeled as they swim around, meeting here and there to perform their tasks. A great deal is known about the mechanics of the immune system: for example, which components bind to which others. Much less is known about the patterns of these components' behavior. It's as though I knew in detail how foragers react to seeds, but knew much less about what foragers accomplish for the colony. Without understanding how foragers contribute to the life of the colony, I would be unable to explain why it matters if one colony forages differently from another. Yet these are the basic medical questions that drive research in immunology: for example, why does it matter if T-cell counts change in older people? How does the failure of a particular receptor contribute to a particular disease?

The SFI workshop was for me a brief and exciting introduction to immunology. I learned that the analogies between immune systems and ant colonies are close enough that we face some similar problems in trying to understand them. I was struck by the immunologists' willingness to accept that there are diverse processes at work, so that different parts of the system require different explanations. Familiarity with an astonishing number of different entities, and a long history of discoveries showing that apparently distinct parts of the system are in fact connected, seems to have taught the field a kind of methodological pluralism. This reminded me how foolish it is for people who study social insects to argue over whether the most recently discovered mechanism or process explains everything we see; of course it won't. Immunology is an old, established area of research, with vast resources and enormous numbers of people; the study of social insects is young, impoverished by comparison to immunology, and has only a few thousand active researchers. If social insects have anything to teach immunologists, it is that the basic questions are the same for any complex biological system: how do the components fit together to produce the dynamics of the whole system?

REFERENCES

[1] Gordon, D. M. "The Organization of Work in Social Insect Colonies." *Nature* **380** (1996): 121–124.

[2] Gordon, D. M. *Ants at Work: How an Insect Society is Organized.* New York: Free Press, Simon and Schuster, 1999.

[3] Gordon, D. M. "Interaction Patterns and Task Allocation in Ant Colonies." In *Information Processing in Social Insects*, edited by J. M. Pasteels, J.-L. Deneubourg, and C. Detrain, 51–67. Berlin: Birkhauser Verlag, 1999.

[4] Gordon, D. M., and N. Mehdiabadi. "Encounter Rate and Task Allocation in Harvester Ants." *Behav. Ecol. & Sociobiol.* **45** (1999): 370–377.

[5] Gordon, D. M., B. Goodwin, and L. E. H. Trainor. "A Parallel Distributed Model of Ant Colony Behaviour." *J. Theor. Biol.* **156** (1992): 293–307.

[6] Pacala, S. W., D. M. Gordon, and H. C. J. Godfray. "Effects of Social Group Size on Information Transfer and Task Allocation." *Evol. Ecol.* **10** (1996): 127–165.

[7] Wagner, D., M., J. F. Brown, P. Broun, W. Cuevas, L. E. Moses, D. L. Chao, and D. M. Gordon. "Task-Related Differences in the Cuticular Hydrocarbon Composition of Harvester Ants, *Pogonomyrmex barbatus*." *J. Chem. Ecol.* **24** (1998): 2021–2037.

Part VI: Applications to Computer Science

Biologically Motivated Distributed Designs for Adaptive Knowledge Management

Luis Mateus Rocha
Johan Bollen

1 HUMAN-COMPUTER INTERACTION AND BIOLOGY

We discuss how distributed designs that draw from biological network metaphors can largely improve the current state of information retrieval and knowledge management of distributed information systems. In particular, two adaptive recommendation systems named *TalkMine* and *@ApWeb* are discussed in more detail. *TalkMine* operates at the semantic level of keywords. It leads different databases to learn new and adapt existing keywords to the categories recognized by its communities of users using distributed algorithms. *@ApWeb* operates at the structural level of information resources, namely citation or hyperlink structure. It relies on collective behavior to adapt such structure to the expectations of users. *TalkMine* and *@ApWeb* are currently being implemented for the research library of the Los Alamos National Laboratory under the Active Recommendation Project. Together they define a biologically motivated information retrieval system, recommending simultaneously at the level of user knowledge categories expressed in keywords, and at the level of individual documents and their associations to other documents. Rather than passive information retrieval, with this system, users obtain an active, evolving interaction with information resources.

Design Principles for Immune System & Other Distributed Autonomous Systems. L. A. Segel and I. R. Cohen, eds. Oxford Univ. Press, 2001

1.1 DISTRIBUTION INFORMATION SYSTEMS AND INFORMATION RETRIEVAL[1]

Distributed information systems (DIS)[2] refer to collections of electronic networked information resources in some kind of interaction with communities of users; examples of such systems are: the Internet, the World Wide Web, corporate intranets, databases, library information retrieval systems, etc. DIS serve large and diverse communities of users by providing access to a large set of heterogeneous electronic information resources. As the complexity and size of both user communities and information resources grows, the fundamental limitations of traditional information retrieval systems have become evident.

Information retrieval (IR) refers to all the methods and processes for searching relevant information out of information systems (e.g., databases) that contain extremely large numbers of documents. Traditional IR systems are based solely on keywords that index (semantically characterize) documents and a query language to retrieve documents from centralized databases according to these keywords. This setup leads to a number of flaws:

- *Passive Environments.* There is no genuine interaction between user and system. The user pulls information from a passive database and therefore needs to know how to query relevant information with appropriate keywords. Furthermore, such impersonal interfaces cannot respond to queries in a user-specific fashion because they do not keep user-specific information, or user profiles. The net result is that users must know in advance how to characterize the information they need (with keywords) before pulling it from the environment.
- *Idle Structure.* Structural relationships between documents, keywords, and IR patterns are not utilized. Different kinds of structural relationships are available, but not typically used, for different DIS: e.g., citation structure in scientific library databases, the hyperlink structure in the WWW, the clustering of keyword relationships into different meanings of keywords, temporal patterns of user retrieval, etc.
- *Fixed Semantics.* Keywords are initially provided by document authors (or publishers, librarians, and indexers), and do not necessarily reflect the evolving semantic expectations of users.
- *Isolated Information Resources.* No relationships are created and no information is exchanged among documents and/or keywords in different information resources such as databases, web sites, etc. Each resource is accessed with its own set of keywords and query language.

---

[1] This subsection draws from ongoing collaboration with Cliff Joslyn at the Los Alamos National Laboratory. Many of the ideas here presented are undoubtedly due to him.

[2] The main abbreviations used this article in addition to DIS are: IR (Information Retrieval), SA (Spreading Activation), and ARP (Active Recommendation Project).

These flaws prevent traditional IR processes in DIS to achieve any kind of interesting coupling with users. No system-user evolution or learning can be achieved because of the following fundamental limitations:

- There is no *recommendation*. Because of passive environments and idle structure, IR systems cannot proactively push relevant information to its users about related topics that they may be unaware of.
- There is no *conversation* between users and information resources, between information resources, and between users. Because of passive environments and isolated information resources there is no mechanism to exchange knowledge, or crossover of relevant information.
- There is no *creativity*. Because of fixed semantics, isolated information resources, idle structure, and passive environments, there is no mechanism to recombine knowledge in different information resources to infer new categories of keywords used by different communities of users.

1.2 DRAWING FROM BIOLOGY

The limitations of traditional IR and DIS are even more dramatic when contrasted with biological distributed systems such as immune, neural, insect, and social networks. Biological networks function largely in a distributed manner, without recourse to central controllers, while achieving tremendous ability to respond in concerted ways to different environmental necessities. In particular, they are typically endowed with the ability to elicit appropriate responses to specific demands, to transfer and process relevant information across the network, and to adapt to a changing environment by creating novel behaviors (often from recombination of existing ones). These abilities are precisely what has been lacking in IR, in which context they become ways to surmount the recommendation, conversation and creativity limitations described above.

Biological networks effectively evolve in an open-ended manner; we would like to endow DIS with a similar open-ended capacity to evolve with their users—to achieve an open-ended semiosis with them [52]. In biology, open-ended evolution originates from the existence of material building blocks that self-organize nonlinearly (e.g., Kauffman [27]) and are combined via a specification control, such as the genetic system, which nonetheless does not precisely describe or program the dynamical outcome [2, 40, 41, 46, 49]. In contrast, computer systems were precisely constructed with building blocks constrained in such a way as to allow minimum dynamic self-organization and maximum programmability, which results in no inherent evolvability [13]. Therefore, to attain any evolvability in current digital computer systems, we need to program in some building blocks that can be used to realize the kind of dynamical richness we encounter in biological systems.

Biological systems possess an enabling chemistry (the building blocks) leading to fluid evolvability, as the possible interactions between a biological agent and its environment are open-ended. For instance, Gordon [18] shows how different biochemical profiles of ants with different roles in their colonies may be a reflection of their embodied interaction with the environment, and not necessarily a consequence of genetic differences. The gender of the Mississippi alligator too, rather than being genetically programmed, is environmentally regulated by the temperature the eggs encounter in the nest [17]. At all levels of biological systems we find this dynamic agent-environment coupling (or embodiment [11]) coexisting with the specification or loose programmability of the initial conditions for arrangements of dynamic building blocks, which then self-organize to produce phenotypes, behaviors, organizations, etc. [41, 42, 53]. The programmability can be genetic, immune, cognitive, or social.[3] Indeed, biological systems combine a small amount of programmability with rich dynamic building blocks to produce an unbounded set of self-organizing behaviors that can be picked up by natural selection [51].

Computer systems possess the description or programmability part, what they now need is an amount of dynamic agent-environment coupling, which is distributed and therefore not under complete control from a programming center. Mitchell [35], describing her Copycat system, suggests that in order to construct distributed, bottom-up systems capable of solving complicated cognitive tasks that are not explicitly programmed, one needs to endow computer systems with enabling *relationship packages.* In other words, there is a need for an *enabling substrate* to achieve dynamic agent-environment couplings with a smaller degree of programmability and a higher degree of self-organization.

The inherent material dynamics that permeates biology, "comes for free" [36] for the evolving organism. In contrast, in computer systems, since we relinquished dynamics for full programmability, we need to program in every rule that may allow building blocks to be combined, self-organized, and selected—as if setting up the laws of an artificial physics and biology [54]. Programming in the enabling substrate is, however, very different from programming the ultimate behavior that we wish to obtain. Rather, what is programmed are the lower-level building blocks and rules to relate them, which later self-organize computationally to produce (hopefully open-ended) evolving behaviors which in turn are selected by the demands of an environment or set of tasks we wish to see resolved. The enabling relationship packages are used to combine, re-combine, and transmit building blocks to produce new behavior that is not fully prespecified. This bottom-up design mimics the existence in biology of low programmability and high evolvability.

The success of imbuing computer systems with distributed, bottom-up, designs from biology is apparent in such areas as optimization [21, 34], modeling and simulation of social phenomena and organizations [22, 32, 45], computer security [14] (also see Forrest's article in this volume [15]), Artificial

[3]Clearly these types of programmability of different levels of biological systems are quite distinct. Genetic description is much more clearly understood [46, 49], but each level of biological organization establishes its own sets of constraints which also describe or program the accepted behavior at a given level [40, 55].

Life [31], and even biology itself [56]. We are now interested in improving the limitations of IR in DIS utilizing biologically motivated designs.

The ultimate goal of IR is to produce or recommend relevant information to users. It seems obvious that the foundation of any useful recommendation should be first and foremost based on the identification of users and subject matter. In this sense, the goal of recommendation systems can be seen as similar to that of most biological systems, in particular immune systems: to recognize agents (users) and elicit appropriate responses from components of the distributed information network. Furthermore, the information network should learn and adapt to the community of agents (users) it interacts with—its environment. Naturally, unlike immune systems, the goal is not to be hostile to external agents but rather to produce information they find relevant and desirable: users are not to be treated as pathogens!

Nevertheless, as described in section 1.1, traditional IR does not identify users and classifies subjects only with unchanging keywords. To build more flexible IR, or, more generally, biologically motivated recommendation systems, we need to design the enabling relationship substrate precisely to accommodate the identification of users and their needs, as well as the evolving subjects stored in DIS. This substrate includes:

- A means to recognize *users.*
- A means to characterize *information resources.*
- A two-way means to exchange knowledge between users and information resources: a *conversation* process. As information resources become more and more complex, we cannot expect a simple one-way query to work well. Instead, we need a means to combine the interests of the user with the knowledge specific to each information resource.
- *Adaptation* mechanisms. We also want DIS to adapt to their community of users, as well as to exchange and recombine knowledge leading to evolvability and creativity.

We describe below our efforts to include these biologically motivated design requirements to achieve a useful and more natural knowledge management of DIS. Before that though, we describe other recent efforts to improve IR.

2 ACTIVE RECOMMENDATION SYSTEMS

New approaches to IR have been proposed to address the limitations described in section 1.1. *Active recommendation systems*, also known as *active collaborative filtering* [9], *knowledge mining*, or *knowledge self-organization* [24] are IR systems which rely on active computational environments that interact with and adapt to their users. They effectively push relevant information to users according to previous patterns of IR or individual user profiling.

Recommendation systems are typically based on user-environment interaction mediated by intelligent agents or other decentralized components and come in two varieties [3]:

- In *content-based* recommendation, user profiles are created based on the system's keywords. Documents are recommended to users according to their profiles and some kind of semantic metric obtained from the associations between keywords and documents.
- In *collaborative* recommendation no description of the semantics or content of documents is involved, rather recommendations are issued according to a comparison of the profiles of several users that tend to access the same documents. These user profiles are not based on keywords, but on the actual documents retrieved.

Content-based systems depend on single user profiles, and thus cannot effectively recommend documents about previously unrequested content to a specific user. Conversely, pure collaborative systems, with no content analysis, match only the profiles of users that (to a great extent) have requested exactly the same documents; for instance, different book editions or movie review web sites from different news organizations are considered distinct documents. It is clear that effective recommendation systems require aspects of both approaches.

Hybrid approaches to recommendation usually rely on software agents and a central database. The agents have two distinct roles:

1. To retrieve and collect documents from information resources into a database or router and
2. To select or filter those documents retrieved that match the profile of specific users.

This is the case, for instance, of *Fab* [3] and *Amalthaea* [37]. Systems such as these clearly establish active environments which are capable of recommendation; that is, they push topics that users may have not thought of, rely on user-specific interfaces that enable user identification, and keep track of historical data of the user-DIS interaction. In the terms used above, these systems expand IR beyond passive environments and completely idle structure (they keep track of user-environment interaction).

From the picture of IR depicted in section 1, there is clearly still much more room to improve. The structure and semantics of DIS is still largely idle in these collaborative systems, as they retain their original relations. Indeed these systems can improve considerably by clustering and ranking documents according to the semantics of keyword relationships [26] or the structure of document linkage [29]. Many data-mining and graph-theoretical improvements can and should be used to discover hidden patterns in the structure of DIS, thus achieving a much more powerful recommendation capability.

However, our goal here is to improve recommendation systems by empowering them with biologically motivated conversation and creativity dimensions as described in section 1. Particularly, we want to enable the adaptation of structure and semantics of DIS to users. For this we need to develop more active environments and move beyond fixed semantics, isolated information resources, and mostly idle structure of DIS. In the following, we describe some of the work we have been developing in this direction.

3 THE ACTIVE RECOMMENDATION PROJECT

The *Active Recommendation Project*[4] (ARP), part of the Library Without Walls Project, at the Research Library of the Los Alamos National Laboratory is engaged in research and development of biologically motivated designs to escape the shortcomings of traditional IR and more recent recommendation systems. As discussed in section 1.2, in order to implement any biologically motivated designs we need to define an enabling relationship substrate. In this section we describe how we define such as substrate for our information resources and users.

3.1 INFORMATION RESOURCES: DISTRIBUTED MEMORY

The information resources available to ARP are large databases with academic articles. These databases contain bibliographic, citation, and sometimes abstract information about academic articles. Typical databases are *SciSearch*® and *Biosis*®; the first contains articles from scientific journals from several fields collected by ISI (Institute for Scientific Indexing), while the second contains more biologically oriented publications. We do not manipulate directly the records stored in these information resources; rather, we create a repository of records which point us to documents stored in these databases.

3.1.1 The XML Repository. We store pointers to published documents as XML[5] records. By working with XML records, we gain the ability to change the information associated with their respective documents, which we cannot do with the proprietary databases. Indeed, the XML records should be seen more as dynamic objects rather than static documents. Not only do we gain the ability to change the original keywords and citation information from the respective documents, but also the ability to add annotations, links to other records, associations with other types of media (e.g., sound clips), etc. Furthermore, XML records can even have associated procedures to compute relevant algorithms. We can think of XML records as archival objects, "buckets" of pointers, links, data, and code, which are not affiliated with any one particular information resource, as defined by Nelson et al. [39].

[4] See http://www.c3.lanl.gov/~rocha/lww for more information, results, and test bed.
[5] eXtendable Markup Language.

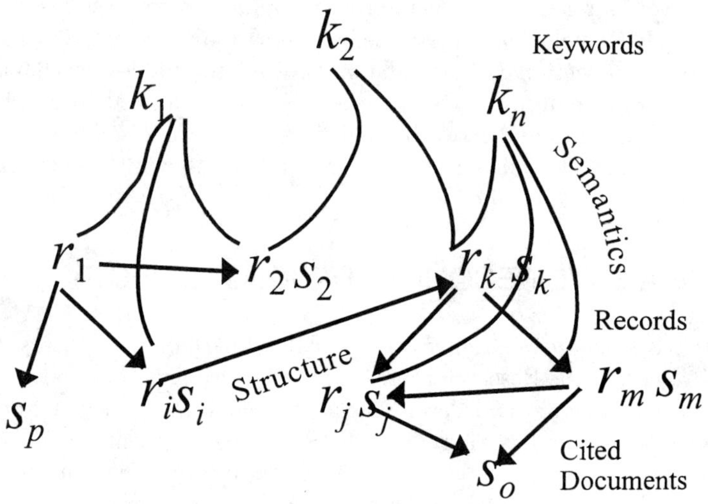

FIGURE 1 Relational Repository. The document Set $D = R \cup S$. Some records are cited, some are not. Some cited documents are records, some are not.

By transforming records from passive documents into active objects, we start our construction of the biologically motivated enabling substrate at the lowest level of information systems: the source data. This is an essential step to set up a distributed design. In centralized systems, documents can be passive since it will be up to a higher level program to decide if a certain document is relevant or not. In contrast, in distributed systems, much of the decision-making is off-loaded to lower-level components, which need to be endowed with computing capabilities. In this sense, records become active objects that store changing information, communicate with other components, and even perform actions (run code) on the information they store.

3.1.2 The Relational Repository. From the XML record repository we can derive relational information between records and keywords and among records: the *semantics* and the *structure*, respectively. This semantic and structural relational repository provides the enabling relationship packages discussed above. They define which record objects are related and how, as well as the semantic tokens (keywords) they are associated with. We can also establish how keywords relate to one another.

From the XML repository we obtain m records $r_j \in \mathbf{R}$, n keywords $k_i \in \mathbf{K}$, and o cited documents $s_m \in \mathbf{S}$. Notice that the cited document set $\mathbf{S}$, is larger than the set of records $\mathbf{R}$, and that these sets overlap only partially, because often records cite documents that are not themselves contained in the XML repository as a record. Furthermore, the two sets are not nested, that is, neither $\mathbf{R} \subseteq \mathbf{S}$ nor $\mathbf{S} \subseteq \mathbf{R}$. For structural analysis we need to create the citation document set $\mathbf{D}$ of all the p documents d_l involved in a citation relation. We can also derive all database semantic information from the relationships between $\mathbf{R}$ and the set of all keywords $\mathbf{K}$. Figure 1 depicts the raw information from the relational repository. We are currently using one information resource from ISI, with data from the years of 1996 to 1999. There are 2,915,258 records and 839,297 keywords. We plan to include another information resource and previous years very soon.

3.1.3 Structural Relations. The structure of an information resource is defined by the relations between documents in the document set $\mathbf{D}$. In academic databases these relations refer to citations, while in the World Wide Web to hyperlinks. In our case, the ISI scientific database, we work with the citation structure. Because we are working with a small interval of years, only less than half of all records (1,111,868) are an element of the set of cited documents $\mathbf{S}$, which contains 8,354,372 documents. We also discovered that many records do not participate in any citation relation (523,804), so the subset of records that participate in a citation relation is $\mathbf{R}'$ (2,391,454). The set of all documents that participate in a citation relation is $\mathbf{D} = \mathbf{R}' \cup \mathbf{S}$ (9,633,958). The citation relations are defined by the *citation matrix* C, a $p \times p$ matrix, of p documents d_l of $\mathbf{D}$. Each entry $c_{i,j}$ in the matrix is Boolean and indicates whether document d_i cites (1) document d_j or not (0). This matrix is not symmetrical and is extremely sparse.

To discern the closeness of documents according to citation structure, we define measures of proximity between any two documents. The *inwards structural proximity matrix* P^{in} is a square matrix of dimension p. For two documents d_i and d_j, it is their direct *co-citation* [57], that is, the number of documents that cite d_i *and* d_j, over the number of documents that cite either d_i or d_j. Documents that cite d_i are referred to as ancestors of d_i. The inwards proximity varies in the unit interval and is defined by:

$$\begin{aligned} p^{\text{in}}(d_i, d_j) &= \frac{\sum_{k=1}^{p} (c_{k,i} \wedge c_{k,j})}{\sum_{k=1}^{p} (c_{k,i} \vee c_{k,j})} = \frac{N^{\text{in}} \cap (d_i, d_j)}{N^{\text{in}} \cup (d_i, d_j)} \\ &= \frac{N^{\text{in}} \cap (d_i, d_j)}{N^{\text{in}}(d_i) + N^{\text{in}}(d_j) - N^{\text{in}} \cap (d_i, d_j)} . \end{aligned} \tag{1}$$

$N^{\text{in}}(d_i)$ is the number of documents that cite document d_i, and $N^{\text{in}} \cap (d_i, d_j)$ the number of documents that cite both d_i and d_j.

The *outwards structural proximity matrix* P^{out} is a square matrix of dimension p. For two documents d_i and d_j, it is their direct *bibliographic coupling* [28], that is, the number of documents that both d_i *and* d_j cite, over the number of documents that either d_i or d_j cite. Documents that d_i cites are referred to as descendants of d_i. The outwards proximity varies in the unit interval and

TABLE 1 Ten most common (stemmed) keywords and their frequency.

Frequency	Keyword
187705	cell
150795	studi
149594	system
140738	express
127350	protein
124094	model
120215	activ
113740	human
112737	rat
112702	patient

is defined by:

$$\begin{aligned} p^{\text{out}}(d_i, d_j) &= \frac{\sum_{k=1}^{p}(c_{i,k} \wedge c_{j,k})}{\sum_{k=1}^{p}(c_{i,k} \vee c_{j,k})} = \frac{N^{\text{out}} \cap (d_i, d_j)}{N^{\text{out}} \cup (d_i, d_j)} \\ &= \frac{N^{\text{out}} \cap (d_i, d_j)}{N^{\text{out}}(d_i) + N^{\text{out}}(d_j) - N^{\text{out}} \cap (d_i, d_j)} \,. \end{aligned} \tag{2}$$

$N^{\text{out}}(d_i)$ is the number of documents that document d_i cites, and $N^{\text{out}} \cap (d_i, d_j)$ the number of documents that both d_i and d_j cite. These very sparse directed graphs can be combined into a nondirected graph via some linear combination. From this value we can define a neighborhood of a document d_i as the set of documents related to it with proximity greater than $\alpha \in [0, 1]$. Furthermore, we use this structural proximity information to study the relative importance of documents using singular value decomposition [29] as well as standard clustering techniques to obtain clusters of related documents.

3.1.4 Semantic Relations. From the XML record repository we obtain the set of all (2,915,258) records **R** and the set of all (839,297) keywords **K**. The relations between the elements of these sets allow us to infer the semantic value of documents and the interrelations between semantic tokens: the keywords. Naturally, semantics is ultimately only expressed in the brains of users who utilize the documents, but keywords are tokens of this ultimate expression, which we can infer from the relation between **R** and **K**. The sources of keywords are the terms authors and/or editors chose to qualify documents, as well as title words. The ten most common keywords in our data set are listed in table 1.[6]

The relations between **K** and **R** are formalized by the very sparse *keyword-record matrix* A: $n \times m$ matrix, of n keywords k_i and m records r_j. Each entry $a_{i,j}$ in the matrix is Boolean and indicates whether keyword k_i qualifies (1) record r_j or not (0). To discern the closeness among keywords according to

[6]We considered only keywords which qualify at least two records. For details about our keyword data, consult http://www.c3.lanl.gov/~rocha/lww/keywords.html.

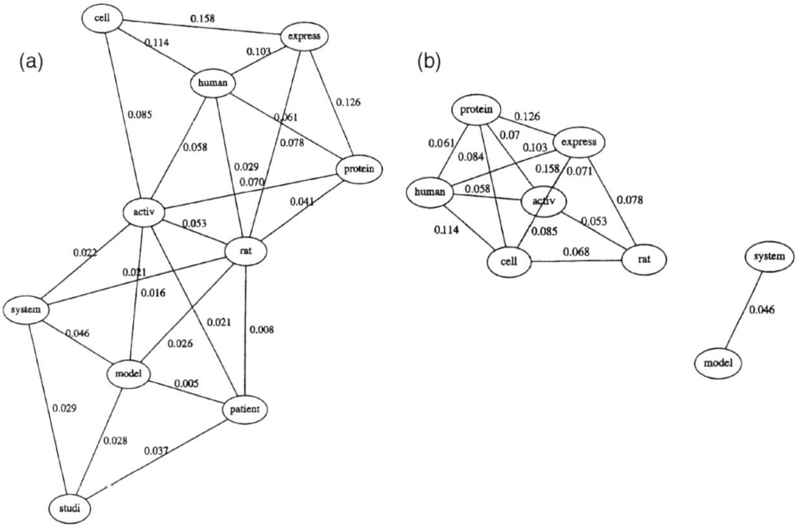

FIGURE 2 Keyword Semantic Proximity for ten most common keywords. (a) Shows the three highest values for each node. (b) Shows all values higher than 0.045.

this relation we compute the *keyword semantic proximity matrix* KSP. It is a sparse square matrix of dimension n. For two keywords k_i and k_j, it is the number of records they both qualify, over the number of records either one qualifies. Proximity varies in the unit interval, and is defined by the following equation:

$$\begin{aligned} ksp(k_i, k_j) &= \frac{\sum_{k=1}^{m}(a_{i,k} \wedge a_{j,k})}{\sum_{k=1}^{m}(a_{i,k} \vee a_{j,k})} = \frac{N\cap(k_i,k_j)}{N\cup(k_i,k_j)} \\ &= \frac{N\cap(k_1,k_j)}{N(k_i)+N(k_j)-N\cap(k_i,k_j)}. \end{aligned} \quad (3)$$

The semantic proximity calculations between two keywords, k_i and k_j, depend on the sets of records qualified by either keyword, and the intersection of these sets. $N(k_i)$ is the number of records keyword k_i qualifies, and $N \cap (k_i, k_j)$ the number of records both keywords qualify. This last quantity is the number of elements in the intersection of the sets of records that each keyword qualifies. Thus, two keywords are near if they tend to qualify many of the same records. Table 2 presents the values of KSP for the ten most common keywords, and figure 2 depicts the same information in graphical form.

Conversely, to discern the closeness of records according to relation A, we compute the *record semantic proximity matrix* RSP. It is a sparse square matrix of dimension m. For two records r_i and r_j, it is the number of keywords that qualify both, over the number of keywords that qualify either one. It

TABLE 2 Keyword Semantic Proximity for the ten most frequent keywords.

	cell	studi	system	express	protein	model	active	human	rat	patient
cell	1.000	0.022	0.019	0.158	0.084	0.017	0.085	0.114	0.068	0.032
studi	0.022	1.000	0.029	0.013	0.017	0.028	0.020	0.020	0.020	0.037
system	0.019	0.029	1.000	0.020	0.017	0.046	0.022	0.014	0.021	0.014
express	0.158	0.013	0.020	1.000	0.126	0.011	0.071	0.103	0.078	0.020
protein	0.084	0.017	0.017	0.126	1.000	0.013	0.070	0.061	0.041	0.014
model	0.017	0.028	0.046	0.011	0.013	1.000	0.016	0.016	0.026	0.005
activ	0.085	0.020	0.022	0.071	0.070	0.016	1.000	0.058	0.053	0.021
human	0.114	0.020	0.014	0.103	0.061	0.016	0.058	1.000	0.029	0.021
rat	0.068	0.020	0.021	0.078	0.041	0.026	0.053	0.029	1.000	0.008
patient	0.032	0.037	0.014	0.020	0.014	0.005	0.021	0.021	0.008	1.000

varies in the unit interval, and it is defined by the following equation:

$$\begin{aligned} rsp(r_i, r_j) &= \frac{\sum_{k=1}^{m}(a_{k,i} \wedge a_{k,j})}{\sum_{k=1}^{m}(a_{k,i} \vee a_{k,j})} = \frac{N\cap(r_i,r_j)}{N\cup(r_i,r_j)} \\ &= \frac{N\cap(r_i,r_j)}{n(r_i)+N(r_j)-N\cap(r_i,r_j)} \,. \end{aligned} \tag{4}$$

The semantic proximity calculations between two records, r_i and r_j, depend on the sets of keywords qualifying either record, and the intersection of these sets. $N(r_i)$ is the number of keywords that qualify record r_i, and $N \cap (r_i, r_j)$ the number of keywords that qualify both records. Thus, two records are near if they tend to be qualified by many of the same keywords.

From the inverse of these very sparse matrices, we can obtain a measure of distance between keywords and between records. These distances are not Euclidean metrics because they do not observe the triangle inequality. This means that the shortest distance between two keywords or records may not be the direct link but rather an indirect pathway. Such measures of distance are referred to as semi-metrics [16]. We are currently investigating if the characteristics of metricity can function as an indication of related semantic topics. The semantic side of the relational repository also allows us to conduct other IR techniques such as latent semantic indexing [4, 26] as well as semantic proximity clustering.

3.1.5 Knowledge Contexts. Each information resource (e.g., a database) is characterized by the relational information described in sections 3.1.2 through 3.1.4, which is obtained from the record objects of section 3.1.1. The collection of this relational information associated with an information resource is an expression of the particular knowledge it conveys to its community of users. Notice that most information resources share a very large set of keywords and documents pointed to by records. However, these are organized differently in each resource, leading to different collections of relational information. Indeed, each resource is tailored to a particular community of users, with a distinct history of utilization and deployment of information by its authors and users.

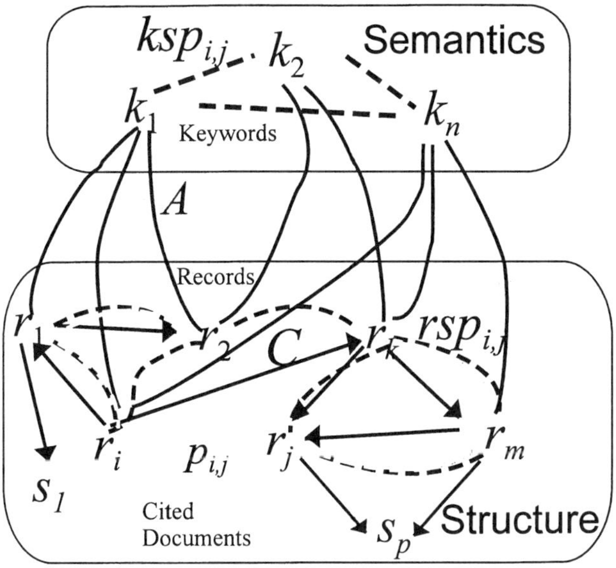

FIGURE 3 Generic knowledge context, with structure and semantic levels, of an information resource.

The same keywords will be related differently in different resources. Therefore, we refer to the relational information of each information resource as a *knowledge context* (fig. 3).

With this name we do not mean to imply that such computational structures possess cognitive abilities. Rather, we note that the way records are organized in information resources is an expression of the knowledge of its community of users. Records and keywords are only tokens of the knowledge that is ultimately expressed in the brains of users. A knowledge context simply mirrors the collective knowledge relations and distinctions of a community of users.

Notice that none of the proximity relations that define a knowledge context exist explicitly in traditional databases. Building this infrastructure is essential as an enabling relationship substrate for biologically motivated designs such as those described below in sections 4 and 5. Our object records and the proximity relations between them function as a metaphor for the material components and allowable interactions among components of biological distributed systems. Based on this substrate, we can now move to adaptive biological designs.

3.2 USERS

The information resources and respective knowledge contexts interact with users whose behavior we will use below to adapt the associative knowledge stored in the proximity measures. But before discussing this interaction, we need to define the capabilities of users: our agents. The following capabilities are implemented in enhanced "browsers" or centralized services that users have access to.

1. *Present Interests.* Described by a set of keywords $\{k_1, \ldots, k_i\}$.
2. *History of IR.* This history is also organized as a knowledge context as described in section 3.1.5, containing the records the user has previously accessed, the keywords associated with them, as well as the structure of this set of records. This way, we treat users themselves as information resources with their own specific knowledge context defined by its own proximity information.
3. *Communication Protocol.* Users need a two-way means to communicate with other information resources in order to retrieve relevant information, and to send signals leading to changes in all parties involved in the exchange.

The collective interaction of users defined by these capabilities, and a set of knowledge contexts from information resources of a DIS is depicted in figure 4. The knowledge contexts defined for information resources and users establish the necessary enabling substrate to set up biologically motivated designs which we describe in sections 4 and 5.

4 TALKMINE: CATEGORIZATION THROUGH CONVERSATION IN DIS

Given the enabling substrate defined in section 3, to accomplish the goals expressed in section 1, we need a mechanism to enable the communication between users/agents and information resources, leading to information exchange, adaptation and recombination. *TalkMine* is a system designed especially for that. It is both a content-based and collaborative recommendation system based on a model of cognitive categories [50], which are created from the conversation between users and information resources and used to recombine knowledge as well as adapt it to users [52].

4.1 THE DISTRIBUTED MEMORY STRUCTURE

The proximity information of knowledge contexts is abstracted from the record-keyword (A) and record-record (C) relations and is not stored as such in the record repository. There is a parallel here to connectionist devices. Clark [10]

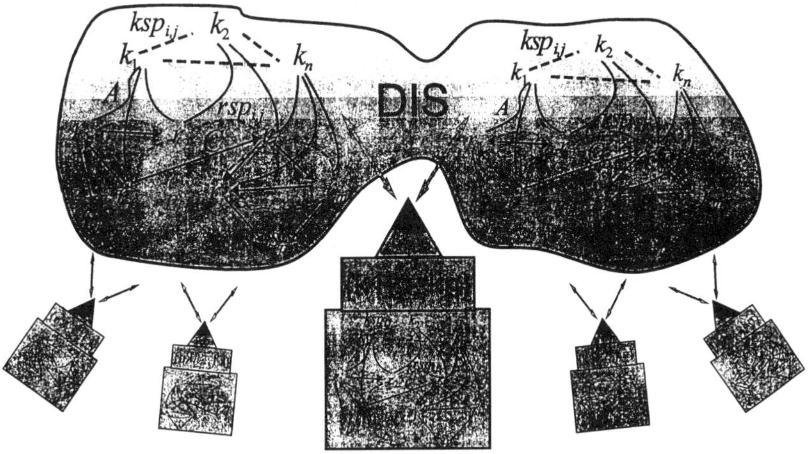

FIGURE 4 A collection of users interacts with two knowledge contexts of a DIS.

proposed that connectionist memory devices work by producing metrics that relate the knowledge they store (our enabling substrate). These metrics and the knowledge tokens they relate are not stored locally in the nodes of a connectionist network, but rather nonlinearly superposed over its weights [58].

Our knowledge context is not a connectionist structure in a strong sense since keywords and records can be identified in particular nodes in the network. However, the same keyword qualifies many records, the same record is qualified by many keywords, and the same record typically is engaged in a citation relation with many other records. Losing or adding a few records or keywords does not affect significantly the derived semantic and structural proximity measures (as defined in section 3) of a large network. In this sense, the knowledge conveyed by such proximity measures is distributed over the entire network of records and keywords in a highly redundant manner, as required of sparse distributed memory models [25]. Below we discuss how such distributed knowledge adapts to users (the environment) with Hebbian-type learning.

In the *TalkMine* system, we use the keyword semantic proximity measure (eq. 3) from the knowledge context, which we regard as the long-term memory banks of an information resource. This proximity measure is unique, reflecting the semantic relationships obtained from the set of records stored, which in turn echo the knowledge of the resource's community of users and authors. Because we use a keywords proximity, *TalkMine* is a content-based recommendation system (section 2). Next we describe how it is also collaborative by integrating the user patterns of IR.

4.2 SHORT-TERM CATEGORIZATION THROUGH CONVERSATION

TalkMine uses a set structure named evidence set [47, 48, 50], an extension of a fuzzy set [59], as a model of cognitive categories. Evidence sets are used to quantify the relative interest of users in each of the available knowledge contexts from several information resources. *TalkMine* is based on a question-answering process that integrates the user's present interests (a set of keywords) with the long-term distributed memory of the intervening knowledge contexts (including the users'). In a sense, this is done by projecting the user's interests onto the keyword proximity measures (eq. 3) of the available information resources. The result of this nonlinear integration is a category, implemented as an evidence set of keywords. This way, each user interacts with several information resources simultaneously, engaging in a multiway conversation process.

The conversation between user and information resources is an extension of Nakamura and Iwai's [38] question-answering IR system (for a single information resource), using uncertainty measures [48] and the evidence set operations of intersection and union [50]. The algorithm of this conversation process is defined elsewhere [47, 52]. It constructs neighborhood functions from the semantic distances of the intervening knowledge contexts, and integrates these into an evidence set with a question-answering process that relies on (evidence set) union and intersection operations. The questions are used to reduce the uncertainty content of intermediate evidence sets, and are answered either by the user or her associated knowledge context.[7] At the end of this process an evidence set of keywords is obtained, which we regard as a knowledge category that contains the interests of the user as "seen" by the intervening information resources.

It is important to notice that the evidence set categories constructed with the question-answering algorithm, are not stored in any location in the distributed memory. They are temporarily constructed by integration of long-term knowledge from several information resources (the enabling substrate) and the present interests of the user. These constructed categories are therefore temporary containers of knowledge nonlinearly integrated from and relevant to the user and the collection of information resources. They model Clark's [10] "on the hoof" categories. Such "on the hoof" construction of categories, triggered by interaction with users, allows several unrelated information resources to be searched simultaneously, temporarily generating categories that are not really stored in any location.

After construction of this final category, *TalkMine* returns relevant records to the user. The records returned are those that are qualified to a high degree by many of the keywords contained in the respective evidence set. De-

---

[7] Users possess a browser with their IR history stored as a knowledge context (section 3.2). They can set up their browsers to respond to every question themselves or allow the browser to do it automatically given the past learned experience. Users can choose an intermediate value of answering between these two extremes.

tails of the actual operations used to choose relevant records are presented elsewhere [50].

4.3 ADAPTATION OF LONG-TERM MEMORY TO USERS BY SHORT-TERM CATEGORIZATION

The final component of *TalkMine* is the adaptation of the long-term distributed memory to the community of users of this system. Given the original relations of records in information resources, the derived semantic proximity measures may fail to construct associations between keywords that their users find relevant. Furthermore, the documents pointed to by records in a given information resource do not change (e.g., scientific articles), producing a fixed semantics as discussed in section 1. In contrast, the semantics of users changes with time as new keywords and associations between keywords are constantly being created and changed. Therefore, an effective recommendation system for DIS needs to adapt its knowledge contexts to the evolving semantics of its users.

The Hebbian reinforcement scheme used to implement this adaptation is very simple: the more certain keywords are combined with each other, by often being simultaneously included in the final categories, the more the distance between them is reduced. Conversely, if certain keywords are not frequently associated with one another, the distance between them is increased (details in Rocha [47, 50, 52]). This implements an adaptation of the distributed memory of information resources to their users according to repeated inclusion of keywords in categories constructed in conversation with users. This adaptation leads the semantic proximity measures involved to increasingly match the expectations of the community of users with whom information resources interact. In other words, the distributed memory is selected by the community of users.

Furthermore, when keywords in the final category are not present in one of the information resources that are combined, they are added to the information resource that does not contain them. If the association in knowledge categories of the same keywords keeps occurring, then an information resource that did not previously contain a certain keyword will have its presence progressively strengthened, even though such a keyword does not really qualify any records in this information resource.

4.4 EVOLVING KNOWLEDGE SYSTEMS VIA CATEGORICAL RECOMBINATION

Besides adapting independent information resources to users, *TalkMine* implements a kind of knowledge recombination that leads to evolving knowledge systems. The short-term categories bridge together a number of possibly highly unrelated contexts, which in turn creates new keyword associations in the respective information resources that would never occur within their own limited context.

Consider the following example. Two distinct information resources (databases) are going to be searched using the system described above. One database contains records of documents (books, articles, etc.) of an institution devoted to the study of computational complex adaptive systems (e.g., the library of the Santa Fe Institute), and the other the documents of a Philosophy of Biology department. A group of users is interested in the keywords "genetics" and "natural selection." If they were to conduct this search a number of times, due to their own interests and history, the final category obtained would certainly contain other keywords such as "adaptive computation," "genetic algorithms," etc. Let us assume that the keyword "genetic algorithms" does not initially exist in the Philosophy of Biology digital library. After these users conduct this search a number of times, the keyword "genetic algorithms" is created in this database, even though it does not contain any records about this topic. However, with these users' continuing to perform this search over and over again, the keyword "genetic algorithms" becomes highly associated with "genetics" and "natural selection," introducing a new perspective of these keywords. From this point on, users of the Philosophy of Biology library, by entering the keyword "genetic algorithms" would have their own data retrieval system point them to other information resources such as the library of the Santa Fe Institute or/and recommend documents ranging from *The Origin of Species* to treatises on Neo-Darwinism which in the meantime would have become associated with "genetic algorithms"—at which point they might rethink external access to their networked database!

TalkMine's learned categories, implemented as evidence sets, integrate the knowledge of a set of information resources with user interests through conversation. This integration is in effect a temporary recombination of knowledge. If many users tend to produce similar categories, this recombination becomes fixed in the long-term distributed memory. Therefore, categorization functions as a recombination mechanism to obtain new knowledge, which can become fixed via adaptation. This is in effect a variation and selection mechanism: short-term categories provide variation of stored knowledge, while selection is implemented by the community of users. High fitness of a category corresponds to many users producing similar categories from their conversations with networked information resources. In this sense, short-term categorization not only adapts existing information resources to users, but effectively creates new knowledge in different, otherwise independent, information resources, solely by virtue of its temporary construction of categories. This open-ended semiosis of *TalkMine* is discussed in Rocha [52].

There are obvious parallels between the user-driven evolution of knowledge systems achieved by *TalkMine* and social insect models [19]. In a sense, the knowledge categories that users create function as insect trails, the more similar categories are created the more users will be attracted to them because of reinforced proximity values among their constituent keywords. In the IR world, we refer to this organization design as collaborative recommendation. As we can see, *TalkMine* is both content-based (it organizes keyword proximity measures) and collaborative (its organization is driven by user interaction). Notice, however, that *TalkMine* by itself does not adapt the structure of knowledge contexts—it works solely on the semantic proximity measures. In section 5 we present another collaborative system we use in ARP to adapt the citation structure of DIS starting from *TalkMine*'s user-derived keyword categories.

5 ADAPTIVE ASSOCIATION AND SPREADING ACTIVATION

In section 3 we arrived at a general relational structure for information resources: a knowledge context. The relations of this structure are obtained from information contained in published documents, such as keywords and citations. Notice that this information is designed by the authors and editors of these documents. Therefore, the knowledge contexts we obtain reflect the associations that authors and editors deemed relevant. The quality of these relational structures rests then on the assumption that authors and editors generate logical semantic (keywords) and structural (citation) associations with their documents. Indeed, these associations might be different from those that users may find relevant.

In section 4 we described the *TalkMine* system which in time can precisely adapt the *semantics* (the keyword proximity measures) of the original knowledge contexts to particular communities of users. In other words, *TalkMine* recommends documents according to the original knowledge derived from author and editor keyword associations, but it changes this original knowledge base according to user choices leading to an on-going evolution of the knowledge bases accessed. In this section, we describe a system used in ARP to adapt the original (e.g., citation) *structure* of knowledge contexts to users. This system, named *@ApWeb*, provides another biologically motivated means of obtaining an enabling relationship substrate on which recommendation can be issued. We also describe how another biologically motivated algorithm, *spreading activation* (SA), can be used for IR of the original knowledge contexts or those obtained from *@ApWeb* and *TalkMine*.

5.1 @APWEB

@ApWeb is an adaptive system of obtaining a relational structure between documents of an information resource with user retrieval patterns. The knowledge contexts obtained from this system are user-determined, implicit, and collective. Its basic assumption is that the sequential retrieval of two records implies a measure of relevancy of the link or relation between them. The amount of use for a given relation between two records, is taken as an expression of its strength.

The background of this technique lies on the organization of hypertext networks such as the Web. Measuring user traversal frequencies for hyperlinks [44] in a hypertext network has been proven to be quite a successful technique not only in predicting future user link preferences [23] but also in the interactive shaping of the structure of existing hypertext networks [6].

In the general description of an information resource of section 3, links refer to the structural associations between records. On the web these are hyperlinks, while in academic databases they are citations. *@ApWeb* starts from the same original structure (the Citation Matrix C of section 3.1.3), to compute a *traversal proximity measure* T, a square matrix of $p \times p$ documents d_l of the document set $\mathbf{D}$ (see section 3), that takes values in the unit interval. This measure is then used to complement the structural proximity measures (eqs. 1 and 2) and the record semantic proximity (eq. 4) in a recommendation process that in this way issues recommendations according to both user patterns and author defined relations.

The original *@ApWeb* experiments [5], were conducted on a randomly initiated structure that adapted to users in real time. In these experiments, users accessing a web page were initially shown a random set of possible links. As the adaptation process took hold, the set of links given to a user of a web page adapted to the expectations of users. Here, we describe a design[8] that relies on a fixed structure (C) to produce traversal proximity. In this case, when users retrieve a document, they are always shown the same set of related (by C) documents which they can also retrieve, namely the documents cited by the first. Only posteriorly do we use the obtained traversal proximity to recommend relevant documents not necessarily associated in C.

The algorithm for obtaining T follows[9]:

1. Initialize T. $\forall_{i,j=1\ldots p}\ t_{i,j} = 0$.
2. Obtain the n user paths. A user path is the 3-tuple $\pi(d_i, d_j, d_k)$ of three documents retrieved sequentially by the same user.
3. For each path $\pi(d_i, d_j, d_k)$ apply the following *learning rules*:
   (a) *Frequency*: $t_{i,j} = t_{i,j} + r_{\text{freq}}$ and $t_{j,k} = t_{j,k} + r_{\text{freq}}$. This rule implements a form of Hebbian learning in that the proximity between two documents that have been retrieved sequentially by the same user is reinforced with reward $r_{\text{freq}} = 1/n$.
   (b) *Symmetry*: $t_{j,i} = t_{j,i} + r_{\text{symm}}$ and $t_{k,j} = t_{k,j} + r_{\text{symm}}$. This rule instantiates a partially symmetric proximity. If the proximity between d_i and d_j increases by r_{freq} with the frequency rule, the proximity between d_j and d_i increases by $r_{\text{symm}} < r_{\text{freq}}$ with the symmetry rule. In our experiments we have used $r_{\text{symm}} = 0.3 r_{\text{freq}}$.
   (c) *Transitivity*: $t_{i,k} = t_{i,k} + r_{\text{trans}}$. This rule instantiates a partially transitive proximity. If the proximity values between d_i and d_j and d_j and d_k increase by r_{freq} with the frequency rule, the proximity between

[8]More details in Bollen et al. [8] and Bollen and Vandesompel [7].

[9]Note that other normalization schemes are possible; the three learning rules are the important part.

d_i and d_k increases by $r_{\text{trans}} < r_{\text{freq}}$ with the transitivity rule. In our experiments we have used $r_{\text{trans}} = 0.5 r_{\text{freq}}$.

After computing T for a large number of user traversals, when a document is retrieved, we can recommend more documents than those it initially cited. We describe this process in section 5.2, but first let us discuss how we collect user paths.

Any recommendation system will eventually produce a list of relevant documents to users. A traditional IR system (section 1) may obtain such a list from simple keyword lexical matching, while a system like *TalkMine* obtains it from the keyword categorization process described in section 4. But ultimately, we obtain a list of documents. Once a user selects one of these documents, he begins a browsing path that we can use for *@ApWeb*. Each document selected, which we store in the object records (section 3.1.1), is associated with a set of other documents in the particular structure of each information resource. Initially, this set may be solely the cited documents and the derived proximity measures (eqs. 1, 2, and 4). The browsing path follows this structure and is stored in logs subsequently used for *@ApWeb*.

We have used *@ApWeb* to produce a user traversal proximity for the 423 web pages comprising the Principia Cybernetica Project web site[10] [8], as well as producing a proximity measure for the approximately 800 academic journals of the documents in the ARP database [7]. In the latter case, every time a user retrieves a document, its respective journal was recorded, therefore obtaining journal traversal paths which were fed to the *@ApWeb* algorithm. The journal proximity obtained provides an associative network of related journals. Below we describe how this information is used for recommendation.

5.2 RECOMMENDATION WITH SPREADING ACTIVATION

The technique of spreading activation (SA) is based on a model of facilitated retrieval [33] from human memory [1, 12] and has also been implemented for the analysis of hypertext network structure [43]. The model assumes that the coding format of human memory is an associative network in which the most similar memory items have strongest connections [20, 30]. SA works by activating a set of cue nodes in an associative network which spreads out to all other related nodes modulated by the network connection weights. The nodes that directly or indirectly accumulate most (or above a certain threshold) activation energy are considered relevant to the set of initial cue nodes. The algorithm itself works with simple linear algebra by iteratively multiplying an activation vector of all network nodes by the matrix of associative weights between nodes [8].

This IR method is ideal for networks such as those defined by the proximity measures of section 3 (eqs. 1 to 4), as well as the traversal proximity of

---

[10] http://pespmc1.vub.ac.be mirrored at http://pcp.lanl.gov.

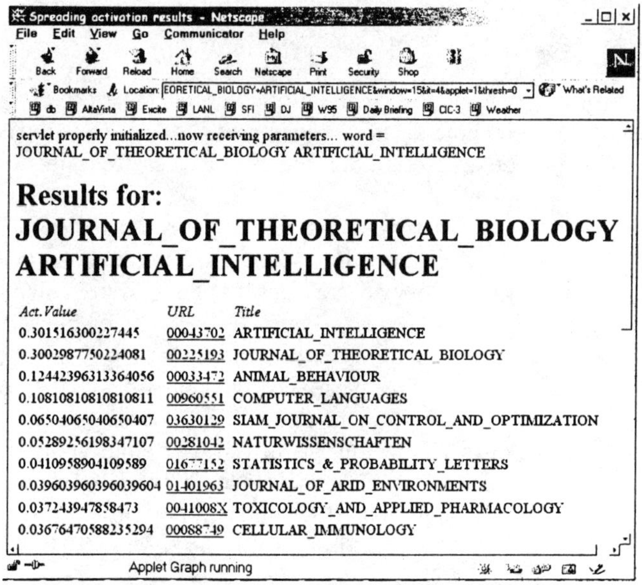

FIGURE 5 Results of spreading activation on the associative networks (traversal proximity) obtained by *@ApWeb* for journal titles.

section 5.1. Several SA utilities built on these proximity measures are available on ARP's web site 1. The main advantage of SA is that it does not depend on keyword lexical matching, but rather it exploits associative knowledge contexts built from relevant information—our enabling substrate. In other words, SA defines a process of context-dependent, knowledge-driven IR [8].

We have applied SA to the network of journal titles obtained from *@ApWeb* described in section 5.1 with very positive results [7]. An example of a recommendation from this utility (available in our web site) is shown in figure 5. In this example, the initial query is *Journal of Theoretical Biology* and *Artificial Intelligence.* A traditional IR search engine, such as those we typically use on the web, would perform a keyword match and return all other journal titles that include the words "biology," "intelligence," "theoretical" "artificial," etc. But with the combination of SA with a *@ApWeb*-organized associative network, we obtain a much more interesting recommendation of journals that do not match syntactically the query's keywords, but are in effect semantically related, e.g., *Statistics and Probability Letters*, *Cellular Immunology*, and *Naturwissenschaften.*

Because the associative network was organized via the *@ApWeb* algorithm, these recommendations reflect the interests of the community of researchers at the Los Alamos National Laboratory. *@ApWeb* and SA define

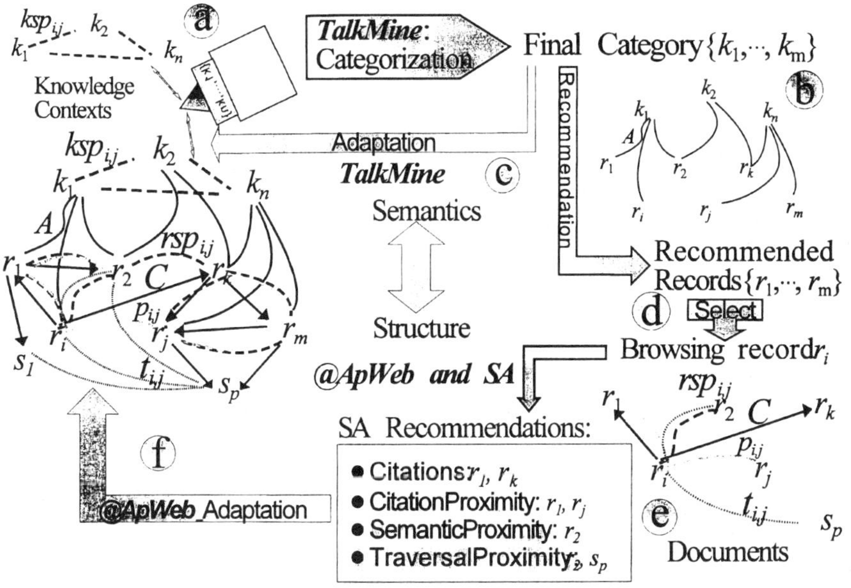

FIGURE 6 The interaction of *TalkMine* and *@ApWeb* in ARP.

another instance of collaborative recommendation. Without access to semantic tokens such as keywords, the structure of knowledge contexts is adapted to the collective semantics of its users. Again, as discussed for *TalkMine*, this user-driven knowledge organization is based on processes similar to social insect organization. But to be able to produce such recommendation processes in DIS, we need to collect an enabling substrate of relationship packages in DIS. With the construction of the knowledge contexts of section 3, we are capable of deploying simple algorithms such as SA, while adapting them with *TalkMine* and *@ApWeb*. In the next section we describe how we integrate *TalkMine* with *@ApWeb* in ARP.

6 ACTIVE RECOMMENDATION PROJECT AND THE FUTURE

With ARP we are interested in porting some of the evolvability of biological systems into IR distributed designs. As discussed in section 2, we also insist on recommendations systems that are both content-based and collaborative. The two systems described in sections 4 (*TalkMine*) and 5 (*@ApWeb*) operate at distinct levels of the knowledge contexts of information resources:

semantics and structure respectively. This allows users to search information resources with keywords, while adapting both keywords and structural links. Their interaction in ARP is depicted in figure 6, which we describe here.

Users search several information resources using *TalkMine* to interact with their knowledge contexts (a). Users express their interests in terms of a set of keywords, which *TalkMine* integrates with the keyword proximity measures of the information resources, utilizing the question-answering categorization process described in section 4. This leads to a final category representing the interests of the user. The category is used to recommend a set records (b) and to adapt the keyword proximity measures of the intervening parties (c). The user then selects a record from the recommended set (d). The document stored in this record is shown to the user, together with a recommendation of other related documents (e). This recommendation implemented with SA (section 5.2) utilizes the citation structure, the structural proximity measures (eqs. 1–2s), the record semantic proximity measure (eq. 4), and the traversal proximity from *@ApWeb* (section 5.1). The browsing information defined by the user's document choices, the document selection paths, are then used as the *@ApWeb* adaptation signals back to the structure of the initial information resources (f).

TalkMine was initially developed as a prototype application for personal computers [47, 50], while *@ApWeb* was designed for adaptive hypertext [5, 8]. *TalkMine* is an adaptive recommendation system which is both content-based and collaborative, while *@ApWeb* with SA is strictly collaborative. The ARP test bed, where both these systems are being integrated, tackles the flaws of information retrieval in DIS as depicted in section 1 in the following manner:

- It establishes an *active environment* of user-system interaction capable of recommending information relevant to the particular users and the expectations of the overall community of users.
- It explores *structural relationships* in the document structure with proximity measures, which are now adaptive via *@ApWeb*.
- It establishes an *evolving semantics* as keyword associations adapt to the expectations of users and new keywords are introduced from the crossover of information among multiple information resources and users with *TalkMine*.
- It establishes *linked information resources* as users can search several resources simultaneously and establish all-way information exchanges.

Therefore, *TalkMine* and *@ApWeb* in ARP overcome the limitations of information retrieval outlined in section 1:

- There is *recommendation* as the system proactively pushes relevant documents to users about related topics that they may have been unaware of. This is achieved because of the structural and semantic proximity measures of knowledge contexts, how they are integrated with user-specific information in the categorization and adaptation processes, and finally by the document retrieval operations of *TalkMine* and SA.

- There is *conversation* between users and information resources and among information resources (and indirectly among users) as a mechanism to exchange or crossover knowledge among them is established.
- There is *creativity* as new semantic and structural associations are set up by *TalkMine* and *@ApWeb*. The categorization process brings together knowledge from the different information resources. This not only adapts existing knowledge, but combines knowledge not locally available to individual information resources. In this sense, because of the conversation process, information resources gain new knowledge previously unavailable.

For all of these characteristics, ARP is establishing an open-ended human-machine symbiosis, based on biologically motivated distributed designs. This design is used in the automatic, adaptive, organization of knowledge in DIS such as library databases or the Internet, facilitating the rapid dissemination of relevant information and the discovery of new knowledge.

ACKNOWLEDGMENTS

We would like to thank Cliff Joslyn for his involvement in ARP and the development of the research presented here, Herbert Van de Sompel for his contributions for our research on Spreading Activation, and Rick Luce for making this research possible at the Los Alamos National Laboratory.

REFERENCES

[1] Anderson, John R. "A Spreading Activation Theory of Memory." *J. Verbal Learn. and Verbal Behav.* **22** (1983): 261–295.

[2] Atlan, H., and M. Koppel. "The Cellular Computer DNA: Program or Data." *Bull. Math. Biol.* **52(3)** (1990): 335–348.

[3] Balabanović, M., and Y. Shoham. "Content-Based, Collaborative Recommendation." *Comm. ACM* **40(3)** (1997): 66–72.

[4] Berry, M. W., S. T. Dumais, and G. W. O'Brien. "Using Linear Algebra for Intelligent Information Retrieval." *SIAM Rev.* **37(4)** (1995): 573–595.

[5] Bollen, J., and F. Heylighen. "Algorithms for the Self-Organization of Distributed, Multi-user Networks." In *Proceedings of the 13th European Meeting on Cybernetics and Systems Research*, edited by R. Trappl, 911–917. Vienna: Austrian Society for Cybernetic Studies, 1996.

[6] Bollen, J. and F. Heylighen. "A System to Restructure Hypertext Networks into Valid User Models." *New Rev. Hypermedia and Multimedia* **4** (1998): 189–213.

[7] Bollen, J. and H. Vandesompel. "A Recommendation System Based on Spreading Activation for SFX." In preparation, 2000.

[8] Bollen, J., H. Vandesompel, and L. M. Rocha. "Mining Associative Relations from Website Logs and Their Application to Context-Dependent Retrieval using Spreading Activation." In *Workshop on Organizing Web Space (WOWS)*. ACM Digital Libraries 99, August 1999. Berkeley, CA: ACM, in press.

[9] Chislenko, Alexander. "Collaborative Information Filtering and Semantic Transports." 1998. ⟨http://www.lucifer.com/~sasha/articles/ACF.html⟩.

[10] Clark, Andy. *Associative Engines: Connectionism, Concepts, and Representational Change.* Cambridge, MA: MIT Press, 1993.

[11] Clark, Andy. *Being There: Putting Brain, Body, and World Together Again.* Cambridge, MA: MIT Press, 1997.

[12] Collins, A. M., and E. F. Loftus. "A Spreading Activation Theory of Semantic Processing." *Psych. Rev.* **82** (1975): 407–428.

[13] Conrad, Michael. "The Geometry of Evolutions." *BioSystems* **24** (1990): 61–81.

[14] Forrest, S., S. Hofmeyr, and A. Somayaji. "Computer Immunology." *Comm. ACM* **40(10)** (1997): 88–96.

[15] Forrest, S., and S. A. Hofmeyr. "Immunology as Information Processing." This volume.

[16] Galvin, F., and S. D. Shore. "Distance Functions and Topologies." *Amer. Math. Month.* **98(7)** (1991): 620–623.

[17] Goodwin, B. *How the Leopard Changed its Spots: The Evolution of Complexity.* New York: Charles Scribner's Sons, 1994.

[18] Gordon, D. "Task Allocation in Ant Colonies." This volume.

[19] Heylighen, Francis. "Collective Intelligence and Its Implementation on the Web: Algorithms to Develop a Collective Mental Map." *Comp. & Math. Org. Theory* **5(3)** (1999): 253–280.

[20] Hinton, G., and J. R. Anderson. *Parallel Models of Associative Memory.* New Jersey: Hillsdale, 1981.

[21] Holland, J. H. *Hidden Order: How Adaptation Builds Complexity.* Reading, MA: Addison-Wesley, 1995.

[22] Hutchins, E., and B. Hazlehurst. "Learning in the Cultural Process." In *Artificial Life II*, edited by C. Langton, C. Taylor, J. D. Farmer, and S. Rasmussen, 689–706. Santa Fe Institute Studies in the Sciences of Complexity. Reading, MA: Addison-Wesley, 1991.

[23] Joachims, T., D. Freitag, and T. Mitchell. "Web-Watcher: A Tour Guide for the World Wide Web." In *15th International Joint Conference on Artificial Intelligence (IJCAI–97)*. Nagaya, Aichi, Japan, 1997.

[24] Johnson, N., S. Rasmussen, C. Joslyn, L. Rocha, S. Smith, and M. Kantor. "Symbiotic Intelligence: Self-Organizing Knowledge on Distributed Networks, Driven by Human Interaction." In *Proceedings of the 6th Inter-*

national Conference on Artificial Life, edited by C. Adami, R. K. Belew, H. Kitano, and C. E. Taylor, 403–407. Cambridge, MA: MIT Press, 1998.

[25] Kanerva, P. *Sparse Distributed Memory.* Cambridge, MA: MIT Press, 1988.

[26] Kannan, R., and S. Vempala. "Real-Time Clustering and Ranking of Documents on the Web." Unpublished manuscript, 1999.

[27] Kauffman, S. *The Origins of Order: Self-Organization and Selection in Evolution.* New York: Oxford University, 1993.

[28] Kessler, M. M. "Bibliographic Coupling between Scientific Papers." *Am. Document.* **14** (1963): 10–25.

[29] Kleinberg, J. M. "Authoritative Sources in a Hyperlinked Environment." In *Proc. of the 9th ACM-SIAM Symposium on Discrete Algorithms*, 668–677. Berkeley, CA: ACM, 1998.

[30] Klimesch, W. *The Structure of Long Term Memory: A Connectivity Model of Semantic Processing.* Hillsdale, NJ: Lawrence Erlbaum, 1994.

[31] Langton, C. G. "Artificial Life." In *Artificial Life*, edited by C. Langton. Reading, MA: Addison-Wesley, 1989.

[32] Lindgren, K. "Evolutionary Phenomena in Simple Dynamics." In *Artificial Life II*, edited by C. Langton, C. Taylor, J. D. Farmer, and S. Rasmussen, 295–312. Santa Fe Institute Studies in the Sciences of Complexity. Reading, MA: Addison- Wesley, 1991.

[33] Meyer, D. E., and R. W. Schvaneveldt. "Facilitation in Recognition Pairs of Words: Evidence of a Dependence Between Retrieval Operations." *J. Exp. Psych.* **90** (1971): 227–234.

[34] Mitchell, Melanie. *An Introduction to Genetic Algorithms.* Cambridge, MA: MIT Press, 1996.

[35] Mitchell, Melanie. "Analogy Making as a Complex Adaptive System." This volume.

[36] Moreno, A., A. Etxeberria, and J. Umerez."Universality Without Matter?" In *Artificial Life IV*, edited by R. Brooks and P. Maes, 406–410. Cambridge, MA: MIT Press, 1994.

[37] Moukas, A., and P. Maes. "Amalthaea: An Evolving Multi-Agent Information Filtering and Discovery Systems for the WWW." *Auton. Agents and Multi-agent Sys.* **1** (1998): 59–88.

[38] Nakamura, K., and S. Iwai. "A Representation of Analogical Inference by Fuzzy Sets and Its Application to Information Retrieval Systems." In *Fuzzy Information and Decision Processes*, edited by M. M. Gupta and E. Sanchez, 373–386. New York: North-Holland, 1982.

[39] Nelson, M. L., K. Maly, N. T. Shen, and M. Zubair. "Buckets: Aggressive, Intelligent Agents for Publishing." *Web-Net J.* **1(1)** (1999): 58–66.

[40] Pattee, Howard H., ed. *Hierarchy Theory: The Challenge of Complex Systems.* New York: George Braziller, 1973.

[41] Pattee, Howard H. "Cell Psychology: An Evolutionary Approach to the Symbol-Matter Problem." *Cognition & Brain Theory* **5(4)** (1982): 191–200.

[42] Pattee, Howard H. "Evolving Self-Reference: Matter, Symbols, and Semantic Closure." *Comm. & Cognition—AI* **12(1–2)** (1995): 9–27.

[43] Pirolli, P., J. Pitkow, and R. Rao. "Silk from a Sow's Ear: Extracting Usable Structure from the Web." In *Proceedings of CHI'96 (ACM), Human Factors in Computing Systems*, Vancouver, Canada, April, 1996. Berkeley, CA: ACM, 1996.

[44] Pitkow, J. *Proceedings of the Sixth International WWW Conference.* Santa Clara, CA, 1997.

[45] Richards, D., B. D. McKay, and W. A. Richards. "Collective Choice and Mutual Knowledge Structures." *Adv. Comp. Sys.* **1** (1998): 221–236.

[46] Rocha, Luis M. "Eigenbehavior and Symbols." *Sys. Resh.* **13(3)** (1996): 371–384.

[47] Rocha, Luis M. "Evidence Sets and Contextual Genetic Algorithms: Exploring Uncertainty, Context and Embodiment in Cognitive and Biological Systems." Ph.D. Dissertation. State University of New York at Binghamton, 1997.

[48] Rocha, Luis M. "Relative Uncertainty and Evidence Sets: A Constructivist Framework." *Intl. J. Gen. Sys.* **26(1–2)** (1997): 35–61.

[49] Rocha, Luis M. "Selected Self-Organization and the Semiotics of Evolutionary Systems." In *Evolutionary Systems: Biological and Epistemological Perspectives on Selection and Self-Organization*, edited by S. Salthe, G. Van de Vijver, and M. Delpos, 341–358. The Netherlands: Kluwer Academic Publishers, 1998.

[50] Rocha, Luis M. "Evidence Sets: Modeling Subjective Categories." *Intl. J. Gen. Sys.* **27** (1999): 457–494.

[51] Rocha, Luis M. "Syntactic Autonomy: Or Why There is No Autonomy without Symbols and How Self-Organizing Systems Might Evolve Them." In *Closure: Emergent Organizations and Their Dynamics*, edited by J. L. R. Chandler and G. Vande Vijves, vol. 901, 207–223. New York: New York Academy of Sciences, in press.

[52] Rocha, Luis M. "Adaptive Recommendation and Open-Ended Semiosis." *Intl. J. Human-Comp. Stud.*: in press.

[53] Rocha, Luis M., and Wim Hordijk. "Representations and Emergent Symbol Systems." *Cognitive Science* (submitted).

[54] Rocha, Luis M., and Cliff Joslyn. "Models of Embodied, Evolving, Semiosis in Artificial Environments." In *Proceedings of the Virtual Worlds and Simulation Conference*, edited by C. Landauer and K. L. Bellman, 233–238. San Diego, CA: The Society for Computer Simulation, 1998.

[55] Salthe, Stanley N. *Development and Evolution: Complexity and Changes in Biology.* Cambridge, MA: MIT Press, 1993.

[56] Schuster, P. "Artificial Life and Molecular Evolutionary Biology." In *Advances in Artificial Life*, edited by F. Moran, A. Moreno, J. J. Merelo, and P. Chacon, 3–19. New York: Springer, 1995.

[57] Small, H. "Co-Citation in the Scientific Literature: A New Measure of the Relationship between Documents." *J. Am. Soc. Info. Sci.* **42** (1973): 676–684.
[58] van Gelder, Tim. "What is the 'D' in 'PDP': A Survey of the Concept of Distribution." In *Philosophy and Connectionist Theory*, edited by W. Ramsey et al. Hillsdale, NJ: Lawrence Erlbaum, 1991.
[59] Zadeh, Lofti A. "Fuzzy Sets." *Info. & Control* **8** (1965): 338–353.

Analogy Making as a Complex Adaptive System

Melanie Mitchell

1 INTRODUCTION

This chapter describes a computer program, called Copycat, that models how people make analogies. It might seem odd to include such a topic in a collection of chapters mostly on the immune system. However, the immune system is one of many systems in nature in which a very large collection of relatively simple agents, operating with no central control and limited communication among themselves, collectively produce highly complex, coordinated, and adaptive behavior. Other such systems include the brain, colonies of social insects, economies, and ecologies. The general study of how such emergent adaptive behavior comes about has been called the study of "complex adaptive systems."

The Copycat program is meant to model human cognition, and its major contribution is to show how a central aspect of cognition can be modeled as the kind of decentralized, distributed complex system described above. In doing so it proposes principles that I believe are common to all complex adaptive systems, and that are particularly relevant to the study of immunology.

Copycat was developed by Douglas Hofstadter and myself, and has been described previously [6, 10, 11, 16, 17]. This chapter summarizes these earlier works, and makes explicit links to the immune system.

2 ANALOGY MAKING AND COGNITION

Analogy making can be defined as "the perception of two or more nonidentical objects or situations as being the 'same' at some abstract level." We chose to focus on analogy making because of its centrality to every aspect of cognition.

Analogy making is at the core of recognition and categorization. For example, children learn to recognize instances of categories such as "dog" or "cat." Even though different dogs look very different, children perceive some essential sameness at an abstract level and can differentiate a dog from a cat. Likewise, children learn to recognize cats and dogs in books as well as in real life, even though on the surface such images are very different from one another and from the corresponding real-life creatures. Hofstadter has pointed out that even the ability to recognize the letter *A* in many different typefaces and handwriting styles requires a highly sophisticated analogy-making ability [4]. For example, the collection of *A*'s given in figure 1, taken from a typeface catalog, illustrates the ease with which people can recognize vastly different shapes as instances of *A* because of some essential abstract similarity. Hofstadter points out that "no single feature, such as having a pointed top or a horizontal crossbar (or even a crossbar at all) is reliable" as an indicator of being an *A*. (Hofstadter [5], p. 242.)

At a more abstract level, people can easily recognize styles of music—"That sounds like Mozart"—or familiar music transported to a less familiar idiom—"Hey, that's a muzak version of 'Hey Jude.'" When you think about it, these are examples of sophisticated analogy making as well. Any two pieces by Mozart are superficially very different, but at some level we perceive a common essence. Likewise, the Beatles rendition of "Hey Jude" and the version you might hear in the supermarket have little in common in terms of instrumentation, tempo, vocals, and other readily apparent musical features, but we can easily recognize it as "the same song."

People make analogies all the time, both consciously and unconsciously. You have probably had the experience of hearing a friend's story about how her flight from Boston to San Francisco was delayed for four hours, and how her four pieces of luggage were lost. You exclaim "The same thing happened to me," thinking of your flight from Adelaide to Perth and how it was delayed for two hours and how two of your three pieces of luggage were lost. Not exactly the same thing, but close. Or you might have read about a war waged by the Soviets in Afghanistan in which their superior military power was thwarted again and again by the determination of a small Western-supported army fighting on its own soil, and been instantly reminded of a war the Americans waged during the 1960s and 1970s in Asia against a small Soviet-supported army fighting on its own soil. You might have even thought, "This is another Vietnam." Again, it is not exactly, but basically, the same. It is that "basically" where analogy lies. Anytime you recognize something, are reminded of something, or see a similar essence in two different situations, you are making an analogy. (For an excellent discussion of analogies, conscious and uncon-

FIGURE 1 A page of *A*'s in different typefaces from a recent Letraset catalog. From *Metamagical Themas: Questing for the Essence of Mind* by Douglas Hofstadter [5]. Copyright © 1985 by Basic Books, Inc. Printed with permission by Basic Books, a member of Perseus Books, L.L.C.

scious, in human thought, see Hofstadter [6].) Such abstract analogies come about by what might be called *high-level perception*, in which objects, pieces of music, memories, or complex situations are viewed in the mind's eye and found to be similar to one another.

It should be clear from all these examples that in making analogies, elements of one situation are fluidly mapped to another situation. A four-hour flight delay in Boston is easily mapped to a two-hour flight delay in Adelaide or perhaps even a six-hour train stopover in Providence. The parallel diagonal crossbar in the Tintoretto face (third row, second column of fig. 1) is easily seen to correspond to the curved and striped crossbar/foot of the Stripes face (seventh row, sixth column of fig. 1). The lilt and clean lines of Mozart's *Eine Kleine Nachtmusik* is easily seen to be similar to the style of his *Divertimento in D.* The ability for concepts to "slip" from situation to situation in this fluid way is a hallmark of human thought and is one of the salient differences between human intelligence and the rigid literality of computers. Our goals in the Copycat project are to understand how human concepts attain this fluidity and how to impart such fluidity to computers.

3 IDEALIZING ANALOGY MAKING

As a first step in modeling analogy making in a computer, Hofstadter devised a "microworld" consisting of analogies between strings of letters [6]. This microworld captures many of the features of analogy making described above, in an idealized fashion.

For example, consider the following problem: if **abc** changes to **abd**, what is the analogous change to **ijk**? Most people describe the change as something like "Replace the rightmost letter by its alphabetic successor," and answer **ijl**. But clearly there are many other possible answers, among them:

- **ijd** ("Replace the rightmost letter by a d"),
- **ijk** ("Replace all c's by d's; there are no c's in **ijk**"), and
- **abd** ("Replace any string by **abd**").

There are, of course, an infinity of other, even less plausible answers, such as **ijxx** ("Replace all c's by d's and all k's by two x's"), and so on, but almost everyone immediately views **ijl** as the best answer. This being an abstract domain with no practical consequences, I may not be able to convince you that **ijl** is a better answer than, say, **ijd** if you really believe the latter is better. However, it seems that humans have evolved in such a way as to make analogies in the real world that affect their survival and reproduction, and their analogy-making ability seems to carry over into abstract domains as well. This means that almost all of us will, at heart, agree that there is a certain level of abstraction that is "most appropriate," and here it yields the answer **ijl**. Those people who truly believe that **ijd** is a better answer would probably, if alive during the Pleistocene, have been eaten by tigers, which explains why there are not many such people around today.

The knowledge available to an analogy maker in this microworld is fairly limited. The 26 letters are known, but only as members of a Platonic linear sequence; shapes of letters, capital versus lower case, sounds, words, and all other linguistic and graphic facts are unknown. The only relations explicitly known are predecessor and successor relations between immediate neighbors in the alphabet. Ordinal positions in the alphabet (e.g., the fact that *S* is the 19th letter) are not known. (In this chapter I denote "Platonic" letters by italic capitals—"*S* is the 19th letter." I denote their instances by nonitalic lower case—"the a in **abc** is the leftmost letter in its string." Strings of letters appearing in analogy problems are in boldface.)

A and *Z*, being alphabetic extremities, are salient landmarks of equal importance. The alphabet is not circular; that is, *A* has no predecessor and *Z* has no successor. The alphabet is known equally well backward and forward (i.e., the fact that *N* is the letter before *O* is as accessible as the fact that *P* is the letter after *O*). In addition, strings (such as **abc** or **kkjjii**) can be parsed equally well from left to right and from right to left. The analogy maker can count, but is reluctant to count above three or so, and has a commonsense notion of grouping by sameness or by alphabetical adjacency (forward or backward with equal ease).

With these restrictions in mind, let us proceed to a second analogy problem: if **abc** changes to **abd**, what is the analogous change to **iijjkk**? The **abc** ⇒ **abd** change can again be described as "Replace the rightmost letter by its alphabetic successor," but if this rule is applied literally to **iijjkk** it yields answer **iijjkl**, which does not take into account the double-letter structure of **iijjkk**. Most people will answer **iijjll**, implicitly using the rule "Replace the rightmost *group of letters* by its alphabetic successor," letting the concept *letter* of **abc** slip into the concept *group of letters* for **iijjkk**.

Another kind of conceptual slippage can be seen in the problem

abc	⇒	**abd**
kji	⇒	?

A literal application of the rule "Replace the rightmost letter by its alphabetic successor" yields answer **kjj**, but this ignores the reverse structure of **kji**, in which the increasing alphabetic sequence goes from right to left rather than from left to right. This puts pressure on the concept *rightmost* in **abc** to slip to *leftmost* in **kji**, which makes the new rule "Replace the *leftmost* letter by its alphabetic successor," yielding answer **lji**. This is the answer given by most people. Some people prefer the answer **kjh**, in which the sequence **kji** is seen as going from left to right but decreasing in the alphabet. This entails a slippage from "alphabetic successor" to "alphabetic predecessor," and the new rule is "Replace the rightmost letter by its alphabetic *predecessor*."

It should be clear by now that the key to analogy making in this microworld (as well as in the real world) is what I am calling *conceptual slippage.* Finding appropriate conceptual slippages given the context at hand is the essence of finding a good analogy. The Copycat program is a model of how concepts slip in response to pressures brought about by ongoing perception of a situation. Copycat is the successor to two previous programs that modeled high-level perception and conceptual slippage: Jumbo, which produced English-like anagrams [8], and Seek Whence, which searched for patterns underlying numerical sequences [9, 15].

As a prelude to developing Copycat, we created thousands of letter-string analogy problems to explore what kinds of slippages come about in response to different kinds of perceptual pressures. Two more examples will be instructive.

Consider

abc	⇒	**abd**
mrrjjj	⇒	?

You want to make use of the salient fact that **abc** is an alphabetically increasing sequence, but how? This internal "fabric" of **abc** is a very appealing and seemingly central aspect of the string, but at first glance no such fabric seems to weave **mrrjjj** together. So either (like most people) you settle for **mrrkkk** (or possibly **mrrjjk**), or you look more deeply. The interesting thing about this problem is that there happens to be an aspect of **mrrjjj** lurking beneath the surface that, once recognized, yields what many people feel is a more satisfying answer. If you ignore the *letters* in **mrrjjj** and look instead at *group lengths*, the desired successorship fabric is found: the lengths of groups increase as "1-2-3." Once this connection between **abc** and **mrrjjj** is discovered, the rule describing **abc** ⇒ **abd** can be adapted to **mrrjjj** as "Replace the rightmost *group of letters* by its *length* successor," which yields "1-2-4" at the abstract level, or, more concretely, **mrrjjjj**.

Finally, consider

abc	⇒	**abd**
xyz	⇒	?

At first glance this problem is essentially the same as the problem with target string **ijk** given above, but there is a snag: *Z* has no successor. Most people answer **xya**, but in Copycat's microworld the alphabet is not circular and thus the program could not come up with this answer. We intentionally excluded it because one of the goals of the project is to model the process by which people deal with impasses. This problem forces an impasse that requires analogy makers to restructure their initial view, possibly making conceptual slippages that were not initially considered, and thus to discover a different way of understanding the situation.

People give a number of different responses to this problem, including **xy** ("Replace the z by nothing at all"), **xyd** ("Replace the rightmost letter by a d"; given the impasse, this answer seems less rigid and more reasonable than did **ijd** for the first problem above), **xyy** ("If you can not take the z's *successor*, then the next best thing is to take its *predecessor*"), and several other answers. However, there is one particular way of viewing the problem that, to many people, seems like a genuine insight, whether or not they come up with it themselves. The essential idea is that **abc** and **xyz** are "mirror images"—**xyz** is wedged against the end of the alphabet, and **abc** is similarly wedged against the beginning. Thus the z in **xyz** and the a in **abc** can be seen to correspond, and then one naturally feels that the x and the c correspond as well. Underlying these object correspondences is a set of slippages that are conceptually parallel: *alphabetic-first* ⇒ *alphabetic-last*, *rightmost* ⇒ *leftmost*, and *successor* ⇒ *predecessor*. Taken together, these slippages convert the original rule into a rule adapted to the target string **xyz**: "Replace the *leftmost* letter by its *predecessor*." This yields a surprising but strong answer: **wyz**.

It is important to emphasize once again that the goal of this project is not to model specifically how people solve these letter-string analogy problems (it is clear that the microworld involves only a very small fraction of what people know about letters and what knowledge they might use in solving these problems), but rather to propose and model mechanisms for high-level perception and analogy making in general. Analogy making can be characterized very broadly as distilling the *essence* of one situation and *adapting* it (via conceptual slippage) to fit another situation. The letter-string analogy problems were designed to isolate and make very clear some of the mental abilities that are required for this process of understanding and perceiving similarity between situations. These abilities include the following (which, though listed separately, are of course strongly interrelated):

- Mentally constructing a coherently structured whole out of initially unattached parts.
- Describing objects, relations, and events at the appropriate level of abstraction.
- Grouping certain elements of a situation while viewing others individually.
- Focusing on relevant aspects and ignoring irrelevant or superficial aspects of situations.
- Taking certain descriptions literally and letting others slip when perceiving correspondences between aspects of two situations.
- Exploring many avenues of possible interpretations while avoiding a search through a combinatorial explosion of possibilities.

The rest of this chapter describes how these abilities are modeled in Copycat.

4 DYNAMICS OF EXPLORING WAYS OF UNDERSTANDING SITUATIONS

When given a situation with many components and potential relations among components, be it a visual scene, a friend's story, or a scientific problem, how does a person (or how might a computer program) mentally explore the typically intractably huge number of possible ways of understanding what is going on and possible similarities to other situations?

The following are two opposite and equally implausible strategies, both to be rejected:

1. Some possibilities are *a priori* absolutely excluded from being explored. For example, after an initial scan of **mrrjjj**, make a list of candidate concepts to explore (e.g., *letter*, *group of letters*, *successor*, *predecessor*, *rightmost*) and rigidly stick to it. The problem with this strategy, of course, is that it gives up flexibility. One or more concepts not immediately apparent as relevant to the situation (e.g., *group length*) might emerge later as being central.
2. All possibilities are equally available and easy to explore, so one can do a "full width" search through all concepts and possible relationships that would ever be relevant in any situation. The problem with this strategy is that in real life there are always too many possibilities, and it is not even clear ahead of time what might constitute a possible concept or relationship for a given situation. If you hear a funny clacking noise in your engine and then your car won't start, you might give equal weight to the possibilities that (a) the timing belt has accidentally come off its bearings or (b) the timing belt is old and has broken. If, for no special reason, you give equal weight to the third possibility that your next-door neighbor has furtively cut your timing belt, you are a bit paranoid. If, for no special reason, you also give equal weight to the fourth possibility that the atoms making up your timing belt have quantum-tunneled into a parallel universe, you are a bit of a crackpot. If you continue and give equal weight to every other possibility...well, you just can't, not with a finite brain. But, on the other hand, there is some chance you might be right about the malicious neighbor, and the quantum-tunneling possibility shouldn't be forever excluded from your cognitive capacities or you risk missing a Nobel Prize.

The moral is that all possibilities have to be potentially available, but they can't all be equally available. Counterintuitive possibilities (e.g., your malicious neighbor; quantum tunneling) must be potentially available but must require significant pressure to be considered (e.g., you've heard complaints about your neighbor; you've just installed a quantum-tunneling device in your car; every other possibility that you have explored has turned out to be wrong).

The problem of finding an exploration strategy that achieves this goal has been solved many times in nature. One example is the way ant colonies forage for food. Many ants wander in random directions away from the nest. When one locates a food source (by sight or smell), it picks up some of the food and returns to the nest, leaving a pheremone trail. Other ants follow such trails when they encounter them, following the trails to the food sources and reinforcing the trails with more pheremone. In this way, the shortest trails leading to the best food sources attain the strongest scent, and increasing numbers of ants follow these trails. But, at any given time, some ants are still following weaker, less-plausible trails, and some ants are still foraging randomly, allowing for the possibility of new food sources to be found.

This is an example of what John Holland has called the balance between "exploration and exploitation" [12]. When promising possibilities are identified, they should be exploited at a rate and intensity related to their estimated promise, which is continually being updated. But at all times exploration for new possibilities should continue. The problem is how to allocate limited resources—be they ants or thoughts—to different possibilities in a dynamic way that takes new information into account as it is obtained. Ant colonies have solved this problem by having large numbers of ants follow a combination of two strategies: continual random foraging combined with a simple feedback mechanism of preferentially following trails scented with pheremone and laying down additional pheremone while doing so.

The immune system also seems to maintain a near-optimal balance between exploration and exploitation. At any time, large numbers of B lymphocytes with different receptors are available for matching potential antigens; these different receptor types are formed via random combinations of genetic material in B-cell precursors. In this way the immune system uses randomness to attain the potential for responding to virtually any antigen it encounters. This potential is realized when an antigen activates a particular B cell and triggers the proliferation of that cell and the production of antibodies with increasing specificity for the antigen in question. Thus the immune system exploits the information it encounters in the form of antigens by allocating much of its resources toward targeting those antigens that are actually found to be present. But it always continues to explore additional possibilities that it might encounter by maintaining its huge repertoire of different B cells. Like ant colonies, the immune system combines randomness with highly directed behavior based on feedback.

Holland formalized the exploitation-versus-exploration balance in terms of a "multiarmed bandit" and proved some theorems regarding the optimal allocation of resources in uncertain environments in which information is continually being obtained. He proposed these as candidate general principles for all adaptive systems [12]. Hofstadter proposed a similar, more specific scheme for exploring such environments: the "parallel terraced scan" [8]. In this scheme many possibilities are explored in parallel, each being allocated resources according to feedback about its current promise, whose estimation is updated continually as new information is obtained. Like in an ant colony or the immune system, all possibilities have the potential to be explored, but at any given time only some are being actively explored, and not with equal resources. When a person (or ant colony, or immune system) has little information about the situation facing it, the exploration of possibilities starts out being very random, highly parallel (many possibilities being considered at once), and "bottom up": there is no pressure to explore any particular possibility more vigorously than any other. As more and more information is obtained, exploration gradually becomes more focused (increasing resources are concentrated on a smaller number of possibilities), less random, and more "top down": possibilities that have already been identified as promising are exploited. As in ant colonies and the immune system, in Copycat such an exploration strategy emerges from myriad interactions among simple, autonomous, and interacting components.

5 OVERVIEW OF THE COPYCAT PROGRAM

Copycat's task is to use the concepts it possesses to build perceptual structures—descriptions of objects, links between objects in the same string, groupings of objects in a string, and correspondences between objects in different strings—on top of the three "raw," unprocessed strings given to it in each problem. The structures that the program builds represent its understanding of the problem and allow it to formulate a solution. Since, for every problem, the program starts out from exactly the same state with exactly the same set of concepts, its concepts have to be adaptable, in terms of their relevance and their associations with one another, to different situations. In a given problem, as the representation of a situation is constructed, associations arise and are considered in a probabilistic fashion according to a parallel terraced scan in which many routes toward understanding the situation are tested in parallel, each at a rate and to a depth reflecting ongoing evaluations of its promise.

Copycat's solution of letter-string analogy problems involves the interaction of the following components:

- The *Slipnet*: A network of concepts, each of which consists of a central node surrounded by potential associations and slippages. A picture of some of the concepts and relationships in the current version of the program is given in figure 2. Each node in the Slipnet has a dynamic *activation* value which gives its current perceived relevance to the analogy problem at hand, and which therefore changes as the program runs. Activation also spreads from a node to its conceptual neighbors, and decays if not reinforced. Each link has a dynamic *resistance* value which gives its current resistance to slippage. This also changes as the program runs. The resistance of a link is inversely proportional to the activation of the node naming the link. For example, when *opposite* is highly active, the resistance to slippage between

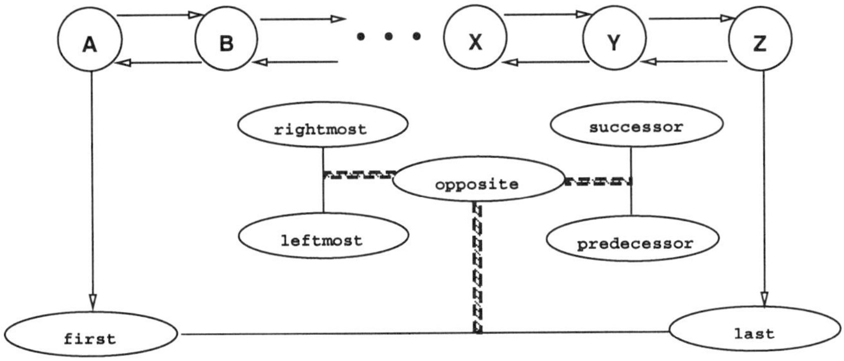

FIGURE 2 Part of Copycat's Slipnet. Each node is labeled with the concept it represents (e.g., *A–Z*, *rightmost*, *successor*, etc.). Solid-line arrows represent directed links between concepts. Some links between nodes (e.g., *rightmost–leftmost*) are connected to a label node giving the link's relationship (e.g., *opposite*). Each node has a dynamic activation value (not shown) and spreads activation to neighboring nodes. Activation decays if not reinforced. Each link has an intrinsic resistance to slippage, which decreases when the label node is activated.

nodes linked by opposite links (e.g., *successor* and *predecessor*) is lowered, and the probability of such slippages is increased.

- The *Workspace*: A working area in which the letters composing the analogy problem reside and in which perceptual structures are built on top of the letters.
- *Codelets*: Agents that continually explore possibilities for perceptual structures to build in the Workspace, and, based on their findings, attempt to instantiate such structures. (The term "codelet" is meant to evoke the notion of a "small piece of code," just as the later term "applet" in Java is meant to evoke the notion of a small application program.)
  Teams of codelets cooperate and compete to construct perceptual structures defining relationships between objects (e.g., "b is the successor of a in **abc**," or "the two i's in **iijjkk** form a *group*," or "the b in **abc** corresponds to the group of j's in **iijjkk**," or "the c in **abc** corresponds to the k in **kji**"). Each team considers a particular possibility for structuring part of the world, and the resources (codelet time) allocated to each team depends on the promise of the structure it is trying to build, as assessed dynamically as exploration proceeds. In this way, a parallel terraced scan of possibilities emerges as teams of codelets, via competition and cooperation, gradually build up a hierarchy of structures that defines the program's "understanding" of the situation with which it is faced.
- *Temperature*: This measures the amount of perceptual organization in the system. As in the physical world, high temperature corresponds to disor-

ganization, and low temperature corresponds to a high degree of organization). In Copycat, temperature both measures organization and feeds back to control the degree of randomness with which codelets make decisions. When the temperature is high, reflecting little perceptual organization and little information on which to base decisions, codelets make their decisions more randomly. As perceptual structures are built and more information is obtained about what concepts are relevant and how to structure the perception of objects and relationships in the world, the temperature decreases, reflecting the presence of more information to guide decisions, and codelets make their decisions more deterministically.

6 A RUN OF COPYCAT

The best way to describe how these different components interact in Copycat is to display graphics from an actual run of the program. These graphics are produced in real time as the program runs. This section displays snapshots from a run of the program on **abc** ⇒ **abd**, **mrrjjj** ⇒ ? This is the same run that was described in Hofstadter and Mitchell [11]. For details about the implementation of the program, see Mitchell [16]. The source code for Copycat, written in Common LISP, is publicly available; see http://www.santafe.edu/~mm for instructions on how to get it.

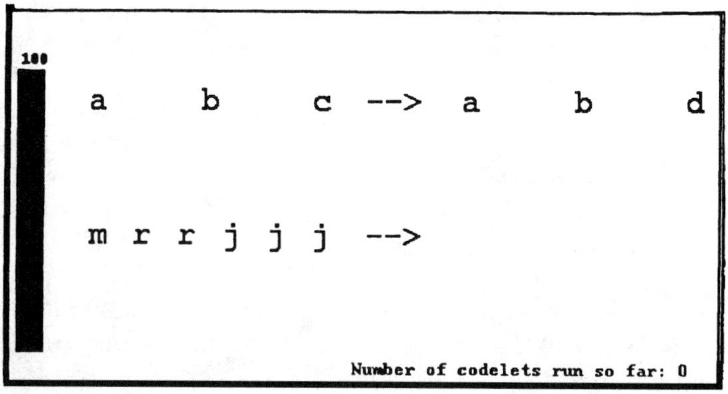

1. The problem is presented. The picture above displays the Workspace (here, the as-yet unstructured letters of the analogy problem); a "thermometer" on the left which gives the current temperature (initially set at 100, its maximum value, reflecting the lack of any perceptual structures); and the number of codelets that have run so far (zero).

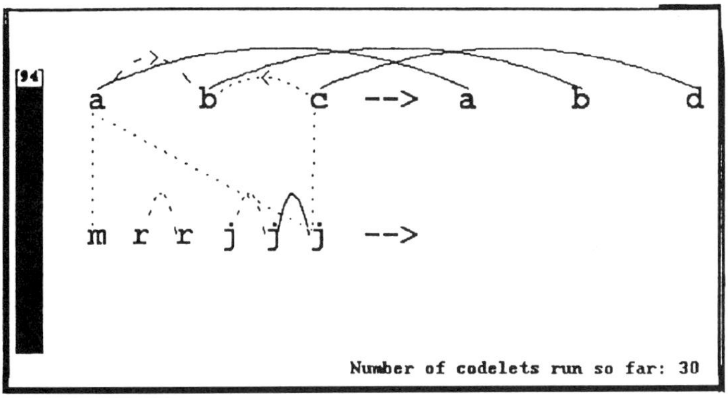

2. Thirty codelets have run and have investigated a variety of s structures. Conceptually, codelets can be thought of as antlike agents, each one probabilistically following a path to explore but being guided by the paths laid down by other codelets. In this case the "paths" correspond to candidate perceptual structures. Candidate structures are proposed by codelets looking around at random for plausible descriptions, relationships, and groupings within strings, and correspondences between strings. A proposed structure becomes stronger as more and more codelets consider it and find it worthwhile. After a certain threshold of strength, the structure is considered to be "built" and can then influence subsequent structure building. In the picture above, dotted lines and arcs represent structures in early stages of consideration; dashed lines and arcs represent structures in more serious stages of consideration; finally, solid lines and arcs represent structures that have been built. The speed at which proposed structures are considered depends on codelets' assessments of the promise of the structure. For example, the codelet that proposed the a–m correspondence rated it as highly promising because both objects are *leftmost* in their respective strings: identity relationships such as *leftmost* ⇒ *leftmost* are always strong. The codelet that proposed the a–j correspondence rated it much more weakly, since the mapping it is based on, *leftmost* ⇒ *rightmost*, is much weaker, especially given that *opposite* is not currently active. Thus the a–m correspondence will be investigated more quickly than the less plausible a–j correspondence.
The temperature has gone down from 100 to 94 in response to the single built structure, the "sameness" link between the rightmost two j's in **mrrjjj**. This sameness link activated the node *same* in the Slipnet (not shown), which creates top-down pressure in the form of specifically targeted codelets to look for instances of sameness elsewhere.

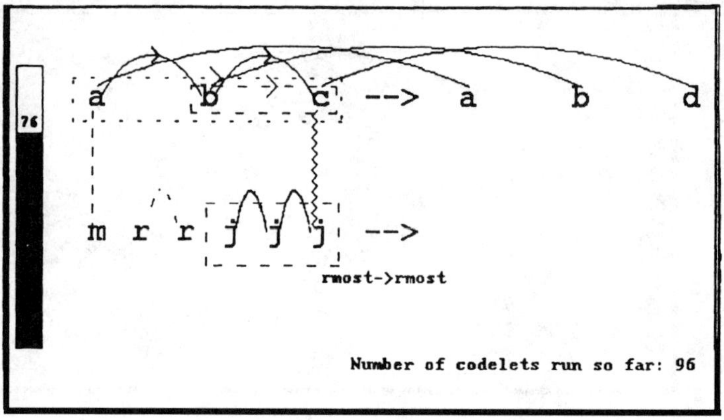

3. Ninety-six codelets have run. The successorship fabric of **abc** has been built. Note that the proposed c-to-b predecessor link of the previous picture has been out-competed by a successor link. The two successor links in **abc** support each other: each is viewed as stronger due to the presence of the other, making rival predecessor links much less likely to destroy the successor links.
   Two rival groups based on successorship links between letters are being considered: **bc** and **abc** (a whole-string group). These are represented by dotted or dashed rectangles around the letters. Although bc got off to an early lead (it is dashed while the latter is only dotted), the group abc covers more objects in the string. This makes it stronger than **bc**—codelets will get around to testing it more quickly and will be more likely to build it than to build **bc**. A strong group, **jjj**, based on sameness is being considered in the bottom string.
   Exploration of the crosswise a–j correspondence (dotted line in the previous picture) has been aborted, since codelets that investigated further found it (probabilistically) too weak to be built. A c–j correspondence has been built (jagged vertical line); the mapping on which it is based (namely, both letters are *rightmost* in their respective strings) is given beneath it.
   Since successor and sameness links have been built, along with an identity mapping (*rightmost* ⇒ *rightmost*), these nodes are highly active in the Slipnet, and are creating top-down pressure in the form of codelets to search explicitly for other instances of these concepts. For example, an identity mapping between the two leftmost letters is being considered.
   In response to the structures that have been built, the temperature has decreased to 76. The lower the temperature, the less random the decisions made by codelets, so less plausible structures such as the **bc** group are even more unlikely to be built.

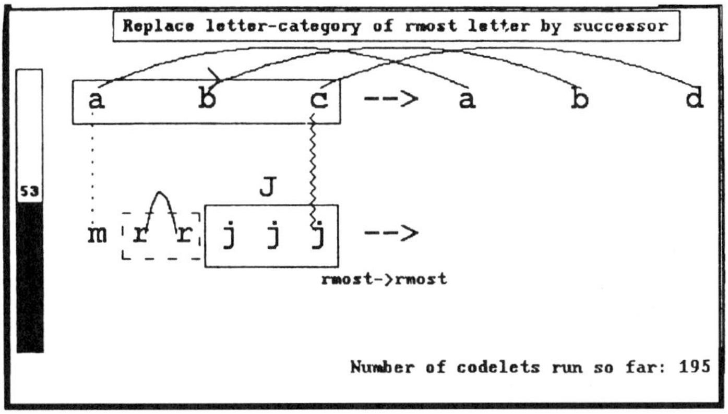

4. The **abc** and **jjj** groups have been built, represented by solid rectangles around the letters. For graphical clarity, the links between letters in a group are not displayed. The existence of these groups creates additional pressure to find new successorship and sameness groups, such as the **rr** sameness group that is being strongly considered. Groups, such as the **jjj** sameness group, become new objects in the string, and can have their own descriptions as well as links and correspondences to other objects. The capital J represents the object consisting of the **jjj** group; the **abc** group likewise is a new object but for clarity a single letter representing it is not displayed. Note that the length of a group is not automatically noticed by the program; it has to be noticed by codelets, just like other attributes of an object. Every time a group node (e.g., *successor group*, *sameness group*) is activated in the Slipnet, it spreads some activation to the node *length*. Thus length is now weakly activated and creating codelets to notice lengths, but these codelets are not urgent compared with others and none so far have run and noticed the lengths of groups.

   A rule describing the **abc** ⇒ **abd** change has been built: "Replace letter-category of rightmost letter by successor." The current version of Copycat assumes that the example change consists of the replacement of exactly one letter, so rule-building codelets fill in the template "Replace ______ by ______," choosing probabilistically from descriptions that the program has attached to the changed letter and its replacement, with a probabilistic bias towards choosing more abstract descriptions (e.g., usually preferring *rightmost letter* to *C*).

   The temperature has fallen to 53, resulting from the increasing perceptual organization reflected in the structures that have been built.

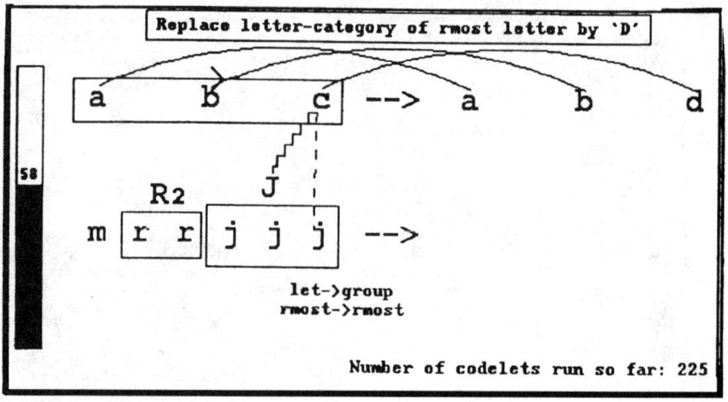

5. Two hundred twenty-five codelets have run. The letter-to-letter c–j correspondence has been defeated by the letter-to-group c–J correspondence. Reflecting this, the *rightmost* ⇒ *rightmost* mapping has been joined by a *letter* ⇒ *group* mapping underlying the correspondence. The c–J correspondence is stronger than the c–j correspondence because the former covers more objects and because the concept *group* is highly active and thus seen as highly relevant to the problem. However, in spite of its relative weakness, the c–j correspondence is again being considered by a new team of codelets.
   Meanwhile, the **rr** group has been built. In addition, its length (represented by the 2 next to the R) has been noticed by a codelet (a probabilistic event). This event activated the node *length*, creating pressure to notice other groups' lengths.
   A new rule, "Replace the letter category of the rightmost letter by 'D'" has replaced the old one at the top of the screen. Although this rule is weaker than the previous one, competitions between rival structures (including rules) are decided probabilistically, and this one simply happened to win. However, its weakness has caused the temperature to increase to 58.
   If the program were to stop now (which is quite unlikely, since a key factor in the program's probabilistic decision of when to stop is the temperature, which is still relatively high), the rule would be adapted for application to the string **mrrjjj** as "Replace the letter category of the rightmost group by 'D'," obeying the slippage *letter* ⇒ *group* spelled out under the c-J correspondence. This yields answer **mrrddd**, an answer that Copycat does indeed produce, though on very rare occasions.
   Codelets that attempt to create an answer run frequently throughout the program (their attempts are not displayed here) but are not likely to succeed unless the temperature is low.

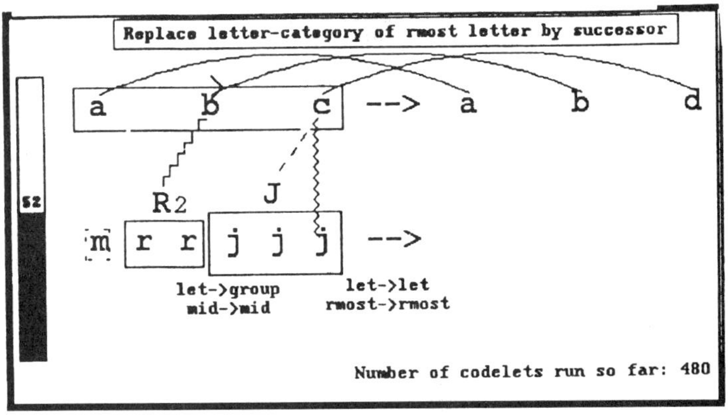

6. Four hundred eighty codelets into the run, the rule "Replace letter-category of rightmost letter by successor" has been restored after it out-competed the previous weaker rule (a probabilistic event). However, the strong c–J correspondence was broken and replaced by its weaker rival, the c–j correspondence (also a probabilistic event). As a consequence, if the program were to stop at this point, its answer would be **mrrjjk**, since the c in **abc** is mapped to a letter, not to a group. Thus the answer-building codelet would ignore the facts that b has been mapped to a group, and that the slippage *letter* ⇒ *group* is now the Workspace. However, the (now) candidate correspondence between the c and the group J is again being strongly considered. It will fight again with the c–j correspondence, and will likely be seen as even stronger than before because of the parallel correspondence between the b and the group R.
   In the Slipnet the activation of *length* has decayed since the length description given to the R group hasn't so far been found to be useful (i.e., it hasn't yet been connected up with any other structures). In the Workspace, the diminished salience of the group R's length description "2" is represented by the fact that the "2" is no longer in boldface.
   The temperature is still fairly high, since the program is having a hard time making a single, coherent structure out of **mrrjjj**, something that it did easily with **abc**. That continuing difficulty, combined with strong top-down pressure from the two sameness groups that have been built inside **mrrjjj**, caused the system to consider the *a priori* very unlikely idea of making a single-letter sameness group. This is represented by the dashed rectangle around the letter m.

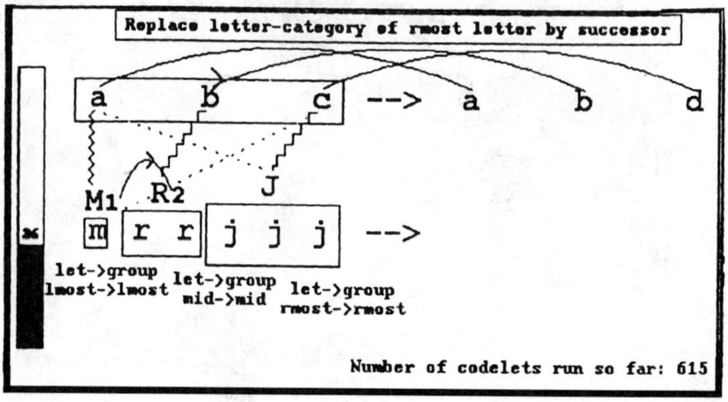

7. As a result of these combined pressures, the M sameness group was built, to parallel the R and J groups in the same string. Its length of 1 has been attached as a description, activating *length*, which makes the program again consider the possibility that group length is relevant for this problem. This activation now more strongly attracts codelets to the objects representing group lengths. Some codelets have already been exploring relations between these objects and, likely due to top-down pressures from **abc** to see successorship relationships, have built a successorship link between the 1 and the 2.
   A consistent trio of *letter* ⇒ *group* correspondences have been built and, as a result of these promising new structures, the temperature has fallen to the relatively low value of 36, which in turn helps to lock in this emerging view.
   If the program were to halt at this point, it would produce the answer **mrrkkk**, which is its most frequent answer (see fig. 3 below).

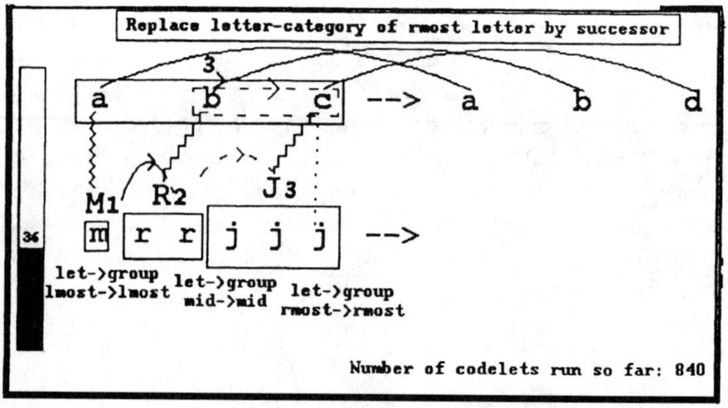

8. As a result of *length*'s continued activity, length descriptions have been attached to the remaining two groups in the problem, **jjj** and **abc**, and a

successorship link between the 2 and the 3 (for which there is much top-down pressure coming from both **abc** and the emerging view of **mrrjjj**) is being considered. Other less likely candidate structures (a **bc** group and a c–j correspondence) continue to be considered, though at considerably less urgency than earlier, now that a coherent perception of the problem is emerging and the temperature is relatively low.

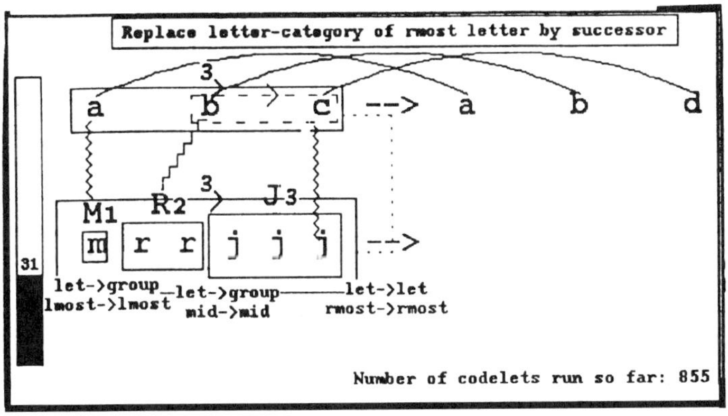

9. The link between the 2 and the 3 was built, which, in conjunction with top-down pressure from the **abc** successor group, allowed codelets to propose and build a whole-string group based on successorship links, here between numbers rather than between letters. This group is represented by a large solid rectangle surrounding the three sameness groups. Also, a correspondence (dotted vertical line to the right of the two strings) is being considered between the two whole-string groups **abc** and **mrrjjj**.
   Ironically, just as these sophisticated ideas seem to be converging, a small renegade codelet, totally unaware of the global movement, has had some good luck: its bid to knock down the c–J correspondence and replace it with a c–j correspondence was successful. Of course, this is a setback on the global level; while the temperature would have gone down significantly because of the strong **mrrjjj** group that was built, its decrease was offset by the now nonparallel set of correspondences linking together the two strings. If the program were forced to stop at this point, it would answer **mrrjjk**, since at this point, as in pictures 4 and 6, the object that changed, the c, is seen as corresponding to the letter j rather than the group J. However, the two other correspondences will continue to put much pressure on the program (in the form of codelets) to go back to the c–J correspondence.

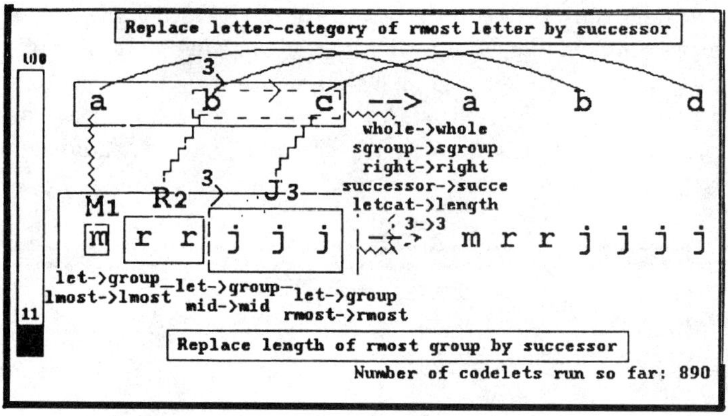

10. Indeed, not much later in the run this happens: the c–j correspondence has been broken and the c–J correspondence has been restored. In addition, the proposed whole-string correspondence between **abc** and **mrrjjj** has been built; underlying it are the mappings *whole* ⇒ *whole*, *successor-group* ⇒ *successor-group*, *right* ⇒ *right* (direction of the links underlying both groups), *successor* ⇒ *successor* (type of links underlying both groups), *letter-category* ⇒ *length*, and *3* ⇒ *3* (size of both groups).
The now very coherent set of perceptual structures built by the program resulted in a very low temperature (11), and (probabilistically) due to this low temperature, a codelet has succeeded in translating the rule according to the slippages present in the Workspace: *letter* ⇒ *group* and *letter-category* ⇒ *length* (all other mappings are identity mappings). The translated rule is "Replace the length of the rightmost group by its successor," and the answer is thus **mrrjjjj**.

It should be clear from the description above that because each run of Copycat is permeated with probabilistic decisions, different answers appear on

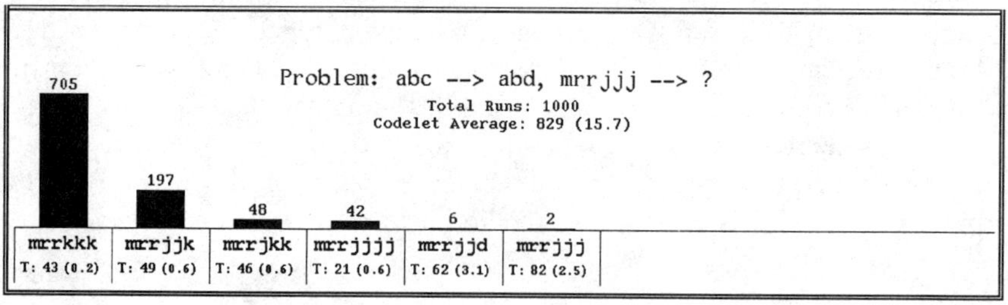

FIGURE 3 A histogram of the different answers Copycat gave over 1000 runs, each starting from a different random number seed.

different runs. Figure 3 displays a histogram of the different answers Copycat gave over 1000 runs, each starting from a different random number seed. Each bar's height gives the relative frequency of the answer it corresponds to, and printed above each bar is the actual number of runs producing that answer. The average final temperature for each answer is also given below each bar's label, with the standard error in parentheses.

The frequency of an answer roughly corresponds to how obvious or immediate it is, given the biases of the program. For example, **mrrkkk**, produced 705 times, is much more immediate to the program than **mrrjjjj**, which was produced only 42 times. However, the average final temperature on runs producing **mrrjjjj** is much lower than on runs producing **mrrkkk** (21 versus 43), indicating that even though the latter is a more immediate answer, the program judges the former to be a better answer, in terms of the strength and coherence of the structures it built to produce each answer.

7 SUMMARY

To summarize, Copycat makes sense of and perceives analogies between situations in a fluid and cognitively plausible way via interaction among three main mechanisms:

- Codelets continually investigating possible structurings in the Workspace and making probabilistic decisions concerning
  - what to look at next;
  - whether to build a structure there (possibly destroying an existing structure);
  - how fast to build it.

  Probabilities are used to insure that no possibilities are ruled out in principle, but that not all possibilities have to be considered.
- The Slipnet, in which concepts
  - become active when instances of them are noticed in the Workspace;
  - feed back to the Workspace by creating top-down pressure, via codelets, to look for further instances of themselves; and
  - spread activation to their neighbors.

  Objects in the Workspace can be mapped onto one another, often requiring slippages between their associated concepts. The Slipnet defines intrinsic resistance to such slippages, but slippages become easier when the concept defining the slippage (e.g., *opposite* for the slippage *successor* $\Rightarrow$ *predecessor*) becomes active.
- Temperature, which starts off high and drops as perceptual structures are built (and rises when they are destroyed). Temperature, in turn, feeds back to codelets by making their decisions more random when temperature is high and more deterministic when temperature is low.

Via these mechanisms, Copycat avoids the Catch 22 of perception: you can't explore everything, but you don't know which possibilities are worth exploring without first exploring them. You have to be open minded, but the territory is too vast to explore everything; you need to use probabilities in order that exploration be fair. In Copycat's strategy, early on there is little information, resulting in high temperature and a high degree of randomness, with lots of parallel explorations. As more and more information is obtained and fitting concepts are found, the temperature falls, and exploration becomes more deterministic and more serial as certain concepts come to dominate. The overall result is that the system gradually changes from a mostly random, parallel, bottom-up mode of processing to a deterministic, serial, top-down mode in which a coherent perception of the situation at hand is gradually discovered and gradually "frozen in." Our claim is that this gradual transition between different modes of processing is a feature common to cognitive systems and to adaptive systems in general.

While Copycat's mechanisms have been shown to be successful in a subset of its letter-string microworld [16], it remains to be demonstrated that such a system will work well on more realistic situations requiring a much larger repertoire of concepts (e.g., visual images). We believe that it will, and this belief is supported by the success of two projects using architectures similar to Copycat's: Tabletop, which makes analogies between objects and relationships on an idealized café table [2], and Letter Spirit, which recognizes and creates letters in different typeface styles on an idealized grid [14]. Copycat has recently been extended to incorporate "self-swatching," in which the program monitors at a high level its own actions [13], an essential component for general high-level perception that was missing in Copycat. Current work in Hofstadter's group includes extending these ideas to the task of solving Bongard problems, a beautiful and open-ended class of visual analogy problems [1, 3]. If successful, this project will go a long way toward the development of a general cognitive architecture for high-level perception and analogy making.

8 EPILOGUE: COPYCAT AND THE IMMUNE SYSTEM

Copycat is one of the "other distributed autonomous systems" referred to in the title of this book. Copycat was not designed with the immune system in mind; ant colonies and cell metabolism (which I didn't discuss here) were the direct biological inspirations. However, there are some interesting parallels between Copycat and the immune system that are worth pointing out. At the most general level, both systems produce global behavior that emerges from the interactions among many simple components with local interactions. At a more specific level, both systems are faced with complex recognition problems that require exploration of many possibilities. They both rely on search controlled by both bottom-up and top-down forces. Like Copycat, the immune system must have the potential to deal with any possibility (i.e., pathogen);

it cannot *a priori* exclude possibilities ahead of time. But the number of possible pathogens is huge and cannot be anticipated. Like Copycat, the immune system uses randomness to avoid this Catch 22. Random combinations are used to construct B-cell receptors from gene libraries, and random mutations allow the system to find increasingly better matches to antigens during affinity maturation. Also like Copycat, the immune system's search is controlled by both bottom-up and top-down aspects. Roughly speaking, bottom-up search consists of a continual patrol of B cells with different receptors, collectively prepared to approximately match any antigen. Top-down search consists of focused B cells, which when activated by a match, create similar B cells to zero in on the particular antigen that has been detected. As in all adaptive systems, maintaining the right balance between these two search modes is essential.

More detailed analogies between Copycat and the immune system are possible. For example, a Slipnet node's activation could be said to correspond to a concentration level of particular B cells or cytokines: Both reflect a particular possibility that the system has deemed worth exploring. Codelets that explore structures might correspond to lymphocytes of various kinds diffusing around in the blood stream and lymph system. Quoting from Hofmeyer's chapter in this volume: "We can abstractly view lymphocytes as mobile, independent *detectors.* There are trillions of these lymphocytes, forming a system of distributed detection, where there is no centralized control, and little, if any, hierarchical control. Detection and elimination of pathogens is a consequence of trillions of cells—detectors—interacting through simple, localized rules." This sounds very much like the way codelets work. Going further, high temperature in Copycat might correspond to fever in the body—the former a signal to Copycat that it must explore more broadly and intensely; the latter a signal to the body that it must increase the intensity of the immune response. (Several of these correspondences were proposed by L. Segel, personal communication.)

Analogies such as these force us to think more broadly about the systems that one is building or trying to understand. If one notices, say, that the role of cytokines in immune signaling is similar to that of codelets that call attention to particular sites in an analogy problem, one is thinking at a general *information-processing* level about the function of a biological entity. Or, if one sees that temperature-like phenomena in the immune system—fever, inflammation—emerge from the joint actions of many agents, one might get some ideas on how to implement an *emergent* temperature in a system like Copycat. The main purposes of this chapter have been to draw the reader's attention to mechanisms of recognition and search that we have proposed as general properties of complex systems, to provoke some thought on how these properties might be implemented in specific complex systems in nature, and to open discussion on how artificial intelligence and cognitive science might additionally benefit from what is being learned about such natural systems.

ACKNOWLEDGMENTS

Many thanks to Lee Segel for inviting me to a very stimulating workshop, and for helpful comments on an earlier version of this chapter.

REFERENCES

[1] Bongard, M. *Pattern Recognition.* Rochell Park, NJ: Hayden Book Co., Spartan Books, 1970.
[2] French, R. M. *The Subtlety of Sameness: A Theory and Computer Model of Analogy Making.* Cambridge, MA: MIT Press, 1995.
[3] Hofstadter, D. R. *Gödel, Escher, Bach: an Eternal Golden Braid.* New York: Basic Books, 1979.
[4] Hofstadter, D. R. "Variations on a Theme as the Crux of Creativity." In *Metamagical Themas*, ch. 12. New York: Basic Books, 1985.
[5] Hofstadter, D. R. *Metamagical Themas: Questing for the Essence of Mind and Pattern.* New York: Basic Books, 1985.
[6] Hofstadter, D. R. "Analogies and Roles in Human and Machine Thinking." In *Metamagical Themas*, ch. 24. New York: Basic Books, 1985.
[7] Hofstadter, D. R. *Fluid Concepts and Creative Analogies.* New York: Basic Books, 1995.
[8] Hofstadter, D. R. "The Architecture of Jumbo." In *Fluid Concepts and Creative Analogies*, ch. 2. New York: Basic Books, 1995.
[9] Hofstadter, D. R. "To Seek Whence Cometh a Sequence." In *Fluid Concepts and Creative Analogies*, ch. 1. New York: Basic Books, 1995.
[10] Hofstadter, D. R., and M. Mitchell. "The Copycat Project: A Model of Mental Fluidity and Analogy-Making." In *Advances in Connectionist and Neural Computation Theory, Volume 2: Analogical Connections*, edited by K. Holyoak and J. Barnden. Norwood, NJ: Ablex, 1984.
[11] Hofstadter, D. R., and M. Mitchell. "The Copycat Project: A Model of Mental Fluidity and Analogy-Making." In *Fluid Concepts and Creative Analogies*, edited by D. R. Hofstadter, ch. 5. New York: Basic Books, 1995.
[12] Holland, J. H. *Adaptation in Natural and Artificial Systems*, 2d ed. Cambridge, MA: MIT Press, 1992.
[13] Marshall, J. B. "Metacat: A Self-Watching Cognitive Architecture for Analogy-Making and High-Level Perception." Ph.D. diss., Bloomington, IN: Indiana Unversity, 1999.
[14] McGraw, G. E., Jr. "Letter Spirit: Emergent High-Level Peception of Letters Using Fluid Concepts." Ph.D. diss., Bloomington, IN: Indiana Unversity, 1995.
[15] Meredith, M. J. "Seek-Whence: A Model of Pattern Perception." Ph.D. diss., Bloomingotn, IN: Indiana University, 1986.

[16] Mitchell, M. *Analogy-Making as Perception: A Computer Model.* Cambridge, MA: MIT Press, 1993.
[17] Mitchell, M., and D. R. Hofstadter. "Perspectives on Copycat: Comparisons with Recent Work." In *Fluid Concepts and Creative Analogies*, edited by D. R. Hofstadter, ch. 6. New York: Basic Books, 1995.

Immunology as Information Processing

Stephanie Forrest
Steven A. Hofmeyr

1 INTRODUCTION

This chapter describes the behavior of the immune system from an information-processing perspective. It reviews a series of projects conducted at the University of New Mexico and the Santa Fe Institute, which have developed and explored the theme "immunology as information processing." The projects cover the spectrum from serious modeling of real immunological phenomena, such as crossreactive responses in animals and the generation of diversity, to computer science applications, especially the attempt to develop an immune system for computers to protect them against viruses, intrusions, and other malicious activities.

In each project, we have used an approach with the following steps: (1) Identify a specific mechanism that appears to be interesting computationally, (2) write a computer program that implements or models the mechanism, (3) study its properties through simulation and mathematical analysis, and (4) demonstrate its capabilities, either by applying the model to a biological question of interest or by showing how it can be used profitably in a computer science setting.

Design Principles for Immune System & Other Distributed Autonomous Systems. L. A. Segel and I. R. Cohen, eds. Oxford Univ. Press, 2001

2 INFORMATION PROCESSING IN THE IMMUNE SYSTEM

The terms "information processing" and "computation" are not easily defined. For the purposes of this chapter we use the term "information" to refer to a spatio-temporal pattern that can be understood and described independently of its physical realization. We will use the words "computation" and "information processing" interchangeably to describe processes that operate on, or transform, information. In the immune system, we believe that such patterns occur in peptides, proteins, and other molecules and that the recognition, learning, storage, communication, and transformation of these patterns governs the behavior of the immune system. This is a strong claim, one that can be contrasted with the more conventional structural view of immunology, which models cells, molecules, and their interactions as mechanical devices. The following sections illustrate how emphasizing the informational properties of the immune system can provide insights which extend our knowledge beyond that provided by the structural view.

The immune system processes peptide patterns using mechanisms that in some cases correspond closely to existing algorithms for processing information (e.g., the genetic algorithm), and it is capable of exquisitely selective and well-coordinated responses, in some cases responding to fewer than ten molecules. Some of the techniques used by the immune system include learning (affinity maturation of B cells, negative selection of B and T cells, and evolved biases in the germline), memory (crossreactivity and the secondary response), massively parallel and distributed computations with highly dynamic components (on the order of 10^8 different varieties of receptors [52] and 10^7 new lymphocytes produced each day [37]), and the use of combinatorics to address the problem of scarce genetic resources (V-region libraries).

It is generally believed that one major function of the immune system is to help protect multicellular organisms from foreign pathogens, especially replicating pathogens such as viruses, bacteria, and parasites. In order to succeed, the immune system must be capable of distinguishing harmful foreign material (which we will refer to as "nonself") from normally behaving constituents of the organism (which we will label "self").[1] That is, it must be able to recognize foreign material (also called "antigen") as foreign—a problem we cast as one of pattern recognition. Detection of pathogens is accomplished by lymphocytes and antibodies, which function as small independent detectors, circulating throughout the body in the blood and lymph systems. Lymphocytes (B cells and T cells) recognize pathogens by forming molecular bonds

[1]For many years, the problem of immunology was described as that of discriminating self from nonself. However, this view is clearly oversimplified. There are, for example, harmless bacteria (nonself) which are tolerated by the immune system. Likewise, the immune system attacks bona fide self cells, for example, in some cancers. Granting these complexities, and without taking a position in the ongoing debate about the immune system's true function, we will use the self/nonself language to refer to the classes of patterns which the immune system tolerates or tries to eliminate, respectively.

between pathogen fragments and receptors on the surface of the lymphocyte. The more complementary the molecular shape and electrostatic surface charge between pathogen and receptor, the stronger the bond (or the higher the affinity). Thus, we say that, in the immune system, pattern recognition is *implemented* as binding. When an immune system detector (including B cells, T cells, or antibodies) binds to a peptide, we say that the immune system has *recognized* the pattern encoded by the peptide.

We have extracted the following informational design principles from our study of immunology: The immune system is *diverse*, which greatly improves robustness, on both a population and individual level, for example, different people are vulnerable to different pathogens; it is *distributed*, consisting of many components which interact locally to provide global protection, so there is no central control, and hence no single point of failure; it is *error tolerant* in that a few mistakes in classification and response are not catastrophic; it is *dynamic*, i.e., individual components are continually created, destroyed, and are circulated throughout the body, which increases the temporal and spatial diversity of the immune system; it is *self-protecting*, i.e., the same mechanisms that protect the body also protect the immune system itself; and it is *adaptable*, i.e., it can learn to recognize and respond to new pathogens, and it can retain a memory of those pathogens to facilitate future responses.

Here, we focus on three examples of the information-processing perspective: How the immune system distributes its detection through negative selection and other mechanisms, the affinity maturation process and its relation to genetic algorithms, and crossreactive memory and its relation to associative memories.

2.1 HOW THE IMMUNE SYSTEM DISTRIBUTES DETECTION AND RESPONSE

The immune system is distributed throughout the body, and this important architectural feature affects nearly everything else. Certainly, any computational model of an immune system must account for distribution, as it constrains many of the details involving detection, learning, memory, and response. One obvious reason for the importance of distribution in the immune system is that the pathogens to which it must respond are themselves distributed throughout the body. Another reason is that distribution makes the system much more robust to attack—it is difficult to neutralize a system which is not in only one place. Distributed systems have always appealed to computer scientists because of their potential efficiency (by distributing workload over multiple locations) and fault-tolerance (robustness to component failures), but these properties are rarely achieved in artificial systems except in highly specialized settings.

How does the immune system achieve highly distributed memory and control? There are three principle mechanisms: (1) negative selection, (2) costimulation and other signaling events, and (3) cell division and death. Negative selection allows the immune system to perform self/nonself discrimination with little or no communication between individual detectors (distributed detection). Costimulation and other signaling occurs locally and controls key processes such as cell proliferation, activation, and death. These processes allow the immune system to allocate resources (cells) dynamically in areas where they are most needed and to prevent many autoimmune reactions. Because each cell carries a complete description of itself, these processes also provide vehicles for distributing the information needed for the immune system to function (distributed autonomous control). We believe that each of these mechanisms is crucial to the highly distributed nature of the immune system. In the remainder of this section, we discuss each of these mechanisms in more detail.

One of the principle ways that the immune system achieves self-tolerance (correct discrimination between self and nonself) is by allowing its detector lymphocytes to mature in isolated settings, the thymus in the case of T cells and the bone marrow for B cells. This is necessary because the binding regions for these cells are created through a pseudorandom genetic process, which could easily lead to self-reactive cells. Focusing on T cells, there are several stages of maturation, including genetic rearrangements, positive selection, and negative selection. Of particular interest is negative selection, in which T cells that bind sufficiently strongly with self-proteins expressed in the thymus are destroyed. In addition to negative selection, there is also positive selection of those T cells that can bind weakly to certain molecules. Negative selection prevents T cells from binding to normal self-proteins, while positive selection ensures that T cells will be able to bind to self-cells that express abnormal peptides. T cells that survive both positive and negative selection are allowed to mature, leave the thymus, and become part of the active immune system. Once in circulation, if a T cell binds to antigen in sufficient concentration, a recognition event is said to have occurred, triggering the complex set of events that leads to elimination of the antigen. These negative-selection processes are called *centralized tolerance* because the cells are censored in a single location.

T-cell censoring can be thought of as defining a set (the set of self-peptides) in terms of its complement (all nonself peptides). We can use this principle to design a distributable change-detection algorithm with interesting properties. Suppose we are given a collection of digital data, which we will call "self," and that we wish to monitor self for changes. The data might be an activity pattern, a static program code, or a file of data. The algorithm works as follows: (1) generate a set of detectors which fail to match self (assuming a "closed world," each detector is then guaranteed to match some portion of nonself); (2) use the detectors to monitor the protected data; and (3) record activated detectors. Whenever a detector is activated, a change is known to have occurred, and the location of the change is known (by examining the pattern which activated the detector). There are several details which must be specified before we have an implementable algorithm (1) How are the detectors represented? (2) How is a match defined? (3) How are detectors generated? (4) How efficient is the algorithm? These topics are explored in detail in Forrest et al. [14] and D'haeseleer [6, 7], but we give highlights here.

There are many possible definitions of self for a computer. Our definitions rely on the idea that any pattern (e.g., a computer program, an execution trace of a program, or the flow of packets through a local area network) can be represented as a finite-length string of symbols. In particular, we represent self as a set of equally sized strings (e.g., by logically segmenting a computer program into equal-length substrings), where the symbol l denotes the length of each string. Similarly, each detector can be defined to be a string of the same length as the substrings.

A perfect *match* between two strings of equal length means that at each location in the string, the symbols are identical. However, perfect matching (perfect binding) is rare in the immune system and improbable between strings of any significant length. Partial matching in symbol strings could be defined in many ways, including Hamming distance or edit distance. However, we typically use a more immunologically plausible rule called r-contiguous bits [38]. This rule looks for r contiguous matches between symbols in corresponding positions. Thus, for any two strings x and y, we say that $match(x, y)$ is true if x and y agree (match) in at least r contiguous locations. The value r is a threshold and determines the specificity of the detector, which is an indication of the number of strings that can be matched (detected) by a single detector. For example, if $r = l$, the matching is completely specific; that is, the detector will detect only a single string (itself). A consequence of a partial matching rule with a threshold, such as r-contiguous bits, is that there is a tradeoff between the number of detectors used, and their specificity. As the specificity of the detectors increases, so the number of detectors required to achieve a certain level of coverage also increases.

Detectors can be generated in several ways. A general method (one that works for any matching rule) is also the one used by the immune system. Simply generate detectors at random, censor them against self, and eliminate those that match self. For the "r-contiguous bits" rule defined above, this generating procedure is inefficient—the number of random strings that must be generated and tested is approximately exponential in the size of self. However, more efficient algorithms based on dynamic programming methods allow us to generate detectors in linear time for the r-contiguous bits rule [6].

The negative-detection algorithm has several interesting properties. First, it can be easily distributed because each detector can function independently of other detectors, that is, without communication between detectors or coordination of multiple detection events. This is because each detector covers part of nonself. A set of detectors can be split up over multiple sites, which will reduce the coverage at any given site but which will provide good syste-mwide coverage. To achieve similar coverage using detectors which match against self would be much more expensive. In this second case (which we call "positive detection"), the system would maintain a description of self and notice when a pattern appeared that failed to match the description. To do this, either a complete set of the detectors (to specify the complete normal pattern) would be needed at every site, resulting in multiple copies of the detection system, or the sites must be in continual communication to coordinate their results. To see why this is true, consider a pattern that fails to match the positive detectors. The match failure could be for two reasons—either the pattern will be matched by a positive detector located at another location, or it is a true anomaly.

A second point about the negative-selection algorithm is that if we assume a closed world and a complete specification of self, it will never report false positives. Depending on how the detector sets are chosen, however, there is a chance of false negatives. Consequently, the algorithm is likely to be more applicable to dynamic or noisy data where perfect discription is difficult to achieve. This is in contract to cryptographic applications, where the data are static, it is important to detect any one-bit change, and efficient change-detection methods already exist. The number of detectors that is required to detect nonself (using the r-contiguous bits matching rule) depends on how the self set is organized, what false-negative rate we are willing to tolerate, and choice of matching rule [6, 14]. For randomly chosen self sets, the number is roughly the same order of magnitude as the size of self, but for nonrandom data the number is often much lower.

A second mechanism that supports distributed processing in the immune system is the concept of a *second signal*, also known as *costimulation*. From the information-processing perspective, this mechanism helps the immune system avoid autoimmune reactions (or false positives), especially in the presence of distributed learning or adaptation (e.g., somatic hypermutation). It also allows the immune system more flexibility in determining tolerance. Rather than a completely centralized tolerance mechanism (e.g., if all self/nonself determination were performed in the thymus), the second signal is one means by which the immune system can determine tolerance in the periphery, by taking advantage of the fact that self-patterns occur much more frequently than nonself patterns.

As Hofmeyr described [26], one example of how a second signal works is given by T-helper lymphocytes. To review, when a B lymphocyte (that is possibly a mutated descendant of an earlier lymphocyte that survived negative selection) binds a foreign peptide (the first signal), it requires additional stimulation by a signal from a T-helper lymphocyte (that has been censored against self in the thymus) in order to trigger an immune response. The second-signal system thus prevents mutating B-lymphocyte cell lines from incorrectly reacting against self. It also helps prevent autoimmunity in T cells, in the following way. Not all peripheral self-proteins are expressed in the thymus. Consequently, T cells emerging will not necessarily be tolerant of all self-proteins. Self-tolerance in these T cells can also be assured through costimulation. In this case, the first signal occurs when binding exceeds the affinity threshold, but the second signal is provided by the cells of the innate immune system, such as macrophages. We call this *frequency tolerization* [24], because a self-reactive T cell is likely to encounter self in the absence of tissue damage with much higher frequency than self in the presence of tissue damage; i.e., self is much more frequent than nonself. A similar situation arises in computer security settings where normal behavior is more frequent than malicious intrusions (see section 4). This general signaling strategy is widespread in the immune system, and it can involve more than two signals.

The processes of negative selection and costimulation help the immune system determine which patterns it should tolerate and which patterns it should eliminate. That is, negative selection and costimulation address the problem of representation and learning of representations. The processes of cell replication and death are a third method by which the immune system achieves its distributed organization. However, here the emphasis is on control. Instead of employing a central process to generate and manage all detectors, immune system control is distributed throughout the body. By control, we refer to the immune system processes that allocate resources, determine which type of immune response will be invoked (effector choice), and know how and when to shut down an immune response. Immune cells are self-replicating (which allows the system to be more autonomous and more adaptable), and their control functions result from the processes of programmed cell death, competition for antigen, and so forth.

How might these ideas be useful to computer science? An example is in computer security, where we are concerned about the problem of protecting an artificial immune system from being attacked. If we implemented protective mechanisms as a distributed collection of self-replicating modules, it would be much more difficult to neutralize the entire defense system. There is a strong analogy here to computer viruses, which are a powerful form of distributed computation (unfortunately, to date mostly harmful). A major problem with computer viruses is uncontrolled replication. At most, they check to see if a file has already been infected before reinfecting it. What is needed is a distributed way to regulate the replication and destruction of distributed agents/detectors. Once again, we can take inspiration from the immune system.

We have already described some of the mechanisms through which B cells can be stimulated to clone themselves. For example, when an antigen is recognized in the presence of T-cell help (indicating that the antigen is not part of self), the B cell is stimulated to divide. Unchecked, this would eventually lead to a disproportionate number of B cells of one type. Thus, each increase in cell division must be balanced out at some point by a concomitant reduction in the rate of reproduction and the removal of excess cells through programmed cell death (apoptosis). The dynamic control of relative cell birth and death rates is an important component of the immune system's distributed control system. The immunology behind this control system is complex and only partially understood (for a review, see Boise and Thompson [1]). However, the basic principles have been identified and provide a reasonable starting point. These include costimulation (discussed earlier) and negative feedback cycles, in which lymphocyte activation promotes immediate survival of cells but also triggers mechanisms that eventually lead to their deletion at the end of the response.

Many different signaling molecules, called cytokines, are believed to participate in the immune response (see Denny [4]), but how they all work together has not been explained systematically [55]. This complex network of signaling molecules and cells apparently has the following properties [41, 42]: (1) every cytokine type affects multiple cells; (2) every function (immune response) is affected by multiple cytokines; (3) immune cells secrete a mixture (vector) of cytokines; (4) signals are molecules, and therefore, distributed (locally) by diffusion; and (5) these signals can be subverted (e.g., viruses can evolve to avoid or interfere with cytokines, perhaps by blocking receptors), so there is an evolutionary pressure toward robust, secure networks. Relevant to computer science, the cytokine signaling networks provide interesting clues about how to design a distributed autonomous control network that is dynamic (both the nodes and connections are changing in time), robust to small perturbations, but responsive to large perturbations.

To summarize, negative selection of detectors provides centralized tolerance and then gives the immune system its ability to distribute detection. Costimulation allows the immune system to distribute its censoring (frequency tolerization). Finally, the processes of cell replication, apoptosis, and the local diffusion of cytokines give the immune system the ability to distribute its resource allocation decisions and control.

2.2 AFFINITY MATURATION AND GENETIC ALGORITHMS

In a *primary response* the immune system uses learning mechanisms similar to biological evolution to design detectors that are specific to a particular antigen. Learning is required if the antigen is "new," that is, if it has not previously been encountered in the lifetime of the organism. Let us consider extracellular pathogens, against which B cells are a primary defense. When a B cell is activated by binding pathogen, it produces many copies of itself (clones that are produced through cell division), in a process called *clonal expansion*. The resulting cells can undergo *somatic hypermutation*, creating daughter B cells with mutated receptors. These new B cells compete for pathogens with their parents and with other clones. The higher the affinity of a B cell for available pathogens, the more likely it is to clone. This results in a Darwinian process of variation and selection, called *affinity maturation*. Affinity maturation enables B cells to adapt rapidly to the specific pathogens present in the body. High-affinity B cells deal with pathogens efficiently by secreting antibody, which can promote pathogen destruction. This is especially important when the immune system is fighting off a replicating pathogen, a situation which is essentially a race between pathogen reproduction and B-cell reproduction. Efficient binding of pathogens is also required to clear infections completely.

The affinity maturation process is reminiscent of a genetic algorithm [10, 16, 27], but there are some important differences. In their traditional form, genetic algorithms process a population of individuals. The population is created randomly in the initial generation. In the simplest case, each individual is a bit string, typically representing a candidate solution to a problem. Variations among individuals in the population result in some individuals being more fit than others (e.g., better problem solutions). These differences are used to bias the selection of a new set of individuals at the next time step, referred to as selection. During selection, a new population is created by making copies of more successful (more fit) individuals and deleting less successful ones. However, the copies are not exact. There is a probability of mutation (random bit flips), crossover (exchange of corresponding substrings between two individuals), or other changes to the bit string during the copy operation. By transforming the previous set of good individuals to a new one, the genetic operations generate a new set of individuals, which have a better than average chance of having high fitness. When this cycle of evaluation, selection, and genetic operations is iterated for many generations, a population of individuals arises, that is biased towards highly fit individuals.

Genetic algorithms can be viewed as a first-order model of the affinity maturation process, as well as a model of change over evolutionary time scales. In the affinity maturation case, each individual in the population can be thought of as a single B cell. High-fitness B cells are those that bind with high affinity to frequently occurring antigen, and they are activated to produce B-cell clones (selection). Somatic hypermutation is the genetic operator, that allows the population of B-cell clones to evolve to be highly specific to the frequently occurring antigen. Over time, a population of B cells is produced that binds much more tightly to the prevalent antigen than before the process started. There are two important differences between affinity maturation in the immune system and conventional genetic algorithms. First, in an immune system undergoing hypermutation, there is no obvious equivalent to crossover. In this case, we can think of the crossover probability being set to zero. A second difference is that in the immune system the amplification (selection) and mutation phases are apparently distinct [31], whereas in the genetic algorithm they are interleaved.

As we will see in section 3.1, the genetic algorithm can be used to study different hypotheses about how the immune system is likely to have evolved. In this case, each individual in the genetic algorithm corresponds to a biological individual (more properly, the immune system genes of a biological individual), fitness corresponds to the survivability of that individual against pathogens, and mutation and crossover correspond to the genetic changes that an individual passes on to his or her offspring.

2.3 CROSSREACTIVITY AND ASSOCIATIVE MEMORIES

A successful immune response results in the proliferation of B cells that have high affinities for the foreign pathogens that caused the response. The information encoded in these B cells constitutes the "memory" of the immune system. Understanding immune memory is problematic because B cells typically live for just a few days, and once an infection is eliminated, it is not well understood what prevents the adapted subpopulation of B cells from dying out. On subsequent encounters with the antigen, however, the immune system responds with a *secondary response* in which the memory cells for the earlier antigen quickly produce large quantities of specific antibodies. These secondary responses are much faster and stronger than primary responses. The distinction between primary and secondary responses is the basis for vaccination, and it is the reason why we get diseases such as measles only once. In vaccination, an attenuated version of the antigen is injected to prime the immune system, so that when the real antigen is encountered, it can produce a response quickly and in large volume.

B-cell receptors do not require an exact match to an antigen in order to be activated. If the immune system is primed with a particular antigen, and then presented with a related antigen (one that is structurally similar but not exactly the same) some of the memory components for the first antigen can be stimulated by the second antigen, producing a secondary immune response. Thus, it is sometimes possible to produce a secondary response to an antigen that the immune system has never seen before. In the field of associative memories, this is called association or generalization [49]. In immunology, such a secondary response is called "crossreactive memory," or "original antigenic sin." Crossreactivity can be beneficial (as in the case of vaccinating with cowpox to protect against the related disease smallpox), and it can be harmful (as in the case where a secondary response is ineffective at eliminating the new antigen but blocks an effective primary response).

3 IMMUNE SYSTEM MODELING

In section 2 we discussed immunology from an informational perspective. Here, we describe how such a perspective can contribute to immune system modeling. The models we describe are all based on a universe in which antigens and detectors are represented by strings over a small alphabet of symbols, and interactions among strings represent molecular binding. In effect, we represent only the receptor region on the surface of a lymphocyte. This approach, first introduced by Farmer et al. [8], is now now widely adopted in the theoretical immunological community. It has been used to study a wide variety of immune system mechanisms. For a recent example see Detours and Perelson [5]. In these "artificial immune systems," binding takes place when an antibody string and an antigen string have similar binary patterns.[2] Binding between idealized antibodies and antigens is defined by a matching function that rewards more specific matches over less specific ones, as we saw earlier with the r-contiguous bits rule. This constraint is related to the immune system's ability to distinguish self from nonself, because recognition of nonself must be fairly specific in order to avoid recognizing self.

Representing binding between antigens and antibodies as simple string matching has advantages and disadvantages. On the one hand, it fails to capture much of what is known and of interest to immunologists—the details of specific molecular and cellular interactions. On the other hand, it provides the ability to model many cells and their interactions, something that we think will be increasingly important over the next several years, both for understanding immune system function and for exploiting immune system principles in computing. Also, it allows us to isolate the informational aspects of immune system processing from other potential factors such as dose, antigenicity, immunocompetence, and virulence and transmissibility, as we will see in section 3.2.

With this basic modeling abstraction in place we can construct a wide variety of model immune systems. For example, affinity maturation can be modeled in such a system by constructing one population of antibodies and one of antigens, each from bit strings. Antigens are "presented" to the antibody population one at a time, and high-affinity antibodies have their fitness increased. The antibody population is then evolved by a genetic algorithm based on its success at matching antigens [13, 50]. Similarly, crossreactive memory can be modeled by constructing populations of B cells, presenting the system with a single antigen type, allowing affinity maturation, then presenting the system with one or more related antigens, and studying the strength of the response [47]. The following subsections describe two such models, one that is concerned with the evolutionary pressures that shaped the immune system and one that is concerned with the question of repeated vaccination against a mutating virus.

3.1 DIVERSITY GENERATING MECHANISMS

One question that can be addressed with models such as these involves the generation of diversity. Several mechanisms have been discovered by which the immune system is able to generate its enormously diverse set of receptors. These include: immune receptor libraries [43], combinatorial rearrangement of entries from multiple libraries [52], junctional diversity [15], and somatic hypermutation and/or gene conversion [54]. We would like to better understand the relative contributions of these different mechanisms, and if possible, what role each mechanism plays in generating the immune repertoire. A re-

[2] A more direct analogy with binding in immunology would be based on complementary binary patterns. In binary alphabets, however, matching rules based on complementary bit patterns are logically equivalent to those based on similarity.

lated question is whether the immune system has evolved its genetics to cover pathogen space randomly or if it has evolved biases that incorporate learning about the pathogenic environment. These questions are difficult to address experimentally in animals.

Ron Hightower's 1996 dissertation [18] studied the mechanisms through which the immune system creates its large number of unique receptors—the multigene families. His work provided insight about how and why the natural immune system evolved as it did [40], showed that robust pattern recognizers can be learned with a surprisingly small amount of information [19], and gave an interesting example of a genotype-to-phenotype map that is nontrivial but still quantifiable. His model inserted an interpretation step between the representation manipulated by evolution (the genetic libraries) and the representation that is operative during an individual's life (the expressed antibodies) [20].

Mihaela Oprea then used Hightower's multigene family model as part of a detailed study of the sources and evolutionary significance of diversity in the immune system [34, 35, 36]. Part of this work was a study of how germline diversity (that is, the antibody gene libraries) contribute to the structure of the antibody repertoire. To study the effect of germline diversity, Oprea used a genetic algorithm to model how the immune system might have evolved. (Note, this is a different use of the genetic algorithm from that described earlier for affinity maturation.) The genetic algorithm employed a population of M individuals, called hosts, which evolved in an environment of hostile pathogens. Each pathogen was represented as a bit string. Each individual in the population consisted of an antibody library, containing A antibodies, where each antibody was represented as a bit string of length L (typically, $L = 16$). Pathogens were also represented as bit strings of length L, and the antibody set was evolved on a fixed, but large (2^9), pathogen set. The A antibodies were concatenated to form a single chromosome. This representation of antibody libraries is reminiscent of the so-called "Pitt" approach to classifier systems [2, 28], in which candidate sets of production rules are represented by concatenating the individual rules to form a single chromosome. Good rule sets (those that solve a given problem well) are then evolved using the genetic algorithm. In the case of Oprea's model, each library (one individual's genome) corresponds to a classifier system if we consider each encoded antibody to take the role of a single classifier rule. This aspect of her representation seems to correspond quite directly to V-region genes in humans.

Oprea then used the model to determine how the fitness of a single individual scales with the amount of diversity (number of entries) and with the matching rule used to determine bond strength. Her results suggest that adding more and more antibodies to the genome-encoded repertoire improves the survival probability of the individual by smaller and smaller amounts, the exact relation being determined by the binding rule. She experimented both with binding rules based on Hamming distance [34] and on free-energy calculations [35]. Theses results suggest an explanation for why the V-region libraries in various species do not seem to number more than approximately one hundred genes. However, if the selection pressure for increasing library size is small, what would keep evolution from producing even smaller libraries than the ones that we observe? One possible explanation is that there is a hard threshold in antibody/pathogen binding, below which recognition will not occur at all. In this case, some minimal number of antibodies would be required to ensure that at least one has minimal affinity for any given pathogen. Alternatively, one can imagine that the pathogen set is structured as a distribution of clusters, such that different antibodies in the library would reflect different clusters of pathogens. Oprea subscribes to the second alternative, conjecturing that the antibody genes encode antibodies which are "strategically" placed in the space of possible receptors, thus providing a form of "coarse graining" of the pathogen space.

3.2 VACCINE DESIGN FOR MUTATING VIRUSES

A second example of how ideas about information processing can be used to better understand the immune system is Derek Smith's 1997 dissertation on the crossreactive immune response and its application to the problem of vaccine design for mutating viruses [48]. Smith developed a model of crossreactive memory in immunology (closely related to Kanerva's Sparse Distributed Memory [30]) and applied the model to the problem of vaccine design for mutating viruses, focusing on influenza [44, 45, 46, 49].

Smith's computational model is concerned with B cells, plasma cells, antibodies, memory B cells, and antigens. The model is *individual based* in the sense that each individual immune cell and antibody is represented in the computer explicitly. (By contrast, many models represent each type of constituent explicitly, together with the concentration in which they exist.) Smith's model is a particularly simple and elegant example of the individual-based style of modeling. Each individual's receptor region is represented as a simple string of twenty symbols, each symbol chosen from a four-letter alphabet (e.g., a, b, c, d). The four symbols are intended to represent generic properties of the binding region; e.g., it might be reasonable to interpret them as polar, nonpolar, large, small. In an earlier study, Smith found that with a binding rule based on the idea of Hamming distance (at how many positions in the string do the symbol strings differ in value?) on strings of length 20 defined over an alphabet of size 4, he obtained crossreactivity patterns that agree well with known biological data [46].

In the model, large populations (10^7) of B cells, each with a randomly generated receptor, are created. The population is then presented with antigen, and the B cells that bind to the antigen with sufficient affinity, under the Hamming rule described earlier, have a chance to be stimulated to divide, undergo somatic hypermutation, and differentiate into a plasma or memory cell. Secreted antibody has a chance to bind antigen, and antigen-antibody complexes are removed from the simulation. Details about how the model is implemented, using a technique called lazy evaluation, are given in Smith et al. [44].

Smith used his model to study the effect of repeated vaccination against a mutating virus [45]. Influenza is an example of such a virus. Antigenic drift causes new but related strains of the virus to circulate through human populations on an annual basis. Influenza vaccines are, therefore, updated regularly to track the antigenic drift. Smith was interested in the case of human populations who are vaccinated annually with the updated strains. Through a series of simulations, based on historical records of actual observed antigenic distances, he observed cases of positive interference between vaccines (where the first year's vaccine reinforced the effect of the second year's vaccine) and negative interference (where the first year's vaccine prevented the second year's vaccine from being effective). He analyzed these cases in detail and concluded that the different outcomes resulted from specific combinations of antigenic distances between the first-year vaccine, the second-year vaccine, and the epidemic strain that actually appears after the second-year vaccine is administered. These different cases are illustrated in figure 1.

Smith then compared his simulation results with the published epidemiological studies on vaccine efficacy (see fig. 2). As the figure shows, Smith's simulation results agree closely with the epidemiological studies, even though his model is exquisitely simple. His results are significant because they provide a parsimonious explanation, based on antigenic distances, for both the cases in which multiyear vaccines are successful and the cases in which they are not. Further, his explanation relies only on the informational properties of the receptors (how closely they are related to each other in sequence space) and not on biological details of how the receptors are implemented, deployed, or on biological properties of the virus. Finally, his dissertation makes testable predictions about how different antigenic strains might interact with one another and with the wild-type strain encountered during an epidemic.

4 IDEAS FROM IMMUNOLOGY APPLIED TO COMPUTER SECURITY

The immune system can also be studied for the purpose of designing better artificial adaptive systems. A natural domain in which to apply immune system mechanisms is computer security, where the analogy between protecting the body and protecting a normally operating computer is evident. A computer security system should ensure the *integrity* of a machine or set of machines, protecting them from unauthorized intruders and foreign code. This is similar in functionality to the immune system protecting the body (*self*) from invasion by harmful microbes (*nonself*). Within this domain, we have studied several problems, including computer virus detection [6, 14], host-based intrusion detection [11, 12, 22, 53], automated response [51], and network intrusion detection [23, 25]. This last project incorporates several different immune sys-

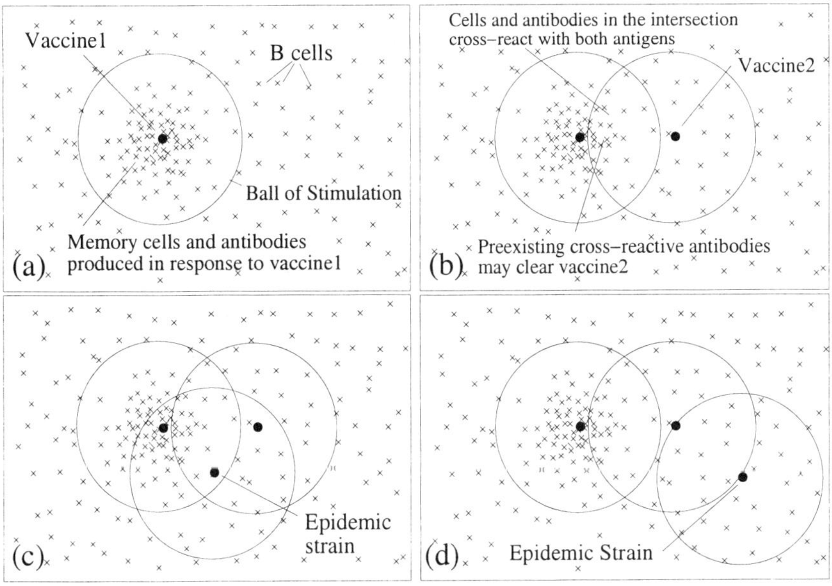

FIGURE 1 An illustration of the antigenic distance hypothesis. *Shape space* diagrams are a way to illustrate the affinities between multiple B cells/antibodies and antigens, and also the antigenic distances between antigens [39]. In these shape space diagrams, the affinity between a B cell or antibody (X) and an antigen (•) is represented by the distance between them. Similarly, the distance between antigens is a measure of how similar they are antigenically. (a) B cells with sufficient affinity to be stimulated by an antigen lie within a *ball of stimulation* centered on the antigen. Thus, a first vaccine (vaccine 1) creates a population of memory B cells and antibodies within its ball of stimulation. (b) Crossreactive antigens have intersecting balls of stimulation, and antibodies and B cells in the intersection of their balls—those with affinity for both antigens—are the crossreactive antibodies and B cells. The antigen in a second vaccine (vaccine 2) will be partially eliminated by preexisting crossreactive antibodies (depending on the amount of antibody in the intersection), and thus the immune response to vaccine 2 will be reduced [3, 9]. (c) If a subsequent epidemic strain is close to vaccine 1, it will be cleared by preexisting antibodies. (d) However, if there is no intersection between vaccine 1 and the epidemic strain, there will be few preexisting crossreactive antibodies to clear the epidemic strain quickly, despite two vaccinations. Note, in the absence of vaccine 1, vaccine 2 would have produced a memory population and antibodies that would hve been protective against both the epidemic strains in (c) and (d). For an antigen with multiple epitopes (such as influenza), there would be a ball of stimulation for each epitope. Printed with permission by *PNAS* **96** (1999). National Academy of Sciences, USA.

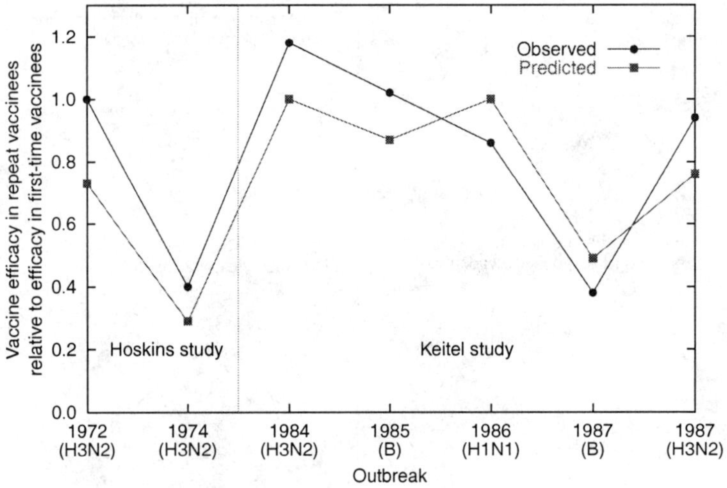

FIGURE 2 Observed vaccine efficacy in repeat vaccinees relative to the efficacy in first-time vaccines, and predicted vaccinees efficacy based on the antigenic distance hypothesis. Printed with permission by *PNAS* **96** (1999). National Academy of Sciences, USA.

tem mechanisms into an integrated system and is the focus of this section, which is largely excerpted from Hofmeyr and Forrest [21].

Hofmeyr's network immune system is situated in a local-area broadcast network (LAN) and is used to protect the LAN from network-based attacks. In contrast with switched networks, broadcast LANs have the convenient property that every location (computer) sees every packet passing through the LAN. In this domain, self is defined to be the set of normal pairwise connections (at the TCP/IP level) between computers, including connections between two computers in the LAN as well as connections between one computer in the LAN and one external computer (fig. 3). A connection is defined in terms of its "data-path triple"—the source IP address, the destination IP address, and the service (or port) by which the computers communicate [17, 33]. This information is compressed to a single 49-bit string that unambiguously defines the connection. Self is then the set of normally occurring connections observed over time on the LAN, each connection being represented by a 49-bit string. Similarly, nonself is also a set of connections (using the same 49-bit representation), the difference being that nonself consists of those connections, potentially an enormous number, that are not normally observed on the LAN.

Natural immune systems consist of many different kinds of cells and molecules. Here, we simplify by introducing one basic type of detector cell which combines useful properties from several different immune cells. This

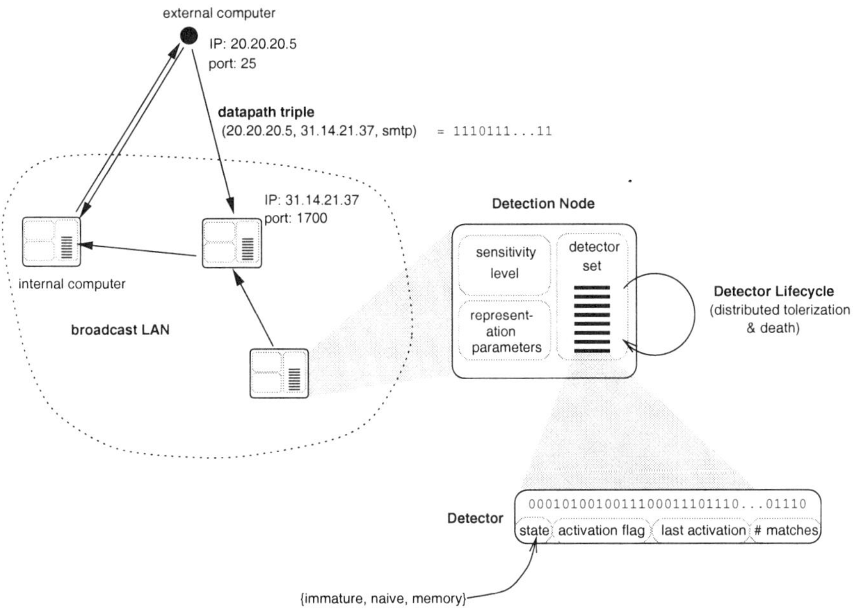

FIGURE 3 Architecture of the artificial immune system.

detector cell has several different possible states, which correspond roughly to thymocytes (immature T lymphocytes undergoing negative selection in the thymus), naïve B lymphocytes (which have never matched foreign material), and memory B lymphocytes (which are long-lived and easily stimulated). The natural immune system also has many different types of effector cells, which implement different immune responses (e.g., macrophages, mast-cells, etc.). This set of features was not included in Hofmeyr's model.

Each detector cell is represented by a single bit string of length $l = 49$ bits, and a small amount of local state (see fig. 3). There are many ways of implementing the detectors; for example, a detector could be a production rule, or a neural network, or an agent. We chose to implement detection (binding) as *string matching*, where each detector is a string d, and detection of a string s occurs when there is a match between s and d, according to a *matching rule.* We use string matching because it is simple and efficient to implement, and easy to analyze and understand. Recall that two strings d and s match under the r-contiguous bits rule if d and s have the same symbols in at least r contiguous bit positions.

The detectors are grouped into sets, one set per machine, or computer, on the LAN; each computer loosely corresponds to a different location in the

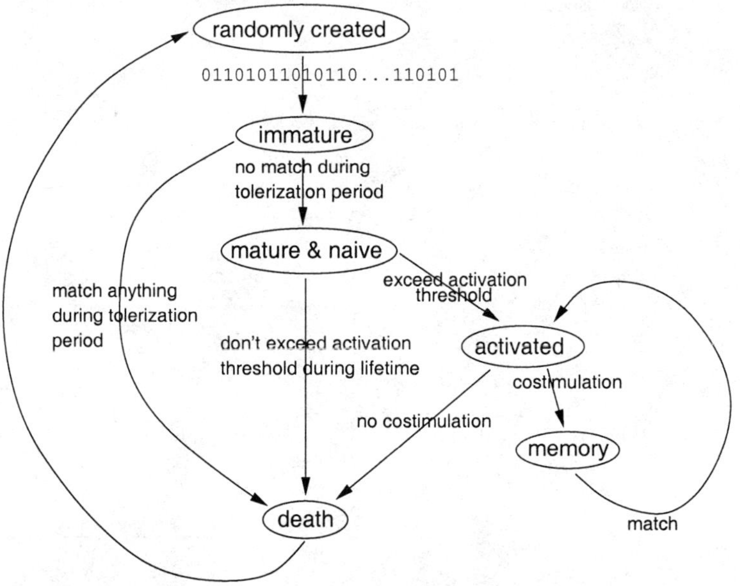

FIGURE 4 Life cycle of a detector. A detector is initially randomly created, and then remains immature for a certain period of time, which is the tolerization period. If the detector matches any string a single time during tolerization, it is replaced by a new randomly generated detector string. If a detector survives immaturity, it will exist for a finite lifetime. At the end of that lifetime it is replaced by a new random detector string, unless it has exceeded its match threshold and becomes a memory detector. If the activation threshold is exceeded for a mature detector, it is activated. If an activated detector does not receive costimulation, it dies (the implicit assumption is that its activation was a false positive). However, if the activated detector receives costimulation, it enters the competition (see above) to become a memory detector with an indefinite lifespan. Memory detectors need only match once to become activated.

body.[3] Because of the broadcast assumption, each detector set is constantly exposed to the current set of connections in the LAN, which it uses as a dynamic definition of self (i.e., the observed connections in a fixed time period are analogous to the set of proteins expressed in the thymus during some period of time). Within each detector set, new detectors, or thymocytes, are created randomly and asynchronously on a continual schedule, similarly to the natural immune system. These new detectors remain *immature* for some period of time, during which they have the opportunity to match any current

[3]The ability of immune system cells to circulate throughout the body is an important part of the immune system that we are currently ignoring. In our system, detectors remain in one location for their lifetime.

network connections. If a detector matches when it is immature, it is killed (deleted). This process, described earlier, is called *negative selection.*

Detectors that survive the initial censoring are promoted to *mature* detectors (analogous to T lymphocytes leaving the thymus and B lymphocytes leaving the bone marrow). Each mature detector can now act independently. If a mature detector d matches a sufficient number of packets (see the discussion of activation thresholds below), an alarm is raised. The time for which d is in the naïve phase can be thought of as a learning phase. At the end of this learning phase, if d has failed to match a packet, it is deleted, but if it has matched a sufficient number of nonself packets then it enters a competition with other activated detectors to become a memory detector. Memory detectors have a greatly extended, potentially infinite, lifetime. Memory detectors have a lower threshold of activation (see below), thus implementing a "secondary response" that is more sensitive and responds more aggressively than naïve detectors to previously seen strings. Although these memory detectors are desirable, a large fraction of naïve detectors must always be present, because the naïve detectors are necessary for the detection of novel foreign packets.

Both the natural immune system and our artificial immune system face the problem of "incomplete self sets." When T lymphocytes undergo negative selection in the thymus, they are exposed to most, but not all, of the proteins in the body. Consequently, the negative selection process is incomplete in the sense that a lymphocyte could survive negative selection but still be reactive against a legitimate self-protein (one that was not presented in the thymus), potentially leading to an autoimmune reaction. In our artificial immune system, such an auto-immune reaction is called a *false positive.* False positives arise if we train the system on an incomplete description of self, and then encounter new but legitimate patterns. We would like the system to be tolerant of such minor, legitimate new patterns, but still detect abnormal activity. We have implemented two methods designed to overcome this problem: activation thresholds and adaptive thresholds. *Activation thresholds* are similar in function to avidity thresholds in lymphocytes. A lymphocyte is covered with many identical receptors, and it is only activated when sufficiently many receptors are bound to pathogens, i.e., when the avidity threshold for binding is exceeded. Analogously, each detector in the artificial immune system must match multiple times before it is activated. Each detector records the number of times it matches, and it raises an alarm only when the number of matches exceeds the activation threshold, which is stored locally for each detector set. Once a detector has raised an alarm, it returns its match count to zero. This mechanism has a time horizon: over time the count of matches slowly returns to zero. Thus, only repeated occurrences of structurally similar and temporally clumped strings will trigger the detection system.

However, some attacks may be launched from many different machines, in which case the first method is unlikely to be successful. To detect such distributed coordinated attacks, we introduce a second method, called *adap-*

tive activation (see fig. 3). Whenever the match count of a detector goes from 0 to 1, the local activation threshold for the set of detectors on a computer is reduced by one. Hence, each different detector that matches for the first time "sensitizes" the detection system, so that all detectors on that machine are more easily activated in future. This mechanism also has a time horizon; over time, the activation threshold gradually returns to its default value. This method will detect diverse activity from many different sources, provided the activity happens within a certain period of time. This mechanism roughly captures the role that inflammation, cytokines, and other molecules play in increasing or decreasing the sensitivity of individual immune system lymphocytes within a physically local region.

Two simple adaptive mechanisms used in our artificial immune system are negative selection and the maturation of naïve cells into memory cells. A third adaptive mechanism has also been incorporated—affinity maturation. In its simple form, detectors compete against one another for foreign packets, just as lymphocytes compete to bind foreign antigen, as described in section 2.2. In the case where two detectors simultaneously match the same packet, the one with the closest match (greatest fitness) wins. This introduces pressure for more specific matching into the system, causing the system to discriminate more precisely between self and nonself. We propose, although we have not yet implemented this, that successful detectors (those that bind many foreign packets) will undergo proliferation (making copies and migrating to other computers) and somatic hypermutation (copying with a high mutation rate).

In our system, we use a human as the second signal. When a detector raises an alarm, there is some chance that it is a false alarm (autoimmune reaction). Before taking action, the artificial immune system waits a fixed amount of time (say, 24 hours) for a costimulatory signal, which in the current implementation is an e-mail message from a human. If the signal is received (confirming the anomaly), the detector enters the competition to become a memory detector with an indefinite lifespan (see above), but, if it loses the competition, it remains naïve and has its match count reset to 0. If the second signal is not received, the artificial immune system assumes that it was a false alarm and destroys the detector, as in the natural immune system. The complete lifecycle of a detector is shown in figure 3.

It might seem more natural to send messages to the artificial immune system in the case of false alarms instead of true anomalies, so that the artificial immune system can adjust itself appropriately by immediately deleting the autoreactive detectors. Unfortunately, this would create a vulnerability, because a malicious adversary could send signals to the artificial immune system, labeling true foreign packets as false alarms, thus tolerizing the artificial immune system against certain forms of attack. The form of costimulation that we have used is much more difficult to subvert. Because false alarms are generally much more frequent than true anomalies, our costimulation method has the additional advantage that action by the human operator is required in the less frequent case.

Each of the mechanisms described above can be implemented with a single detector set running on a single location. We could trivially gain efficiency advantages by distributing the single detector set across all locations on the LAN, thus distributing the computational cost of intrusion detection. Such distribution will give linear speedup, because there are no communication costs (apart from the signaling of alarms and costimulation). However, we take advantage of another immune system feature to implement a more powerful form of distribution.

Molecules of the *major histocompatibility complex* (MHC) play an important role in immune systems, because they transport protein fragments (called peptides) from the interior regions of a cell to its surface, *presenting* these peptides on the cell's surface. This mechanism enables roving immune system cells to detect infections in cells without penetrating the cell membrane. There are many variations of MHC, each of which binds a different class of peptides. Each individual in a population is genetically capable of making a small set of these MHC types (about ten), but the set of MHC types varies in different individuals. Consequently, individuals in a population are capable of recognizing different profiles of peptides, providing an important form of population-level *diversity*.[4] Our artificial immune system uses *permutation masks* to achieve a similar kind of diversity. A permutation mask defines a permutation of the bits in the string representation of the network packets. Each detector set has a different, randomly generated, permutation mask. One limitation of the negative-selection algorithm as originally implemented is that it can result in undetectable abnormal patterns called *holes*, which limit detection rates [6, 7]. Holes can exist for any symmetric, fixed-probability matching rule, but by using permutation masks we effectively change the match rule on each host, and so overcome the hole limitation. Thus, the permutation mask controls how the network packet is presented to the detection system, which is analogous to the way different MHC types present different sets of peptides on the cell surface.

Our network intrusion detection system was empirically tested on actual data collected on a subnet of 50 computers at the Computer Science department at the University of New Mexico. The data consisted of two months of network traffic. This was used as the basis for a simulation of a network of 50 computers. We chose to simulate the environment because we needed to repeat many different runs of the simulation to test out the effects of the various mechanisms. We also collected seven traces of network traffic during real incidents of attempted and successful intrusions (for a description of these intrusions, see Hofmeyr [25]). In the simulation, with each of the 50 computers running with 100 detectors, the false positive rates were on the order of two per day. This is regarded as very low in the intrusion detection community [32]. In addition, the system successfully detected all seven intrusive incidents, in all cases detecting at least 44% of the nonself strings present

[4]For example, there are some viruses, such as the Epstein-Barr virus, that have evolved dominant peptides that cannot be bound by particular MHC types, leaving individuals who have those MHC types vulnerable to the disease [29].

in each trace. The various mechanisms were found to be useful: Activation thresholds reduce false positives by up to a factor of 10; the sensitivity mechanism is useful for detecting distributed coordinated attacks; costimulation reduces false positives by up to a factor of three; memory detectors greatly improve the secondary response; and, diverse permutation masks are useful for detecting anomalies that are similar to the normal traffic.

5 CONCLUSIONS

In this chapter, we emphasized three themes: (1) understanding the distributed memory and control systems of the immune system from an informational perspective, (2) creating models that emphasize the informational properties of the immune system, and (3) building an integrated adaptive immune system that can address an important unsolved problem in computer science.

What is the value of such analogies? In the case of modeling the real immune system, there are several benefits to the information-based approach. As we saw in the influenza example, viewing biological processes as computations can lead to nonobvious predictions that can be tested. It can also allow us to infer global properties of the immune system that are impossible to test experimentally using today's technology. An example of this is Oprea's conclusions about pathogen space coverage, based on her genetic algorithm simulation. Further, simulations allow us the opportunity to perform perturbation experiments and to run controls that may be difficult to do experimentally. Oprea's scaling relations are the result of such an exercise.

In the case of computer science, our study of the immune system has revealed an important set of design principles, many of which are widely appreciated in computer science, but some of which are not. More importantly, we have few examples of working computer systems that illustrate all of these properties in an integrated whole. Hofmeyr's artificial immune system is, however, an important step in this direction. As our computers and the software they run become more complex and interconnected, properties such as robustness, flexibility and adaptability, diversity, self-reliance and autonomy, and scalability can only become more important to computer design.

The research described here stresses the similarities between computers and immunology. Yet, there are also major differences, which it is necessary to respect. The success of all analogies between computing and living systems will ultimately rest on our ability to identify the correct level of abstraction—preserving what is essential from an information-processing perspective and discarding what is not. In the case of immunology, this task is complicated by the fact that real immune systems handle data that are very different from that handled by computers. In principle, a computer vision system or a speech-recognition system would take as input the same data as a human does (e.g., photons or sound waves). In contrast, regardless of how successful we are at constructing a computer immune system, we would never expect or want it to handle pathogens in the form of actual living cells, viruses, or parasites. Thus, the level of abstraction for computational immunology is necessarily higher than that for computer vision or speech, and there are more degrees of freedom in selecting a modeling strategy.

ACKNOWLEDGMENTS

Alan Perelson introduced the authors to immunology and made significant contributions to most of the projects described in this chapter. Lee Segel is an active collaborator in our ongoing efforts to understand the response side of the immune system. Over the past ten years, an unusually talented and dedicated group of students and postdoctoral fellows have comprised the "adaptive" group at UNM. This chapter reflects their achievements and articulates some of the insights that were hammered out in our weekly seminar.

Forrest is affiliated with the Santa Fe Insitute and the University of New Mexico, Department of Computer Science. Hofmeyr conducted the work described in this chapter while he was a graduate student and post-doctoral fellow at the University of New Mexico. The authors gratefully acknowledge the support of the National Science Foundation (grants IRI-9711199 and CDA-9503064), the Office of Naval Research (grant N00014-99-1-0417), and the Intel Corporation.

REFERENCES

[1] Boise, Lawrence H., and Craig B. Thompson. "Hierarchical Control of Lymphocyte Survival." *Science* **274** (1996): 67–68.

[2] Booker, L. B., R. L. Riolo, and J. H. Holland. "Learning and Representation in Classifier Systems." *Art. Intel.* **40** (1989): 235–282.

[3] Davenport, F. M., A. V. Hennessy, and T. Francis. *J. Exp. Med.* **98** (1953): 641–656.

[4] Denny, T. N. "Cytokines: A Common Signaling System for Cell Growth, Inflammation, Immunity, and Differentiation." This volume.

[5] Detours, V., and A. S. Perelson. "Explaining High Alloreactivity as a Quantitative Consequence of Affinity-Driven Thymocyte Selection." *Proc. Natl. Acad. Sci.* **96** (1999): 5153–5158.

[6] D'haeseleer, P., S. Forrest, and P. Helman. "An Immunological Approach to Change Detection: Algorithms, Analysis, and Implications." In *Proceedings of the 1996 IEEE Symposium on Computer Security and Privacy.* Los Alamitos, CA: IEEE Press, 1996.

[7] D'haeseleer, Patrik. "An Immunological Approach to Change Detection: Theoretical Results." In *Proceedings of the 9th IEEE Computer Secu-*

rity Foundations Workshop. Los Alamitos, CA: IEEE Computer Society Press, 1996.

[8] Farmer, J. D., N. H. Packard, and A. S. Perelson. "The Immune System, Adaptation, and Machine Learning." *Physica D* **22** (1986): 187–204.

[9] Fazekas de St. Groth, S., and R. G. Webster. "Disquisitions on Original Antigenic Sin. II. Proof in Lower Creatures." *J. Exp. Med.* **124** (1966): 347–361.

[10] Forrest, S. "Genetic Algorithms: Principles of Adaptation Applied to Computation." *Science* **261** (1993): 872–878.

[11] Forrest, S., S. Hofmeyr, and A. Somayaji. "Computer Immunology." *Comm. ACM* **40(10)** (1997): 88–96.

[12] Forrest, S., S. Hofmeyr, A. Somayaji, and T. Longstaff. "A Sense of Self for Unix Processes." In *Proceedings of the 1996 IEEE Symposium on Computer Security and Privacy.* Los Alamitos, CA: IEEE Press, 1996.

[13] Forrest, S., B. Javornik, R. Smith, and A. Perelson. "Using Genetic Algorithms to Explore Pattern Recognition in the Immune System." *Evol. Comp.* **1(3)** (1993): 191–211.

[14] Forrest, S., A. S. Perelson, L. Allen, and R. Cherukuri. "Self-Nonself Discrimination in a Computer." In *Proceedings of the 1994 IEEE Symposium on Research in Security and Privacy.* Los Alamitos, CA: IEEE Computer Society Press, 1994.

[15] Gilfillan, S., A. Dierich, M. Lemeur, C. Benoist, and D. Mathis. "Mice Lacking TdT: Mature Animals with an Immature Lymphocyte Repertoire." *Science* **261** (1993): 1175–1178.

[16] Goldberg, D. E. *Genetic Algorithms in Search, Optimization, and Machine Learning.* Reading, MA: Addison Wesley, 1989.

[17] Heberlein, L. T., G. V. Dias, K. N. Levitte, B. Mukherjee, J. Wood, and D. Wolber. "A Network Security Monitor." In *Proceedings of the IEEE Symposium on Security and Privacy.* Los Alamitos, CA: IEEE Press, 1990.

[18] Hightower, R. "Computational Aspects of Antibody Gene Families." Ph.D. diss., University of New Mexico, Albuquerque, NM, 1996.

[19] Hightower, R., S. Forrest, and A. S. Perelson. "The Evolution of Emergent Organization in Immune System Gene Libraries." In *GACONF6*, edited by L. J. Eshelman. Los Altos, CA: Morgan-Kaufmann, 1995.

[20] Hightower, R. H., S. Forrest, and A. S. Perelson. "The Baldwin Effect in the Immune System: Learning by Somatic Hypermutation." In *Adaptive Individuals in Evolving Populations*, edited by R. K. Belew and M. Mitchell, 159–167. Santa Fe Institute Studies in the Sciences of Complexity. Reading, MA, Addison-Wesley, 1996.

[21] Hofmeyr, S., and S. Forrest. "Immunity by Design: An Artificial Immune System." In *Proceedings of the Genetic and Evolutionary Computation Conference (GECCO)*, 1289–1296. San Francisco, CA: Morgan-Kaufmann, 1999.

[22] Hofmeyr, S., A. Somayaji, and S. Forrest. "Intrusion Detection using Sequences of System Calls." *J. Comp. Security* **6** (1998): 151–180.

[23] Hofmeyr, S. A., and S. Forrest. "Architecture for an Artificial Immune System." *Evol. Comp. J.* in press.

[24] Hofmeyr, S. A. "Immune System Principles and Mechanisms for Computer Security." Ph.D. diss. proposal, Department of Computer Science, University of New Mexico, 1996.

[25] Hofmeyr, Steven A. "An Immunological Model of Distributed Detection and Its Application to Computer Security." Ph.D. diss., University of New Mexico, Albuquerque, NM, 1999.

[26] Hofmeyr, Steven A. "An Interpretative Introduction to the Immune System." This volume.

[27] Holland, J. H. *Adaptation in Natural and Artificial Systems.* Ann Arbor, MI: The University of Michigan Press, 1975.

[28] Holland, J. H., K. J. Holyoak, R. E. Nisbett, and P. Thagard. *Induction: Processes of Inference, Learning, and Discovery.* Cambridge, MA: MIT Press, 1986.

[29] Janeway, C. A., and P. Travers. *Immunobiology: The Immune System in Health and Disease*, 3d ed. London: Current Biology Ltd., 1996.

[30] Kanerva, Pentti. *Sparse Distributed Memory.* Cambridge, MA: MIT Press, 1988.

[31] Kepler, T. B., and A. S. Perelson. "Cyclic Re-Entry of Germinal Center B Cells and the Efficiency of Affinity Maturation." *Immunol. Today* **14(8)** (1993): 412–415.

[32] Lippman, R. "Lincoln Laboratory Intrusion Detection Evaluation." October 1999. ⟨http://www.ll.mit.edu/IST/ideval/index.html⟩.

[33] Mukherjee, B., L. T. Heberlein, and K. N. Levitt. "Network Intrusion Detection." *IEEE Network* (1994): 26–41.

[34] Oprea, M., and S. Forrest. "Simulated Evolution of Antibody Gene Libraries under Pathogen Selection." In *Proceedings of the 1998 IEEE International Conference on Systems, Man, and Cybernetics.* Los Alamitos, CA: IEEE Press, 1998.

[35] Oprea, M., and S. Forrest. "How the Immune System Generates Diversity: Pathogen Space Coverage with Random and Evolved Antibody Libraries." In *Proceedings of the Genetic and Evolutionary Computation Conference (GECCO)*, 1651–1656. San Francisco, CA: Morgan-Kaufmann, 1999.

[36] Oprea, Mihaela L. "Antibody Repertoires and Pathogen Recognition: The Role of Germline Diversity and Somatic Hypermutation." Ph.D. diss., University of New Mexico, Albuquerque, NM, 1999.

[37] Osmond, D. G. "The Turn-Over of B-cell Populations." *Immunol. Today* **14(1)** (1993): 34–37.

[38] Percus, J., O. Percus, and A. S. Perelson. "Predicting the Size of the Antibody-Combining Region from Consideration of Efficient Self/Non-Self Discrimination." *Proc. Nat. Acad. Sci.* **90** (1993): 1691–1695.

[39] Perelson, A. S., and F. G. Oster. "Theoretical Studies of Clonal Selection: Minimal Antibody Repertoire Size and Reliability of Self-Nonself Discrimination." *J. Theor. Biol.* **81** (1979): 645–670.

[40] Perelson, A., R. Hightower, and S. Forrest. "Evolution (and Learning) of V-Region Genes." *Resh. Immunol.* **147** (1996): 202–208.

[41] Segel, L. A., and R. Lev Bar-Or. "Immunology Viewed as the Study of an Autonomous Decentralized System." In *Artificial Immune Systems and Their Applications*, edited by D. Dasgupta, 65–88. Berlin: Springer-Verlag, 1998.

[42] Segel, L. A., and R. Lev Bar-Or. "On the Role of Feedback in Promoting Conflicting Goals of the Adaptive Immune System." *J. Immunol.* **163** (1999): 1342–1349.

[43] Seidman, J. G., A. Leder, M. H. Edgell, F. Polsky, S. M. Tilghman, D. C. Tiemeier, and P. Leder. "Multiple Related Immunoglobulin Variable Region Genes Identified by Cloning and Sequence Analysis." *Proc. Natl. Acad. Sci. USA* **75** (1978): 3881–3885.

[44] Smith, D. J., S. Forrest, D. H. Ackley, and A. S. Perelson. "Using Lazy Evaluation to Simulate Realistic-Size Repertoires in Models of the Immune System." *Bull. Math. Biol.* **60** (1998): 647–658.

[45] Smith, D. J., S. Forrest, D. H. Ackley, and A. S. Perelson. "Variable Efficacy of Repeated Annual Influenza Vaccination." *Proc. Nat Acad Sci.* **96** (1999): 14001–14006.

[46] Smith, D. J., S. Forrest, R. Hightower, and A. S. Perelson. "Deriving Shape Space Parameters from Immunological Data for a Model of Cross-Reactive Memory." *J. Theor. Biol.* **189** (1997): 141–150.

[47] Smith, Derek. "Towards a Model of Associative Recall in Immunological Memory." Technical Report 94-9, University of New Mexico, Albuquerque, NM, 1994.

[48] Smith, Derek. "The Cross-Reactive Immune Response." Ph.D. diss., University of New Mexico, Albuquerque, NM, 1997.

[49] Smith, Derek, Stephanie Forrest, and Alan Perelson. "Immunological Memory is Associative." In *ICMAS workshop on "Immunity Based Systems,"* 1996.

[50] Smith, R., S. Forrest, and A. S. Perelson. "Searching for Diverse, Cooperative Populations with Genetic Algorithms." *Evol. Comp.* **1(2)** (1993): 127–149.

[51] Somayaji, A., and S. Forrest. "Automated Response Using System-Call Delays." In *Usenix Security Syposium 2000*, submitted.

[52] Tonegawa, S. "Somatic Generation of Antibody Diversity." *Nature* **302** (1983): 575–581.

[53] Warrender, C., S. Forrest, and B. Pearlmutter. "Detecting Intrusions Using System Calls: Alternative Data Models." In *Proceedings of the 1999 IEEE Symposium on Security and Privacy*, 133–145. Los Alamitos, CA: IEEE Computer Society, 1999.

[54] Weigert, M. G., I. M. Cesari, S. J. Yonkovitch, and M. Cohn. "Variability in the Light Chain Sequences of Mouse Antibody." *Nature* **228** (1970): 1045–1047.
[55] Wilson, M., R. Seymour, and B. Henderson. "Bacterial Perturbation of Cytokine Networks." *Infect. & Immunol.* **66** (1998): 2401–2409.

Index

Please note: numbers in boldface indicate terms found in tables and figures, and numbers in italics indicate definitions.

U